Polymer Synthesis

With contributions by
Y. Furusho · Y. Ito · N. Kihara · K. Osakada · M. Suginome
T. Takata · D. Takeuchi

Springer

The series presents critical reviews of the present and future trends in polymer and biopolymer science including chemistry, physical chemistry, physics and material science. It is addressed to all scientists at universities and in industry who wish to keep abreast of advances in the topics covered.

As a rule, contributions are specially commissioned. The editors and publishers will, however, always be pleased to receive suggestions and supplementary information. Papers are accepted for "Advances in Polymer Science" in English.

In references Advances in Polymer Science is abbreviated Adv Polym Sci and is cited as a journal.

The electronic content of APS may be found at http://www.springerLink.com

ISSN 0065-3195
ISBN 3-540-21711-8
DOI 10.1007/b14098
Springer-Verlag Berlin Heidelberg New York

Library of Congress Control Number 2004105254

This work is subject to copyright. All rights are reserved, whether the whole or part of the material is concerned, specifically the rights of translation, re-printing, re-use of illustrations, recitation, broadcasting, reproduction on microfilms or in other ways, and storage in data banks. Duplication of this publication or parts thereof is only permitted under the provisions of the German Copyright Law of September 9, 1965, in its current version, and permission for use must always be obtained from Springer-Verlag. Violations are liable for prosecution under the German Copyright Law.

Springer-Verlag is a part of Springer Science+Business Media

springeronline.com

© Springer-Verlag Berlin Heidelberg 2004
Printed in Germany

The use of registered names, trademarks, etc. in this publication does not imply, even in the absence of a specific statement, that such names are exempt from the relevant protective laws and regulations and therefore free for general use.

Typesetting: Stürtz AG, Würzburg
Cover: Künkellopka GmbH, Heidelberg; design&production GmbH, Heidelberg

Printed on acid-free paper   02/3020/kk – 5 4 3 2 1 0

Editorial Board

Prof. Akihiro Abe
Department of Industrial Chemistry
Tokyo Institute of Polytechnics
1583 Iiyama, Atsugi-shi 243-02, Japan
E-mail: aabe@chem.t-kougei.ac.jp

Prof. A.-C. Albertsson
Department of Polymer Technology
The Royal Institute of Technology
S-10044 Stockholm, Sweden
E-mail: aila@polymer.kth.se

Prof. Ruth Duncan
Welsh School of Pharmacy
Cardiff University
Redwood Building
King Edward VII Avenue
Cardiff CF 10 3XF
United Kingdom
E-mail: duncan@cf.ac.uk

Prof. Karel Dušek
Institute of Macromolecular Chemistry, Czech
Academy of Sciences of the Czech Republic
Heyrovský Sq. 2
16206 Prague 6, Czech Republic
E-mail: dusek@imc.cas.cz

Prof. Dr. W. H. de Jeu
FOM-Institute AMOLF
Kruislaan 407
1098 SJ Amsterdam, The Netherlands
E-mail: dejeu@amolf.nl

Prof. Jean-François Joanny
Institute Charles Sadron
6, rue Boussingault
F-67083 Strasbourg Cedex, France
E-mail: joanny@europe.u-strasbg.fr

Prof. Hans-Henning Kausch
c/o IGC I, Lab. of Polyelectrolytes
and Biomacromolecules
EPFL-Ecublens
CH-1015 Lausanne, Switzerland
E-mail: kausch.cully@bluewin.ch

Prof. S. Kobayashi
Department of Materials Chemistry
Graduate School of Engineering
Kyoto University
Kyoto 615-8510, Japan
E-mail: kobayasi@mat.polym.kyoto-u.ac.jp

Prof. Prof. Kwang-Sup Lee
Department of Polymer Science & Engineering
Hannam University
133 Ojung-Dong
Taejon 300-791, Korea
E-mail: kslee@mail.hannam.ac.kr

Prof. L. Leibler
Matière Molle et Chimie
Ecole Supèrieure de Physique
et Chimie Industrielles (ESPCI)
10 rue Vauquelin
75231 Paris Cedex 05, France
E-mail: ludwik.leibler@espci.fr

Prof. Timothy E. Long
Department of Chemistry and Research Institute
Virginia Tech
2110 Hahn Hall (0344)
Blacksburg, VA 24061, USA
E-mail: telong@vt.edu

Prof. Ian Manners
Department of Chemistry
University of Toronto
80 St. George St.
M5S 3H6 Ontario, Canada
E-mail: imanners@chem.utoronto.ca

Prof. Dr. Martin Möller
Deutsches Wollforschungsinstitut
an der RWTH Aachen e.V.
Veltmanplatz 8
52062 Aachen, Germany
E-mail: moeller@dwi.rwth-aachen.de

Prof. Oskar Nuyken
Lehrstuhl für Makromolekulare Stoffe
TU München
Lichtenbergstr. 4
85747 Garching, Germany
E-mail: oskar.nuyken@ch.tum.de

Dr. E. M. Terentjev
Cavendish Laboratory
Madingley Road
Cambridge CB 3 OHE
United Kingdom
E-mail: emt1000@cam.ac.uk

Prof. Brigitte Voit
Institut für Polymerforschung Dresden
Hohe Straße 6
01069 Dresden, Germany
E-mail: voit@ipfdd.de

Prof. Gerhard Wegner
Max-Planck-Institut für Polymerforschung
Ackermannweg 10
Postfach 3148
55128 Mainz, Germany
E-mail: wegner@mpip-mainz.mpg.de

Advances in Polymer Science
Also Available Electronically

For all customers who have a standing order to Advances in Polymer Science, we offer the electronic version via SpringerLink free of charge. Please contact your librarian who can receive a password for free access to the full articles by registering at:

http://www.springerlink.com

If you do not have a subscription, you can still view the tables of contents of the volumes and the abstract of each article by going to the SpringerLink Homepage, clicking on "Browse by Online Libraries", then "Chemical Sciences", and finally choose Advances in Polymer Science.

You will find information about the

– Editorial Board
– Aims and Scope
– Instructions for Authors
– Sample Contribution

at http://www.springeronline.com using the search function.

Contents

Polyrotaxanes and Polycatenanes: Recent Advances in Syntheses
and Applications of Polymers Comprising of Interlocked Structures
T. Takata · N. Kihara · Y. Furusho 1

Transition Metal-Mediated Polymerization of Isocyanides
M. Suginome · Y. Ito. 77

Coordination Polymerization of Dienes, Allenes,
and Methylenecycloalkanes
K. Osakada · D. Takeuchi 137

Author Index Volumes 101–171 195

Subject Index 211

Polyrotaxanes and Polycatenanes: Recent Advances in Syntheses and Applications of Polymers Comprising of Interlocked Structures

Toshikazu Takata[1] (✉) · Nobuhiro Kihara[2] · Yoshio Furusho[2]

[1] Department of Organic and Polymeric Materials, Tokyo Institute of Technology, 152-8552 Ookayama, Meguro-ku, Tokyo, Japan
ttakata@polymer.titech.ac.jp
[2] Department of Applied Chemistry, Osaka Prefecture University, 599-8531 Gakuen-cho, Sakai, Osaka, Japan

1	Introduction—Chemistry of Polyrotaxanes and Polycatenanes: An Overview	2
1.1	Structures of Interlocked Polymers	3
1.2	Wheel Component	4
1.3	Synthesis of Interlocked Polymers	7
2	Synthesis and Application of Polyrotaxanes	9
2.1	Main Chain-Type Polyrotaxanes Bearing Crown Ethers as The Wheel Components	9
2.1.1	(Pseudo)polyrotaxane Synthesized by Statistical Approach	9
2.1.2	(Pseudo)polyrotaxane Synthesized by "Directed" Approach	11
2.1.2.1	CT Interaction Used for Complexation to Rotaxane Structure	11
2.1.2.2	Hydrogen Bonding Interaction Used for Complexation to Rotaxane Structure	13
2.2	Main-chain Type Polyrotaxanes Having Cyclodextrins as The wheel Components	19
2.2.1	Synthesis of Pseudopolyrotaxane	19
2.2.1.1	Polymerization of Pseudorotaxane	19
2.2.1.2	Complexation with Polymer	21
2.2.2	Synthesis of Polyrotaxane	22
2.2.2.1	Doubly-stranded Pseudopolyrotaxane	26
2.2.3	Application of Cyclodextrin Polyrotaxane	27
2.2.3.1	Molecular Abacus	27
2.2.3.2	Insulated Molecular Wires	27
2.2.3.3	Antenna Molecules	29
2.2.3.4	Biodegradable Polyrotaxanes	30
2.2.3.5	Stimuli-Responsive Polyrotaxanes	31
2.3	Polyrotaxanes Having Cyclodextrin Nanotubes as The Wheel Components	32
2.4	Polyrotaxanes Bearing Miscellaneous Ring Systems as the Wheel Components	33
2.4.1	Cucurbituril	33
2.4.2	Macrocycle Bearing Bidentate Nitrogen Ligand	38
2.4.3	Cyclophane	40
2.4.4	Cyclophane Bearing Bis(4,4'-bipyridinium) Moiety	41
2.4.5	Amide-Type Macrocycle	42
3	Side Chain-Type Polyrotaxanes	44
3.1	Crown Ethers	44

3.2	Cyclodextrin	44
3.3	Cucurbituril	48
4	**Synthesis and Application of Polymers Bearing Interlocked Structures Used for Monomer Linkage**	**49**
4.1	"Topological" Polyrotaxanes	49
4.1.1	Poly[2]rotaxane	49
4.1.2	Poly[3]rotaxane	51
4.1.3	Crosslinked Polyrotaxane	55
4.2	Polycatenanes	58
4.2.1	Poly[2]catenane	58
4.2.2	Poly[n]catenane	62
4.2.3	Polycatenane Network	66
5	**Concluding Remarks**	**68**
References		**68**

Abstract Syntheses and applications of interlocked polymers, polyrotaxanes, and polycatenanes, including corresponding oligomers are reviewed with emphasis on (i) synthesis of interlocked polymers consisting of interlocked structures as the monomer-linking units (genuine "topological" polymers), and (ii) application of the interlocked polymers in both bulk and molecular levels. Further, the review also refers to a few important polyrotaxanes and polycatenane which are still unknown despite many synthetic challenges attempted so far. The review mainly summarizes the recent progress in the chemistry of polyrotaxanes and polycatenanes during this decade, in terms of kind of ring systems.

Keywords Poly(oligo)rotaxane poly(oligo)catenane · Synthesis · Application · Interlocked polymer

1
Introduction—Chemistry of Polyrotaxanes and Polycatenanes: An Overview

Mechanical bonding characteristic of interlocked molecules such as rotaxanes and catenanes assures high freedom and mobility of the whole molecule or its components, as predicted from their unique structures. Meanwhile, complete separation of their components to each other requires energy as high as that for covalent bond breaking. Therefore, the so-called "topological bond" between the components can be regarded as a "soft but strong bond" in comparison with the typical covalent bond. The interlocked molecules having such characteristic features are expected to have special or extraordinary physical and chemical properties.

The chemistry of rotaxanes and catenanes has progressed well in accordance with the interest in their unique structures and the expectation to development as the parts of molecular machines or molecular device, whereas that of polymers comprising these structures as the key repeating units, i.e., interlocked polymers, has progressed less well. The chemistry of [2]rotaxanes and [2]catenanes has recently stressed their applications by utilizing the vast amount of studies as their background, while both synthesis and application of the interlocked polymers have been studied simultaneously.

Although characteristic properties in mechanical and/or rheological aspects have been assigned to the interlocked polymers in the bulk state, the ring size of the wheel component included seems to exert a serious influence, as well as that by the freedom of the component. As a way of looking at the polymer properties, the effect of the entangled polymer chain plays an important role. That is, it can be considered that an elastic property exists based on the interlocked polymer chains as highly interpenetrated ones which may be associated with the properties of rubber and interpenetrating polymers, when a big wheel is used. Meanwhile, when a wheel component is connected with the chain polymers, bonding between the wheel component and the polymer chain results in producing the crosslinked points that can move on the chain. Such a type of "topological crosslinking" can provide special mobility to the polymer, because it is distinguished from both physical and chemical crosslinkings with little mobility.

As mentioned above, studies from the viewpoint of material science and technology of these new type of polymers are progressing, particularly since the start of the twenty-first century, and various unique properties are expected for the polymers characterized by the interlocked structures.

There are reviews including two comprehensive articles of Gibson [1] and Stoddart [2] on the polyrotaxanes and polycatenanes [3–12], in addition to a lot of review articles and books on the rotaxanes and catenanes [13–28]. Short reviews on the applications of polyrotaxanes are also reported [29–38].

1.1
Structures of Interlocked Polymers

Scheme 1 illustrates the simplest structures of rotaxane, catenane, and knot besides polyrotaxane and polycatenane. From the fact that the main chain-type polyrotaxane at the left side is the only interlocked polymer synthesized so far among the three polymers shown at the bottom of the scheme, progress in synthesis of interlocked polymers appears to be sluggish judging from the level of activity in synthetic polymer chemistry in the world.

More detailed general structures of the representative polyrotaxanes and polycatenanes are shown in Scheme 2. Polyrotaxanes can be categorized into two types: one is the polyrotaxanes consisting of the main chains of covalent type as shown in the top four examples (A–D), while the other involves the polyrotaxanes of which monomer linking units are constructed by the rotaxane structure as shown in the following two structures (E, F). The essen-

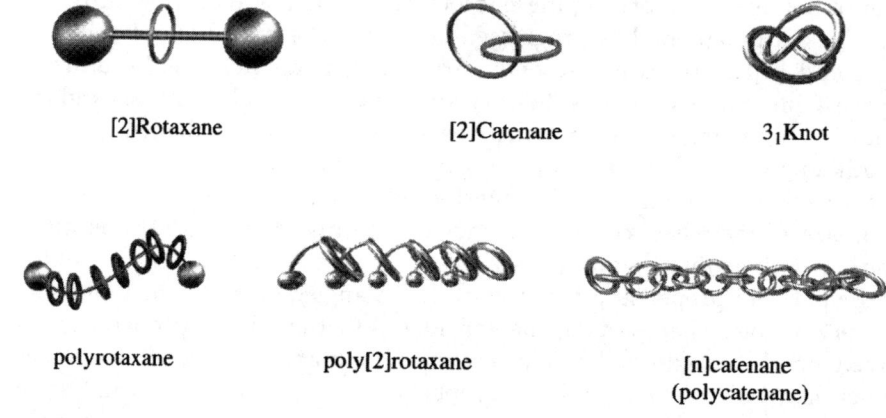

Scheme 1

tial difference in main chain structure between the two types of polyrotaxanes should cause large differences in their physical or mechanical properties. The former polyrotaxanes (A–D) are further divided into main chain-type (A, B) and side chain-type (C, D). As mentioned above, poly[2]rotaxane (E) as one of the latter is an unknown polymer which has still been encouraging the many synthetic challenges made so far aside. Meanwhile, synthesis of the neighboring poly[3]rotaxane (F) has very recently been achieved. Genuine polyrotaxane seems to be one of the polyrotaxanes like the former "topological polyrotaxanes", which may reflect their truly unique structures to their properties.

In addition to three typical structures of poly[2]catenanes (G–I), polycatenane (i.e., [n]catenane) of which the structure is comprised only of wheel components is simply interlinked like a "chain" (J). The polycatenane is one of the most difficult goals in the synthesis of unknown interlocked polymers, like poly[2]rotaxane as already pointed out, although it will be accomplished in the near future because so much effort has been made by synthetic chemists, and this will be continued.

1.2
Wheel Component

It is vital that simple and cheap synthesis of interlocked polymers is achieved in order to make progress in the chemistry of polyrotaxanes and polycatenanes. Since bulk property is essential in polymer science, difficulty in synthesis of interlocked polymers should be avoided, this being different from the case of molecular materials such as molecular devices functioning at a molecular level. Both polyrotaxanes and polycatenanes as well as both rotaxanes and catenanes are becoming easy to synthesize with the progress

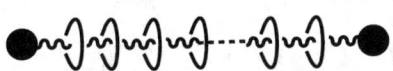

Main Chain-Type Polyrotaxane, **A**

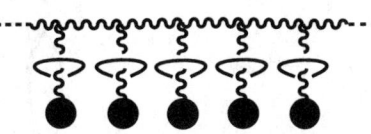

Side Chain-Type Polyrotaxane, **C**

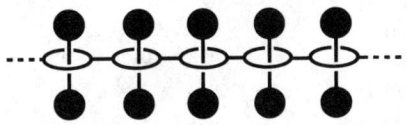

Main Chain-Type Polyrotaxane, **B**

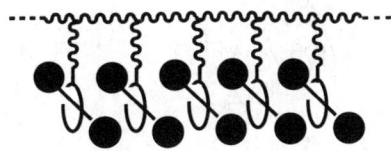

Side Chain-Type Polyrotaxane, **D**

Poly[2]rotaxane, **E**

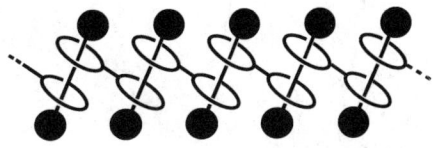

Poly[3]rotaxane, **F**

Typical Structures of Polyrotaxanes

Poly[2]catenane, **G**

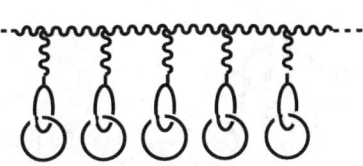

Poly[2]catenane, **I**

Poly[2]catenane, **H**

Polycatenane, [n]Catenane, **J**

Typical Structures of Polycatenanes

Scheme 2

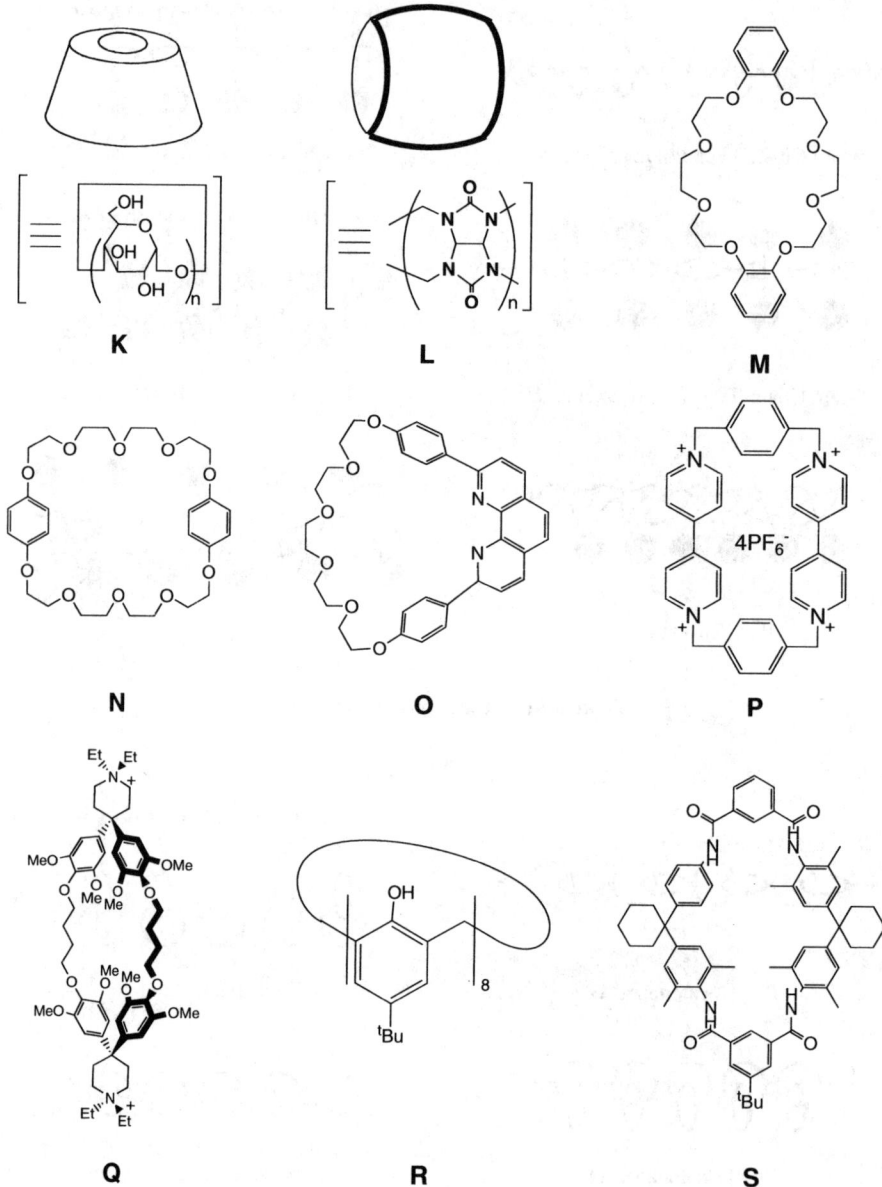

Structures of Representative Wheel Components

Scheme 3

of supramolecular chemistry. Even with such fortunate circumstances in recent times, the biggest problems to be overcome would be the absence of appropriate wheel components. As shown in Scheme 3, there are several examples of wheel components (K–S) which are used in the interlocked polymers among those used in rotaxanes and catenanes. Although these macrocycles are wheels good enough to be interactable with the axle components with each particular interaction, most of them suffer from synthetic difficulty and/or high cost. Creation or development of cheap or easily prepared wheel components is strongly desired.

1.3
Synthesis of Interlocked Polymers

Synthesis of polyrotaxanes and polycatenanes is performed basically by using or applying the synthetic methods for rotaxanes and catenanes.

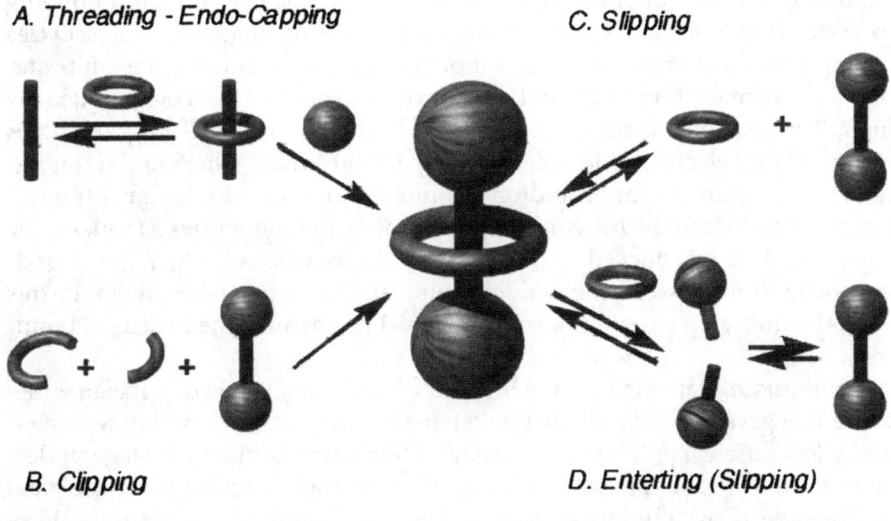

Scheme 4

Scheme 4 summarizes the representative synthetic methods of [2]rotaxanes [17, 21]. Methods A and B are characterized by the kinetically controlled process as the final step to the rotaxane. In particular, method A is the general and most straightforward synthesis: i.e., end-capping of the axle terminal with a bulky group after threading of the axle into the wheel. Most polyrotaxanes are prepared according to this methodology. On the other hand, methods C and D at the left side undergo the thermodynamically controlled process at the equilibrium in the last step of the process. Since the procedures in the two synthetic methods completely differ from each other, the control of the synthetic reaction is also different. The thermodynamic control process has recently attracted much attention from the viewpoint of advantages, not only in yield but in also milder reaction in accordance with progress in supramolecular chemistry.

The above-mentioned "directed" synthesis always requires a certain strong interaction between the wheel and the axle before making the interlocked bond in any method. As for the example of the wheels depicted in Scheme 3, hydrophobic interaction is the major attractive interaction in the cases of cucurbituril and cyclodextrin (**A, B**), thereby resulting in limitation as the wheel-axle complex formation should be done in water. In particular, it is an additional difficulty to employ the complexation in strongly acidic conditions owing to the extremely low solubility of cucurbituril.

In the case of crown ethers (**M, N**) which should have the number of ring members more than ca. 24, since the major attractive interaction is the hydrogen bonding with secondary ammonium ion and/or ion-dipole interaction with cationic species, the complexation should be carried out under conditions capable of encouraging such interactions. Coordination bonding to metal is the attractive interaction in the case of oligoether-macrocycles having a bidentate nitrogen ligand moiety (**F**). The coordination is quite stable with strong "bonding" and therefore the yield of rotaxane is usually high. Paraquat-type cationic host as the wheel component (**G**) requires axles having highly electron-donating property like aromatic ethers and tetrathiafulvalenes, where cation-π and/or CT interactions are the major attractive interactions. Macrocycles consisting of amide functionalities (**J**) make the corresponding interlocked structures with the assistance of hydrogen bonding interaction between the amide groups and the axle components. In this case, the final step reaction to rotaxane need not disturb the hydrogen bonding.

Synthesis of catenane is much more difficult than that of rotaxanes because it always depends on the final step of a ring-closing reaction with generally low efficiency. Namely, macrocycle formation at the final stage undergoes an unfavorable process with regard to entropy, and therefore the yield of catenane is usually low even by a "directed" synthesis, other than those utilizing metal-templated synthesis [16, 23, 27]. Although high yield synthesis is sometimes accessible to [2] and [3]catenanes, no polymer [n]catenane listed in Scheme 2 (**J**) is reported at all. The maximum number of n is 5 at present time. In contrast to polycatenane, poly[2]catenane can be easily prepared by polymerization or copolymerization of [2]catenane with polymerizable groups pre-synthesized through an efficient method.

This review mainly summarizes recent progress in the chemistry of interlocked polymers including oligomers (consisting of more than three components) in this decade. In particular, the review first describes the "genuine" interlocked polymers of which repeating units are linked through the interlocked structures—they can be called "topological polymers"—and also emphasizes the applications of the interlocked polymers. Namely, synthesis and applications of all higher order rotaxanes and catenanes are also the targets of this review, except for the simplest interlocked molecules [2]rotaxane and [2]catenane. Further, it excludes the interlocked molecules which do not have fairly sufficient mobility or freedom of the components, like those in a crystalline state, which seem unlikely to be categorized as an interlocked polymer. Pseudo-interlocked polymers without bulky end-cap groups at the axle terminal as pseudopolyrotaxanes are also described.

2
Synthesis and Application of Polyrotaxanes

2.1
Main Chain-Type Polyrotaxanes Bearing Crown Ethers as The Wheel Components

2.1.1
(Pseudo)polyrotaxane Synthesized by Statistical Approach

One of the most primitive approaches to pseudopolyrotaxanes is the polymerization of certain monomers in the presence of crown ethers. Although this is a statistical approach, either polycondensation, polyaddition, or vinyl polymerization of appropriate monomers using crown ether as a solvent afforded corresponding pseudopolyrotaxane [39]. As a typical example, polycondensation of tetra(ethylene glycol) with methylene diphenyldiisocyanate (MDI) in various ring sizes of crown ethers was carried out [40, 41] (Scheme 5). After the removal of free crown ether from the reaction mixture by repeated precipitation, the resulting polyurethane was analyzed. Every repeating unit of the polyester contained 0.16–0.87 crown ether. Physical prop-

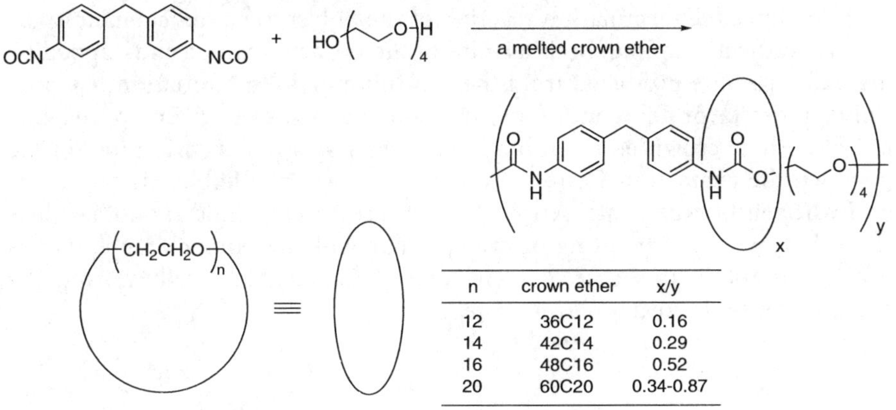

n	crown ether	x/y
12	36C12	0.16
14	42C14	0.29
16	48C16	0.52
20	60C20	0.34–0.87

Scheme 5

erties of this type of pseudopolyrotaxane was studied [42]. In a similar manner, various polyesters having rotaxane structures were prepared [43]. Crown ether statistically threaded the polymer chain to produce pseudopolyrotaxane. It was postulated that hydrogen-bonding interaction between the hydroxy group in tetra(ethylene glycol) or the NH group on the urethane group with crown ether was responsible to the formation of rotaxane structure. However, the authors cast doubt on this explanation because the wheel component in the pseudopolyrotaxane would thread out from the axle component since there is no strongly attractive interaction between the components. The hydrogen-bonding interaction between NH group and crown ether is too weak to maintain the rotaxane structure. It is more plausible that the product was polycatenane rather than pseudopolyrotaxane due to the cyclization that was unavoidably accompanied by linear polymerization. Meanwhile, when a polyurethane was simply mixed with 42C14 or 30C10, a polymer-crown ether complex in which m/n value approached 0.2 was obtained [44]. Since only a very small amount of crown ether was incorporated into the polyester under the same conditions, hydrogen-bonding between the crown ether and the urethane NH groups played a crucial role. However, stability of the complex in the solution was not described. It is very likely that the crown ethers were kinetically captured on the polymer.

When the polymer chain contains the bulky substituent or unit, crown ether can be stably incorporated in the polyrotaxane because these units act as terminators of rotaxane segment [43, 45]. When the polymerization of a diol bearing trityl moiety and a diacid chloride was carried out in molten 30-crown-10 (30C10), a polyrotaxane consisting of the polyester axle and 30C10 wheel was obtained. 30C10 not only acted as a reactant, but also as a solvent. Because of the bulky trityl groups on the main chain, 30C10 could not thread out of the axle. Meanwhile radical polymerization of styrene in molten crown ether may yield pseudopolyrotaxane. To stabilize the polyrotaxane structure, end-capping was investigated by using a bulky azo-initiator [46]. Since the termination reaction of the polystyrene radical mainly occurs via radical coupling, both termini of the polymer can be end-capped by the bulky residue produced from the azo-initiator in the initiation reaction. Under most favorable conditions, the content of crown ether approached 21 wt%. Since crown ether with a small cavity was not incorporated in the polymer, the chain-transfer to the crown ether was negligible. Since there is no hydrogen-bonding interaction between crown ether and monomer, it is clear that hydrogen-bonding of crown ether with monomer or polymer is not important in this case. Crown ether threaded statistically during the polymerization to form polyrotaxane.

2.1.2
(Pseudo)polyrotaxane Synthesized by "Directed" Approach

2.1.2.1
CT Interaction Used for Complexation to Rotaxane Structure

While polyrotaxane can simply be prepared by the statistic method, the polyrotaxanes with controlled structure have been prepared by the directed method. For this purpose, the combination of 4,4'-bipyridinium salt and crown ether having p-dialkoxyphenyl group has been widely used [12]. Not only charge-transfer (CT) interaction between electron deficient 4,4'-bipyridinium moiety and electron-rich benzene ring, but also CH$\cdots$O hydrogen bonding interaction between crown ether and rather acidic hydrogens on the 4,4'-bipyridinium group ensured the stable complexation [47].

Gibson et al. reported that the copolymerization of poly(THF) and a diol-pseudorotaxane consisting of 4,4'-bipyridinium salt and bis-p-phenylene crown ether with diisocyanate afforded the corresponding polyurethane with the interlocked structure [48] (Scheme 6). Although this polyurethane has a pseudopolyrotaxane structure, the interlocked structure is stable because the interaction between 4,4'-bipyridinium salt and bis-p-phenylene crown ether is strong enough to keep the inclusion complex. In this elastic polyurethane, the rotaxane unit acted as a hard segment.

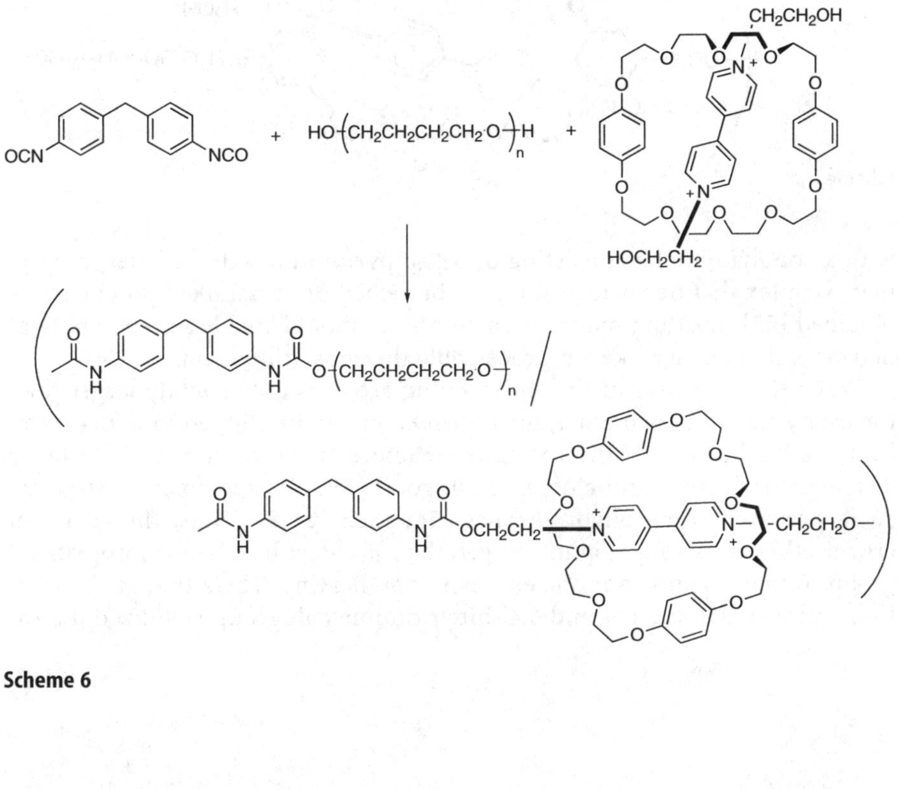

Scheme 6

Since the inclusion complex between 4,4'-bipyridinium salt and bis-*p*-phenylene crown ether is stable, oligorotaxane can be prepared from oligo(4,4'-bipyridinium salt) quantitatively. Stoddart et al. reported that oligorotaxanes were prepared when oligo(4,4'-bipyridinium salt) bearing the pyridine groups at the termini were alkylated by bulky alkylation agent in the presence of bis-*p*-phenylene crown ether [49–51].

Crown ether component can be placed on the axle. Gibson et al. found that a polyester consisting of crown ether formed polypseudorotaxane with 4,4'-bipyridinium salt [52] (Scheme 7). When the same polyester was mixed

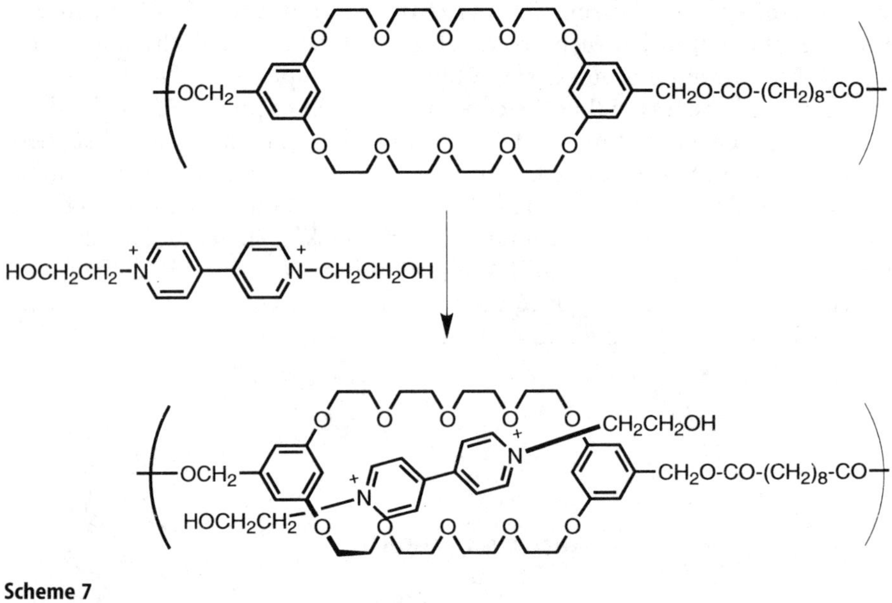

Scheme 7

with a copolyurethane consisting of 4,4'-bipyridinium salt moieties, a polymer complex that behaved as a highly branched or crosslinked polymer was obtained [53]. Loading more interaction sites should enable self-assembly of networks that behave like a reprocessable thermosetting resin.

When the bulkiness of the end-capping group is only slightly larger than the cavity size of crown ether, the crown ether can be slipped in the end-cap by tentative heating. If the rotaxane structure is thermodynamically fairly stable, the rotaxane can be obtained in good yield. In this manner, Stoddart et al. prepared some oligorotaxanes [54] (Scheme 8). Thus, the cavity of crown ether **A** is only slightly larger than 4,4'-*tert*-butyl-4''-isopropyltrityl group so that **A** can slip in the end-cap upon heating. The attractive interaction between crown ether and 4,4'-bipyridinium salt group stabilized the ro-

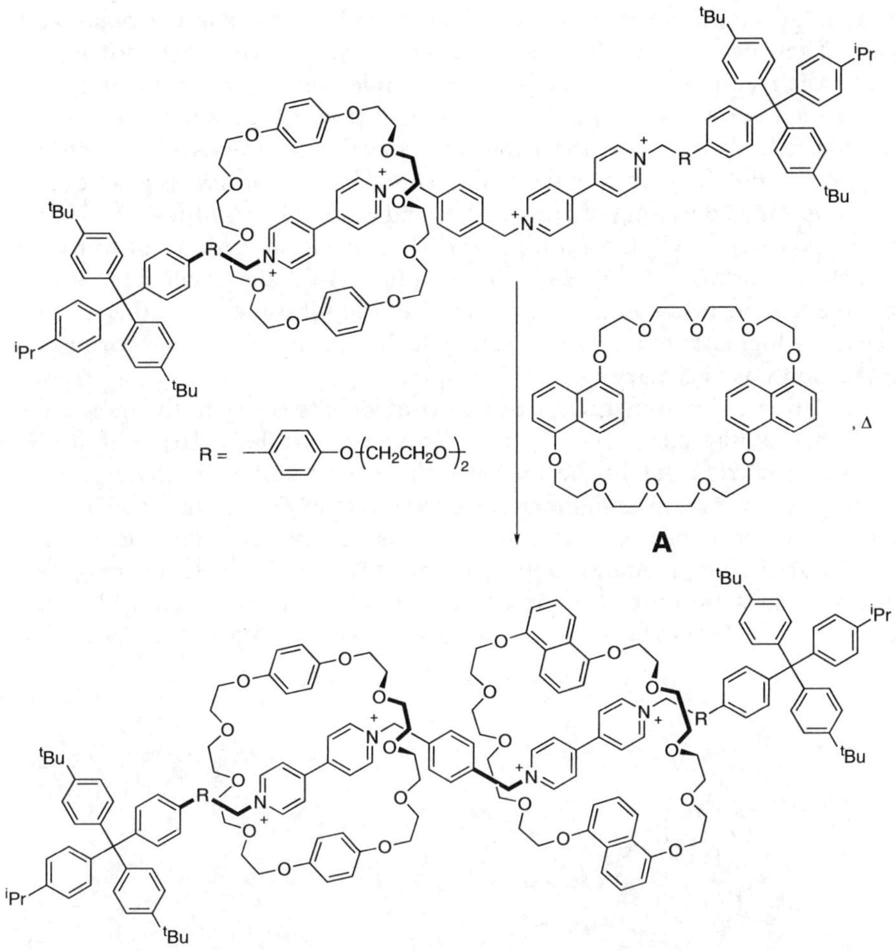

Scheme 8

taxane structure. Branched or dendritic oligorotaxanes can be prepared from branched 4,4′-bipyridinium salt in a similar manner [55, 56].

2.1.2.2
Hydrogen Bonding Interaction Used for Complexation to Rotaxane Structure

The attractive interaction of crown ether with certain secondary ammonium salts is strong enough to prepare various interlocked compounds. The combination of 24-membered crown ether such as dibenzo-24-crown8 (DB24C8) and bis(primary alkyl) or dibenzyl ammonium salt has been widely used for the complex formation. While there are two types of complexes, side-on and inclusion complexes, inclusion complex, which has a pseudorotaxane structure, is generally observed unless the ammonium salt has the bulky endcaps. The rotaxane structure can be fixed by end-capping. It is noteworthy that basic reaction conditions should be avoided during the end-capping because the hydrogen-bonding interaction is lost when the ammonium salt is neutralized. Various reaction conditions have been proposed to construct oligo- and polyrotaxanes without the use of base. The first preparation of oligorotaxane consisting of crown ether and ammonium salt was carried out by Stoddart et al. via the 1,3-dipolar cycloaddition of azide to di-*tert*-butyl acetylenedicarboxylate [57–59]. Although the 1,3-dipolar cycloaddition took place under neutral conditions, the yield of the rotaxane was rather low because the higher reaction temperature which suppressed the pseudorotaxane formation was necessary.

Acylation of hydroxy groups of axle components is one of the most effective end-capping methods of rotaxane synthesis. While acylation of alcohol is usually carried out in the presence of tertiary amines such as triethylamine, the use of amine significantly decreased the yield of the rotaxane. Takata et al. have demonstrated that the use of the combination of tributylphosphine as a catalyst and acid anhydride as an acylation reagent is an excellent solution to this problem [60]. The acid-catalyzed acylation is also effective [61]. The oligorotaxanes were easily prepared in more than

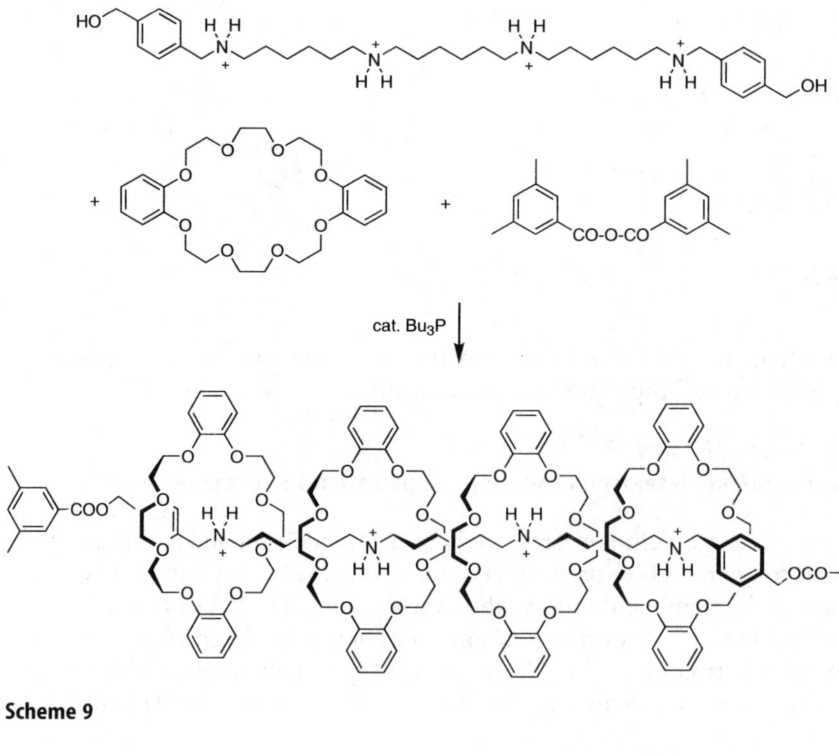

Scheme 9

70% yield by a phosphine-catalyzed acylative end-capping method [62] (Scheme 9).

The homocoupling of [2]pseudorotaxane is one of the most effective methods to prepare [3]rotaxane. The original acylative end-capping method is not effective for this type of oligorotaxane synthesis, however, because of difficulty in obtaining bifunctional acid anhydride. This restriction was solved by the use of an active ester instead of acid anhydride. Actually Takata et al. prepared [3]rotaxane in a good yield by the active ester method [63]. The more effective method for the homocoupling of [2]pseudorotaxane is achieved by the oxidation of mercapto group to form disulfide. The high yielding synthesis of [3]rotaxane was accomplished by Busch et al. by the oxidative coupling of [2]pseudorotaxane bearing the mercapto group at the terminus by iodine [64].

Since disulfide bonding is labile, and the disulfide-exchange reaction is catalyzed by certain nucleophiles such as thiol, a wonderful method to prepare oligorotaxanes was developed by Takata et al. When a bifunctional secondary ammonium salt bearing disulfide linkage and bulky end-caps was mixed with dibenzo-24-crown-8 and a catalytic amount of benzenethiol, crown ether entered into the disulfide linkage to afford oligorotaxane [65] (Scheme 10). The first step of this rotaxane preparation involves the nucleophilic attack of benzenethiol on the disulfide bond. The crown ether rapidly forms the pseudorotaxane complex with the resulting thiol containing *sec*-ammonium group. The second nucleophilic attack of the pseudorotaxane on the starting disulfide gives [2]rotaxane. As a result of the repetition of these processes between the thiol and disulfide species, [2] and [3]rotaxane are obtained. These two rotaxanes were selectively produced in more than 80% yield by choosing the suitable conditions. High polymer was obtained via this procedure when the bifunctional crown ether was used [66].

A pseudorotaxane having a benzyl bromide group at the axle terminus reacted with triphenylphosphine to afford a rotaxane having the phosphonium salt moiety [67]. Since the phosphonium salt undergoes Wittig reaction in the presence of strong base, axle-functionalized rotaxanes with complex structures can be derived from this rotaxane. According to the combination of benzyl bromide and aldehyde, various structures of oligorotaxanes have been prepared by Stoddart et al. besides [2]rotaxanes [68] (Scheme 11). Since the Wittig reaction gave olefin of *cis*- and *trans*-mixture, the products were isolated and characterized after hydrogenation. [3]Catenane and [3]rotaxanes were prepared using this technique [69]. Dendrimer with rotaxane structure was also prepared [70].

Slipping approach is effective for the preparation of oligorotaxane based on crown ether and secondary ammonium salt because the inclusion complex is fairly stable. When dibenzo-30-crown-10 (DB30C10) was used as the wheel component, 3,5-di-*tert*-butylphenyl group is the complementary end-cap to the crown ether. DB30C10 slips over the 3,5-di-*tert*-butylphenyl group

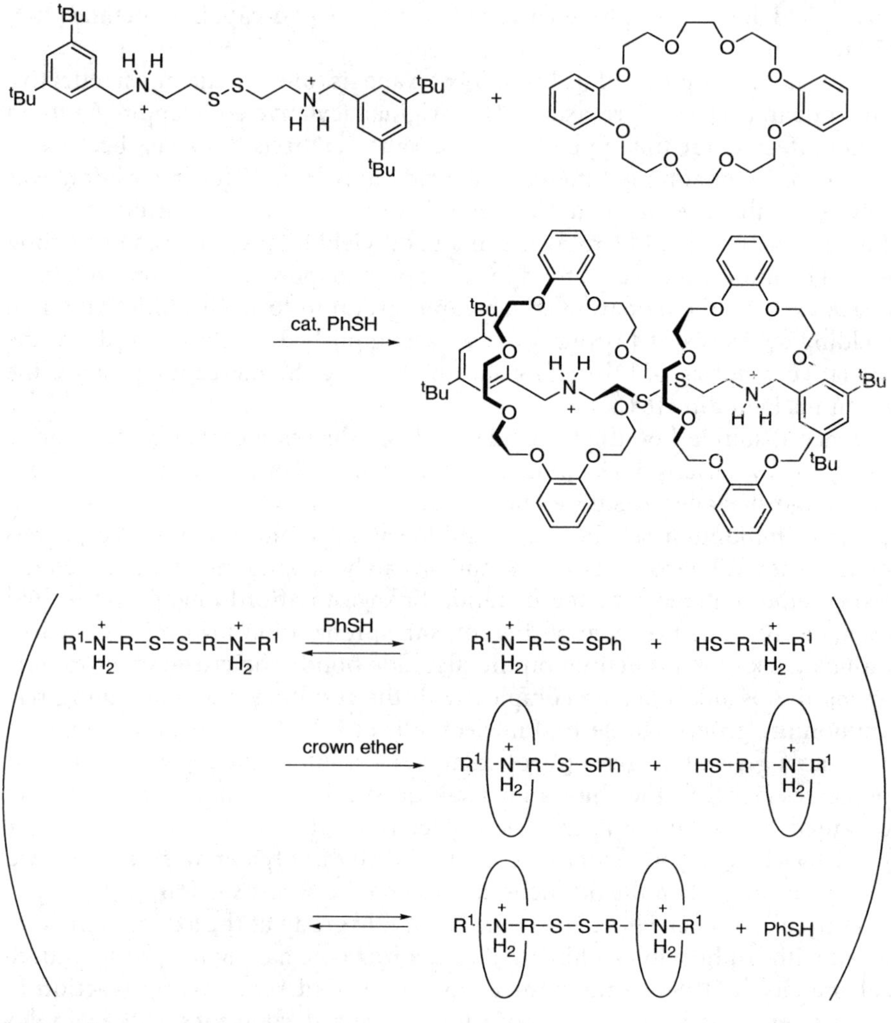

Scheme 10

at elevated temperatures, although it is bulky enough to allow the slipping of DB30C10 at ambient temperature. When the mixture of DB30C10 and a bisammonium salt bearing 3,5-di-*tert*-butylphenyl groups at both termini of the axle was heated, the corresponding [3]rotaxane was obtained [71]. Stoddart et al. demonstrated that for DB24C8 wheel, the cyclohexane group was a proper counterpart to achieve the slipping [72]. A bisammonium salt bearing cyclohexyl groups at the axle termini forms [3]rotaxane with DB24C8 by brief heating. Because of the stability of the inclusion complex, the slipping proceeded quantitatively. Takata et al. used a bis(crown ether) to prepare poly[3]rotaxane [73]. In spite of the fact that the polymerization proceeded

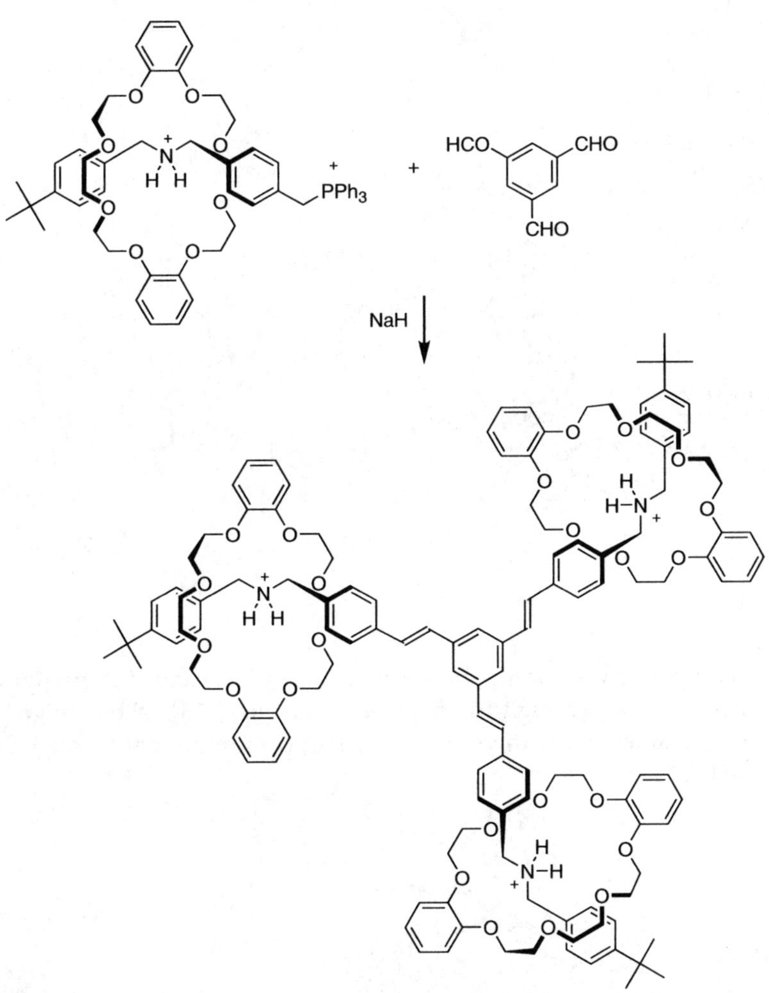

Scheme 11

via a thermodynamic process, high polymer was obtained because the equilibrium was almost completely shifted to the rotaxane formation.

In this type of rotaxane, the intramolecular hydrogen-bonding interaction between ammonium salt and crown ether is too strong to neutralize the ammonium part with usual bases. In fact, Takata et al. found that the rate of proton-exchange between [2]rotaxane and water was far slower than that between pyrrole and water, suggesting the extremely lowered acidity of the ammonium moiety [74]. However, the ammonium group can be quantitatively acylated by the excess amount of electrophile in the presence of excess triethylamine. Some oligorotaxanes can be prepared by this process (Scheme 12).

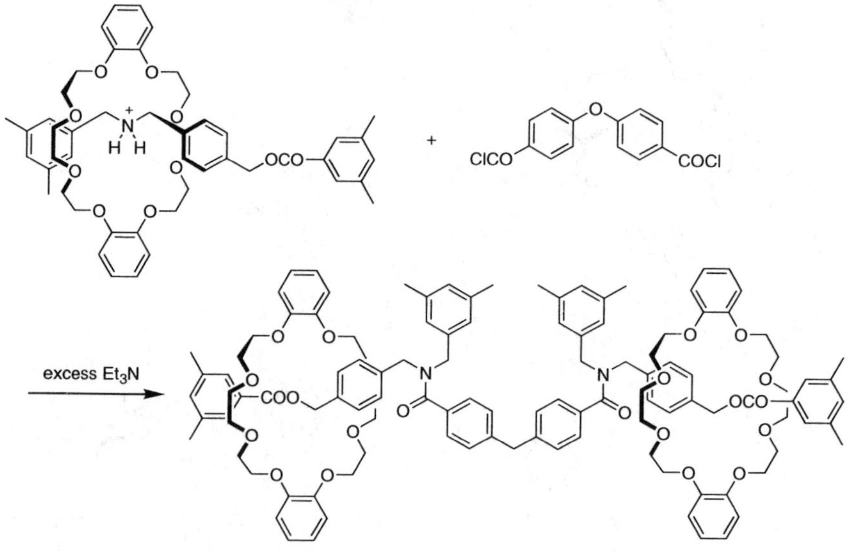

Scheme 12

CH···O hydrogen-bonding between crown ether and 1,2-bis(pyridinium)ethane is strong enough to construct rotaxanes [75]. A few oligorotaxanes were obtained from the oligo 1,2-bis(pyridinium)ethane and DB24C8 [76] (Scheme 13).

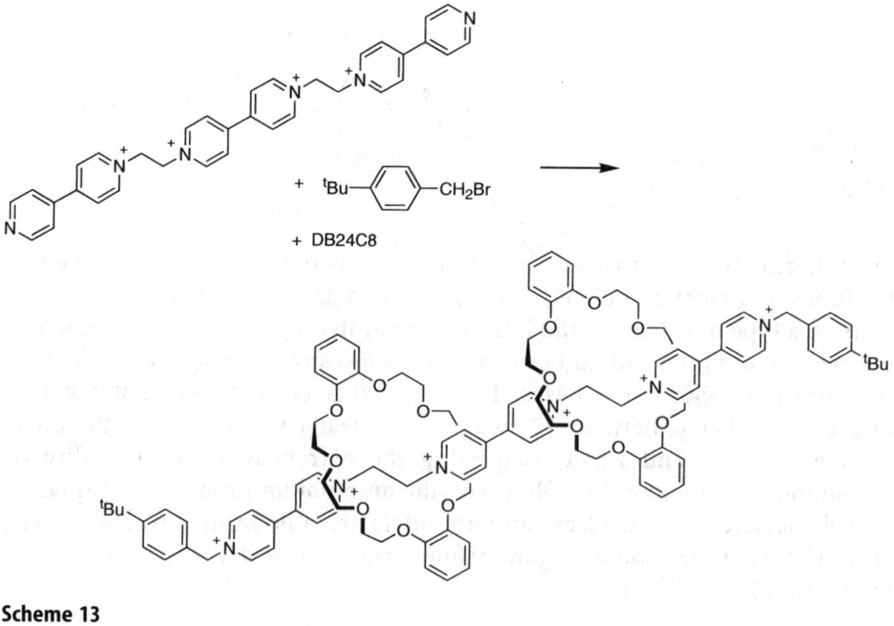

Scheme 13

34-Membered crown ether bearing 2,6-pyridinediyl groups forms the inclusion complex with 4-vinylbenzoic acid via the hydrogen-bonding interaction. Kato et al. reported that the copolymerization of styrene with the complex afforded pseudopolystyrene with polyrotaxane structure [77]. The glass transition temperature of the polymer was significantly lowered compared with that of simple copolymer, and it was deduced that the dimerization of carboxylic acids placed in the polymer backbone was effectively suppressed by rotaxanation.

2.2
Main-chain Type Polyrotaxanes Having Cyclodextrins as The wheel Components

2.2.1
Synthesis of Pseudopolyrotaxane

2.2.1.1
Polymerization of Pseudorotaxane

It was in 1976 that Ogata et al. reported the synthesis of "inclusion polyamides" [78]. When β-cyclodextrin was stirred with aliphatic diamines in water, precipitates were formed and they were characterized as inclusion compounds. Condensation of these inclusion amides with isophthaloyl or terephthaloyl chloride yielded polyamides encircled by many of the cyclodextrins, i.e., the first pseudopolyrotaxanes. Ogata et al. likened them to "a train passing through many tunnels". The solubility and water absorption of the pseudorotaxanes were greater than those of the naked polymer threads. Differential thermal analysis experiments showed that the thermal properties of the pseudorotaxanes were different both from the naked threads and from β-cyclodextrin.

Maciejewski s group reported a series of papers which described several attempts at the preparation of pseudopolyrotaxanes in the solid state [79–83]. For example, the radiation polymerization of the crystalline adduct of vinylidene chloride and β-cyclodextrin yielded a pseudopolyrotaxane, in which one cyclodextrin occupied 2.9 repeat units of vinylidene chloride [80].

Wenz et al. reported the synthesis of pseudopolyrotaxanes by solid-state polycondensation (Scheme 14) [84, 85]. They prepared inclusion complexes between β-cyclodextrin and several α,ω-amino acids, which were obtained as crystalline solids. The X-ray powder diffractograms of the inclusion complexes strongly supported channel-like packings that are required for the successful polycondensation in a solid state. Indeed, the corresponding polyamide, i.e., polypseudorotaxanes, were obtained by annealing inclusion complexes at 150–250 °C under vacuum. To their surprise, these polyamides were highly water-soluble. They attributed the good solubility to the very high coverage of the polyamide chains with β-cyclodextrin.

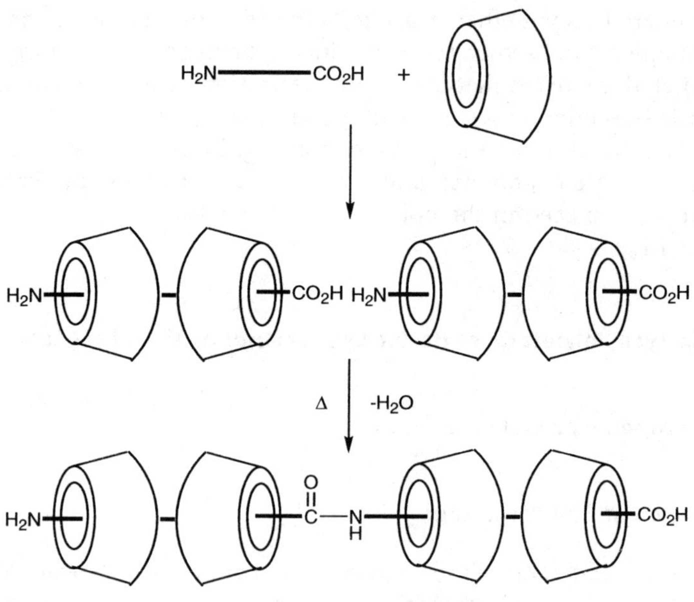

Scheme 14

Urethane-forming reaction between isocyanate and hydroxyl group was utilized by Osakada et al. to prepare polyurethane-cyclodextrin pseudorotaxanes (Scheme 15) [86]. Polyaddition of a diol and MDI in the presence of permethylated α-cyclodextrin or permethylated β-cyclodextrin was carried out in DMF for 20 h at 120 °C to yield the pseudopolyrotaxane. The molar

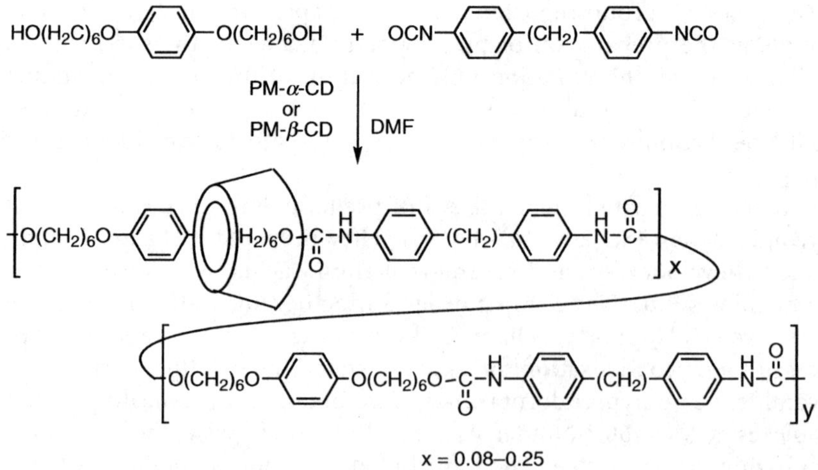

Scheme 15

ratios of the cyclodextrins to the repeating unit of polyurethane chain in the polyrotaxane were estimated to be 0.08–0.25 by ^{1}H NMR and elemental analysis. The polyrotaxane exhibited much lower T_gs than those of the parent polyurethanes. They ascribed the decreases in T_g to the inhibition of intermolecular hydrogen-bonding interaction of the urethane groups by the threaded cyclodextrins. Similarly, Osakada's group utilized ring-opening reaction of carboxylic acids and epoxides to synthesize pseudorotaxanes consisting of an azobenzenepolymer and γ-cyclodextrins [87]. The pseudorotaxane underwent UV-light induced isomerization of *trans*- to *cis*-azobenzene whereas the reverse isomerization did not occur under the usual conditions of irradiation with visible light. This one-way isomerization was explained by stronger hydrogen-bonding interaction between the azo groups and hydroxyl groups in the *cis*-isomer than that in the *trans*-isomer.

2.2.1.2
Complexation with Polymer

In 1990 Harada et al. reported a couple of papers describing how cyclodextrins formed inclusion complexes with poly(ethylene glycol) (PEG) and poly(propylene glycol) (PPG), i.e., pseudopolyrotaxanes [88, 89]. These were the first reports on pseudorotaxane formation between polymer with macrocyclic component, although several examples in which monomers were polymerized in situ within the cavity of cyclodextrin had already been reported by Ogata, Maciejewski, etc., as described above. The molecular recognition in this inclusion complexation is remarkable [88–92]. α-Cyclodextrin forms pseudorotaxane with PEG in an aqueous solution at ambient temperature, while it does not complex with the low molecular weight analogs, ethylene glycol, diethylene glycol, and triethylene glycol under the same conditions. α-Cyclodextrin forms complexes with PEG of molecular weight higher than 200. This finding that a minimum PEG length is required for the formation of stable cyclodextrin complexes illustrates the importance of cooperativity in complexation. The structure of polymer is also critical to the inclusion behavior of cyclodextrins. α-Cyclodextrin forms pseudorotaxane with PEG in an aqueous solution, while it does not complex with PPG at all under the same conditions. On the other hand, β- and γ-cyclodextrins form inclusion complexes with PPG in an aqueous solution, but not with PEG [93].

Harada and coworkers went on to investigate the pseudorotaxane formation behavior of cyclodextrins with various kinds of organic polymers such as poly(methyl vinyl ether), poly(tetrahydrofuran), oligoethylene, and polyesters [94–107]. Wenz et al. reported inclusion complexation behavior between cyclodextrins and some organic polymers [108–110].

In addition, Harada et al. have recently found that β-cyclodextrin and γ-cyclodextrin formed inclusion complexes with poly(dimethylsiloxane) (PDMS), a typical inorganic polymer in aqueous solution [111, 112]. The authors mentioned that these results were the first examples of molecular level "organic inorganic hybrids" using host guest complexation. α-Cyclodextrin does not form complexes with PDMS, apparently because the cavity size of α-cyclodextrin is too small for PDMS to thread in. The yields of the β-cyclodextrin-PDMS inclusion complexes decreased with increasing molecular weight of PDMS and virtually no complexation was observed above 760. In contrast, the yields of the γ-cyclodextrin-PDMS complexes increase with an increase in the molecular weight of PDMS, reach a maximum at around 760, and gradually decrease with increasing molecular weight. The difficulty of β-cyclodextrin to form inclusion complexes was attributed to the steric hindrance between the dimethyl groups of PDMS and the cavity of β-cyclodextrin on the basis of molecular model study. Harada's group has reported recently that β-cyclodextrin and γ-cyclodextrin form inclusion complexes with polydimethylsilane, another inorganic polymer, and they have similar complexation behavior to that of PDMS [113].

2.2.2
Synthesis of Polyrotaxane

In 1992, Harada et al. prepared a compound in which many cyclodextrins are threaded on a single PEG chain and are trapped by end-capping groups at both termini of the main chain (Scheme 16) [111]. The pseudopolyrotaxane formed in water from PEG bisamine (mass 3350) and a large excess of α-cyclodextrin in an aqueous solution was isolated and then subjected to end-capping reaction with 2,4-dinitrofluorobenzene in DMF to yield polyro-

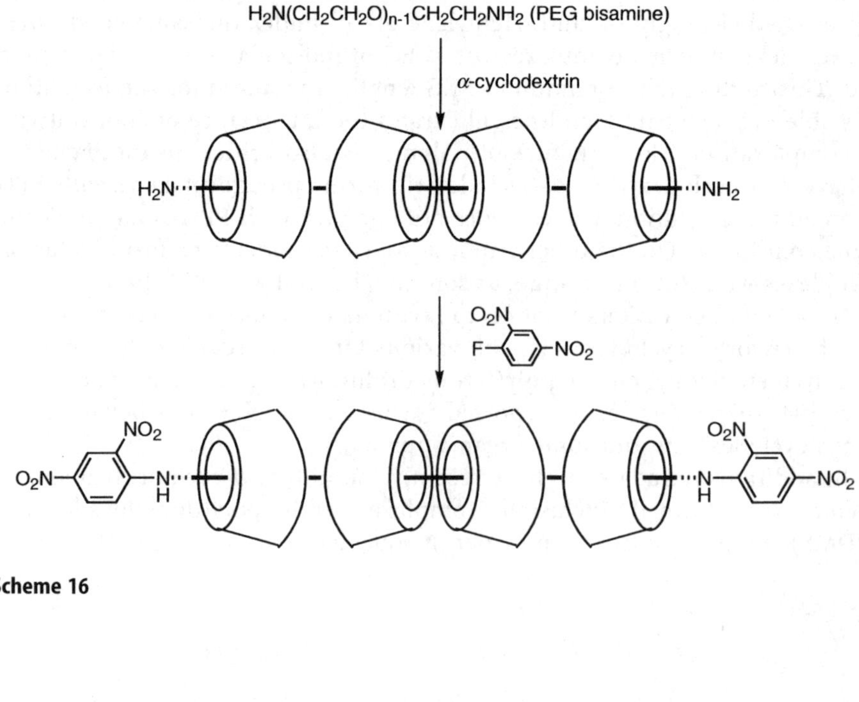

Scheme 16

taxane. The average molecular weight of the polyrotaxane was estimated to be ca. 25,000. Approximately 20–23 cyclodextrins were threaded in one molecule. This is the first example of polyrotaxane (not pseudopolyrotaxane!), which was referred to as a "molecular necklace" by Harada et al. Shortly afterward, Harada and his co-workers prepared polyrotaxanes starting with PEG bisamine of average molecular weight of 2000 and α-cyclodextrin [112]. A polyrotaxane that has about 37 α-cyclodextrins was obtained by the fractionation using GPC. When the monodisperse PEG bisamine ($H_2N(CH_2CH_2O)_{27}CH_2CH_2NH_2$) was employed, the polyrotaxane has 12 α-cyclodextrin rings on each polymer chain, indicating that the cyclodextrin rings are packed closely from one end of the polymer to the other [113].

Simultaneously with the Harada's report on the first polyrotaxane, Wenz reported the preparation of polyrotaxanes from polyamines and α-cyclodextrin in an aqueous solution [108]. α-Cyclodextrin forms inclusion complexes with poly(iminoundecamethylene) and poly(iminotrimethylene-iminodecamethylene) (Scheme 17). The rates of the inclusion of poly(iminoundecamethylene) by α-cyclodextrin are slow on the ^{1}H NMR time scale in D2O at pH<6. The inclusion formation between poly(iminotrimethylene-iminodecamethylene) and α-cyclodextrin took 170 h to reach equilibrium. Wenz

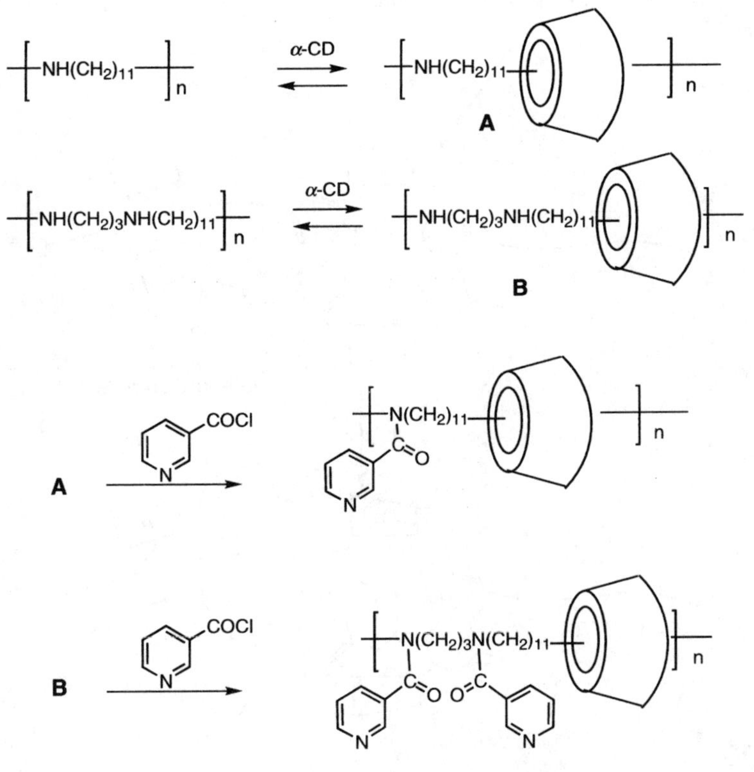

Scheme 17

explained that the cationic groups acted as barriers to the translational motion of the cyclodextrins [108, 109]. Use of heptakis(2,6-di-O-methyl)-β-cyclodextrin having a larger cavity than α-cyclodextrin resulted in virtually instantaneous equilibration upon mixing the components. Nicotinoyl group was chosen to provide blocking groups for the pseudorotaxanes to be converted to polyrotaxane by treatment with nicotinoyl chloride followed by dialysis. When the pseudopolyrotaxanes were treated with nicotinoyl chloride,

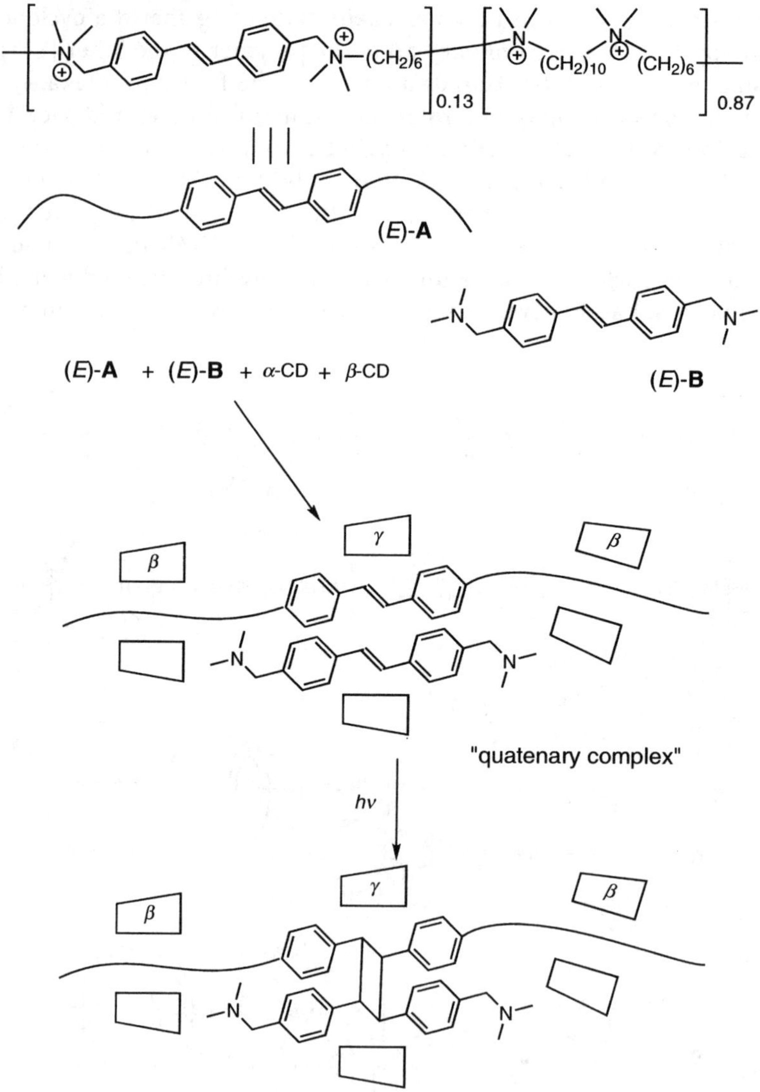

Scheme 18

the N-acylated polyrotaxanes were isolated after dialysis. The bisacylated polyrotaxane had a degree of polymerization of 55±5, with 37 threaded cyclodextrins and an average molecular weight of 55,000±5000, which was determined by the laser scattering method.

Wenz et al. have demonstrated that photoreactions are effective for the construction of polyrotaxanes (Scheme 18) [114]. In the knowledge that stilbenes undergo [2+2]photocycloaddition to yield cyclobutane derivatives by UV irradiation, they prepared a quaternary polymeric inclusion complex from β-cyclodextrin, γ-cyclodextrin, (E)-4,4-bis(dimethylaminomethyl)stilbene (**B**), and (E)-stilbene polymer (**A**). Upon irradiation at 312 nm, the (E)-stilbene units of **A** underwent [2+2]photocycloaddition with **B** by catalysis of γ-cyclodextrin to form the tetraphenylcyclobutane group, which acted as blocking group for β-cyclodextrin. Wenz et al. claimed that "this was the

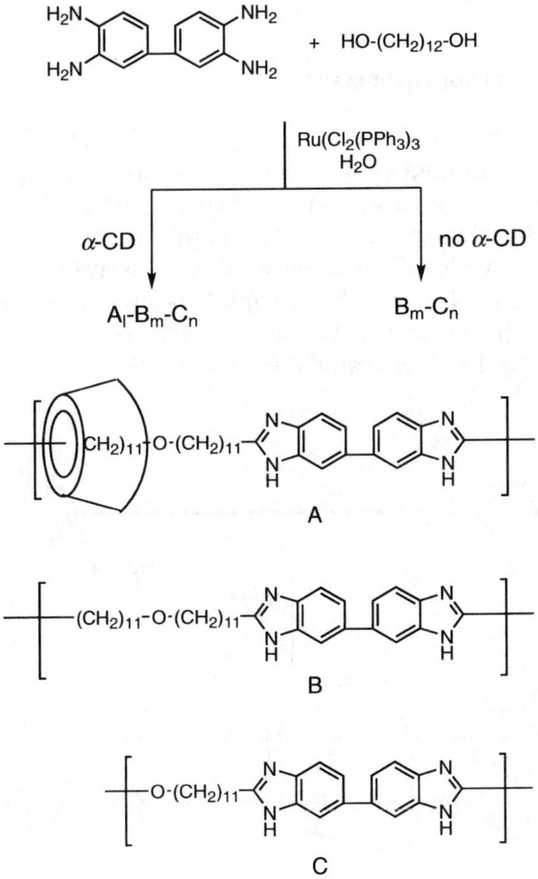

Scheme 19

first demonstration of the supramolecular catalysis of a polymer-analogous conversion".

Osakada et al. prepared a similar kind of polyrotaxane with cyclodextrin rings randomly clipped between "blocking groups" positioned along the polymer chain in 1996 [115]. Their synthetic strategy is quite similar to those employed by Ogata et al. and Maciejewski's group [78–83]. Osakada's group carried out ruthenium-catalyzed polycondensation of 3,3'-diaminobenzidine and 1,12-dodecanediol in the presence of α-cyclodextrin (Scheme 19). The resulting pseudopolyrotaxane (**Al-Bm-Cn**) contained structural units A and B in an 86:14 ratio based on analytical and spectroscopic studies. When 2 mol of the diol per 1 mol of the tetramine was employed, formation of structural unit C took place. Similar reaction in the absence of α-cyclodextrin gave the polymer (**Bm-Cn**) containing structural units B and C in an 80:20 ratio. **Al-Bm-Cn** has a higher T_g than that of **Bm-Cn**. This difference in T_g was attributed to the lower flexibility of the polymer chain of **Al-Bm-Cn** than that of **Bm-Cn**.

2.2.2.1
Doubly-stranded Pseudopolyrotaxane

Although γ-cyclodextrin does not complex with PEG virtually in aqueous solution as mentioned above, it formed inclusion complexes with some PEG derivatives such as bis(3,5-dinitrobenzoyl)-poly(ethylene glycol) and bis(2,4-dinitrophenylamino)-poly(ethylene glycol) to give crystalline compounds in high yields. Fluorescently labeled derivatives, bis(1-naphthylacetyl)-PEG (PEG-1N$_2$) and bis(2-naphthylacetyl)-PEG (PEG-2N$_2$) were prepared to investigate further the nature of the complexes formed between cyclodextrins and PEG derivatives (Scheme 20). PEG-1N$_2$ and PEG-2N$_2$

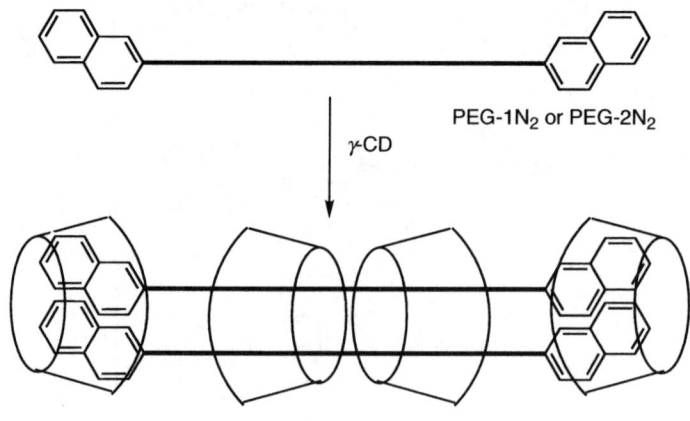

Scheme 20

complexed with γ-cyclodextrin in aqueous solution at a 4:1 ratio, which was confirmed by ^{1}H NMR. Upon irradiation at the naphthyl groups, PEG-2N$_2$-γ-CD complex showed a large contribution from excimers formed from two nearby naphthyl groups, but only a small contribution from monomeric naphthyls. On the other hand, it showed only monomer emission in the absence of γ-cyclodextrin. PEG-1N$_2$ gave the same results. On the basis of the results, Harada et al. concluded that these complexes are composed of double chains of PEGs threaded through the γ-cyclodextrins [116]. The authors have proposed that such complexes, which are reminiscent of DNA polymerases, may be used for the construction of catenanes, in which beads are threaded on double ring.

2.2.3
Application of Cyclodextrin Polyrotaxane

2.2.3.1
Molecular Abacus

A large number of applications of interlocked molecules to molecular machines have been proposed. However, several problems have remained, e.g., how to control and read molecular motions. Shigekawa and Komiyama et al. have nicely demonstrated that the selected α-cyclodextrin ring(s) of one of Harada s polyrotaxanes is(are) reversibly shuttled using scanning tunneling microscope (STM) technique [117]. The group showed three modes of shuttle manipulation, i.e., simple shuttling of a ring, pair-shuttling of two adjacent rings, and bending. This STM manipulation of the polyrotaxane was likened to calculation using a "molecular abacus" by Shigekawa et al.

2.2.3.2
Insulated Molecular Wires

The concept of use of polyrotaxane structures as insulated molecular wires was introduced by Anderson in 1996 [118, 119]. He demonstrated an approach to insulated molecular wires using a [3]rotaxane consisting of water-soluble macrocycles and a conjugated dumbbell. The presence of the cyclophane rings in the [3]rotaxane increased the fluorescence of the conjugated dumbbell by hindering quenching and by increasing the kinetic stability of the excited state.

Ito and co-workers prepared pseudopolyrotaxanes from polyaniline with emeraldine base and β-cyclodextrin, which were studied by frequency-domain electric birefringence (FEB) spectroscopy in a solution of N-methyl-2-pyrrolidone (NMP), scanning tunneling microscopy, etc. [120, 121]. The FEB results showed that polyaniline in the solution with cyclodextrin changed its conformation from coil to rod at low temperature below 275 K. Some rod-like structures were observed on a substrate by STM. Thus, the pseudorotaxane formation between polyaniline and β-cyclodextrin was confirmed. When polyaniline was oxidized by iodine solution, the solution color changed from blue to violet. However, little change in color was observed at 273 K, when the pseudorotaxane was treated with an iodine solution. Thus, the chemical oxidation of the polyaniline was prevented by the cyclodextrin rings, i.e., the cyclodextrin rings "insulated" the polyaniline wire. Shortly after, Yamaguchi et al. reported the synthesis of pseudorotaxanes from polypyridine and β-cyclodextrin [122]. The fluorescence efficiency of the polypyridine chain was reportedly increased to a great extent by the insulation effect of the β-cyclodextrin rings.

Anderson's group reported the synthesis of cyclodextrin-encapsulated conjugated polyrotaxanes that were the first examples of conjugated polyrotaxanes, i.e., conjugated polymers threaded through macrocyclic units with bulky stoppers (Scheme 21) [123]. The conjugated polyrotaxane was pre-

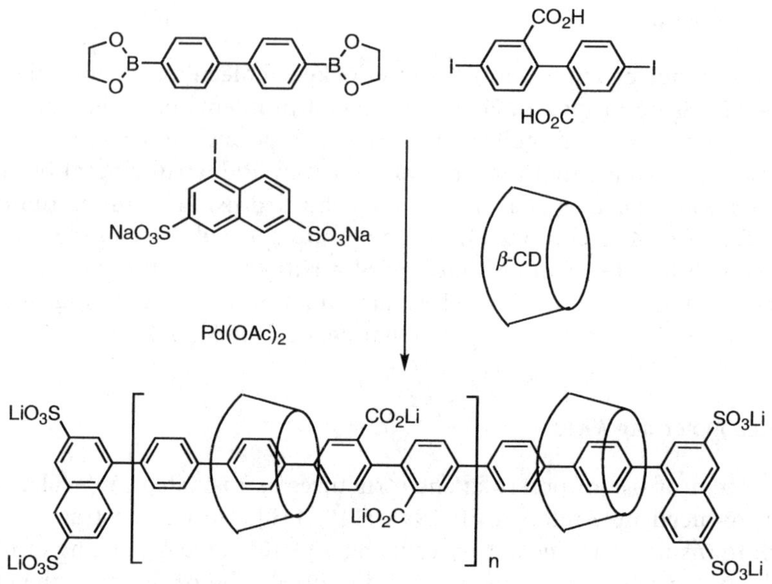

Scheme 21

pared by Suzuki coupling-based polymerization of 4,4'-biphenyldiboronic ester, 4,4'-diiodobiphenyl-2,2'-dicarboxylic acid, and disodium 1-iodonaphthalene-3,6-disulfonate in the presence of β-cyclodextrin. The polyrotaxane was isolated in 45% yield after precipitation and extensive dialysis. The average degree of polymerization was estimated to be 8±2 by ^{1}H NMR. The fluorescence efficiency of **23** was again increased by the insulation effect of the β-cyclodextrin rings. The Oxford group, in collaboration with English and German groups, prepared light-emitting diodes (LEDs) by spin-coating films of the polyrotaxane and analogous polyrotaxanes onto indium tin oxide (ITO) anodes and using thermally evaporated calcium or aluminum thin films as cathodes [124].

2.2.3.3
Antenna Molecules

Ueno and co-workers utilized cyclodextrin-based polyrotaxanes as platforms for antenna molecules [125–127]. For example, a series of α-cyclodextrin-based polyrotaxanes having two anthracene rings as end-capping groups were synthesized at several ratios of α-cyclodextrin units to naphthalene-modified α-cyclodextrin units (Scheme 22) [127]. Upon irradiation, photons captured by the naphthalene rings were transferred to the terminal anthracene rings and emitted as fluorescence. The antenna effect became more marked with increasing ratio of the naphthalene-modified α-cyclodextrin units in the polyrotaxanes. However, the energy transfer efficiency from naphthalene to anthracene decreased with an increase in the ratio of α-cyclodextrin.

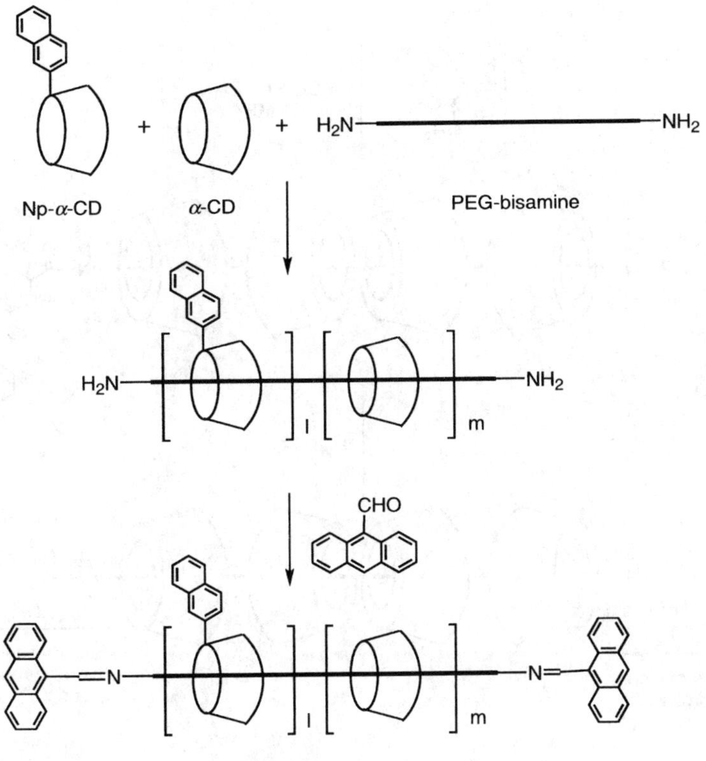

Scheme 22

2.2.3.4
Biodegradable Polyrotaxanes

Yui's group has investigated biodegradable polymers based on the cyclodextrin-based polyrotaxanes over the course of the last decade [128–139]. First, they prepared polyrotaxanes from α-cyclodextrin and PEG bisamine (Scheme 23) [128]. For example, L-phenylalanine was employed as an enzymatically hydrolyzable endcap. The in vitro degradation experiments using papain showed that α-cyclodextrin was indeed released only when the terminal peptide linkages were hydrolyzed. They have prepared various kinds of polyrotaxanes based on cyclodextrin and have demonstrated that these polyrotaxanes are effective as drug delivery systems.

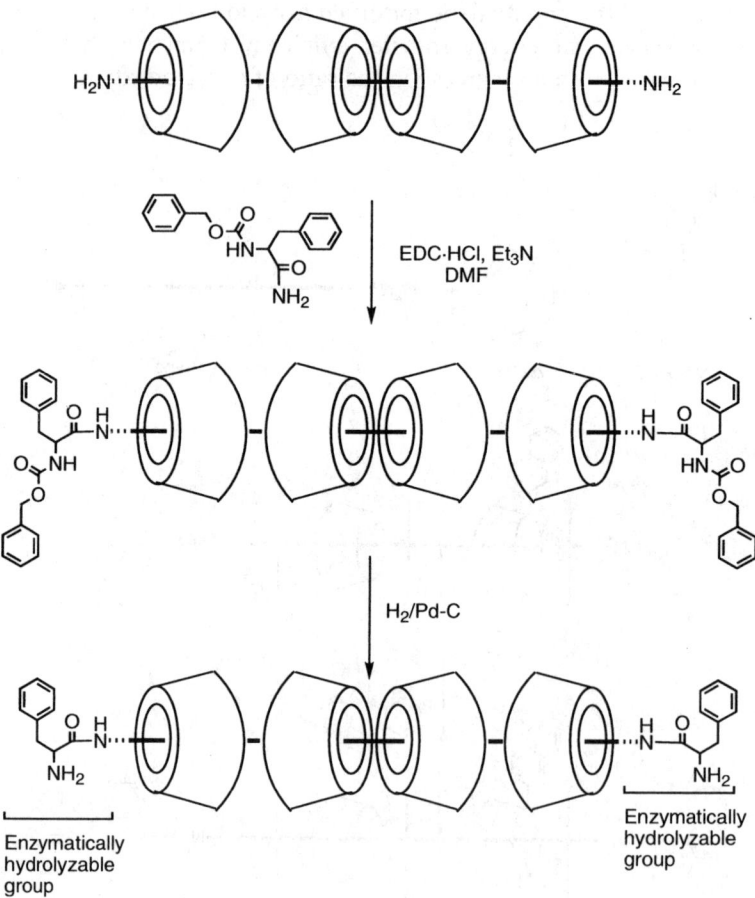

Scheme 23

2.2.3.5
Stimuli-Responsive Polyrotaxanes

Yui's group has also developed stimuli-responsive polyrotaxane materials based on cyclodextrins in the last decade [140–155]. They synthesized polyrotaxanes, in which many β-cyclodextrins are threaded onto a triblock copolymer PEG-*co*-PPG-*co*-PEG and trapped with fluorescein (Scheme 24)

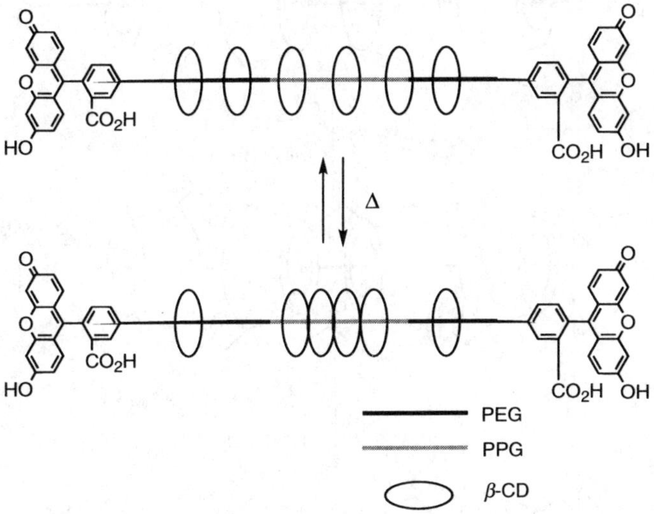

Scheme 24

[140]. The interaction of the β-cyclodextrins with the terminal fluorescein moieties in the polyrotaxane was significantly observed at low temperature [142]. However, the interaction of the β-cyclodextrins with the PPG segment was observed with increasing temperature. Yui concluded on the basis of these results that the majority of the β-cyclodextrins moved toward the PPG segment with an increase in temperature although some β-cyclodextrins might be located on the PEG segment.

Yui et al. found that α-cyclodextrin formed inclusion complexes with poly(ε-lysine) in aqueous solution [151]. The inclusion complexation is controllable by changing several parameters, such as the molar composition between poly(ε-lysine) and α-cyclodextrin, the pH and the ionic strength of aqueous media [152]. Yui s group is actively developing new polyrotaxane drugs utilizing multi-valent interactions [153–155].

2.3
Polyrotaxanes Having Cyclodextrin Nanotubes as The Wheel Components

Harada et al. constructed a "molecular tube" out of α-cyclodextrin-based polyrotaxane (Scheme 25) [156]. The polyrotaxane was synthesized from a

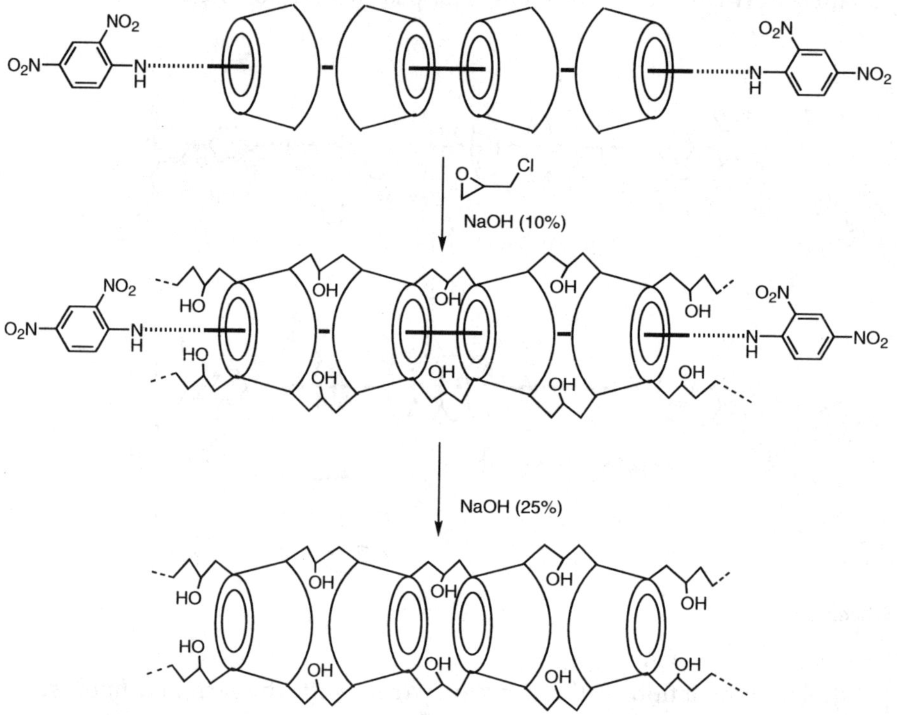

Scheme 25

monodisperse poly(ethylene glycol) bisamine (mass of 1450) used as a template. Epichlorohydrin was reacted with the hydroxyl groups of the threaded α-cyclodextrins in an alkaline solution to link up the cyclodextrin beads. The stopper groups of the resulting macromolecular [2]rotaxane were removed by treatment with a stronger alkaline solution. The molecular tube was isolated in 92% yield and its average molecular weight was estimated to be ca. 20,000 by GPC. The signals in the ^{1}H NMR spectrum of the molecular tube were broadened, probably due to the random substitution of the cyclodextrin. Harada clearly showed that the molecular tube could accommodate I_3^- ion in an aqueous solution of KI–I_2 efficiently, whereas α-cyclodextrin did not.

Yui's group analyzed the thermodynamics on the inclusion complexation between the α-cyclodextrin-based nanotube and sodium alkyl sulfonate [148]. They prepared a supramolecular hydrogel utilizing enthalpy-driven complexation between the molecular tube and an amphiphilic molecule [147]. They carried out the thermodynamic analysis of inclusion complexation between α-cyclodextrin-based molecular tube and poly(ethylene oxide)-*block*-poly(tetrahydrofuran)-*block*-poly(ethylene oxide) triblock copolymer in terms of isothermal titration calorimetry [157]. Furthermore, they incorporated the tube into gels that could recognize the length of alkyl chain [158].

Liu et al. synthesized a bis(molecular tube) from organoselenium-bridged β-cyclodextrin–Pt(IV) complex and propylene glycol, utilizing Harada's protocol and they characterized the tube by TEM and AFM [159].

Ito and co-workers investigated experimentally the inclusion behavior between the molecular tubes and poly(ethylene glycol)s in aqueous solutions. It was found that the amount of polymer chains included into nanotubes increased with increasing polymer length. The results were in good accordance with a theory based on the Flory-Huggins lattice model [160, 161]. The temperature dependence of inclusion-dissociation behavior between molecular tubes and linear polymers was also disclosed by the same group [162]. The association ratio of the molecular tube and linear polymers decreased with an increase in temperature, as expected theoretically. The molecular nanotube complexed with a star polymer to form a self-assembling dendritic supramolecular architecture, as observed by STM [163].

An insulated molecular wire was prepared by Ito's group from the α-cyclodextrin-based molecular tube and polyaniline with emeraldine base [164]. They found that the insulated wire was moved and cut off by manipulating with the cantilever tip of AFM [165].

2.4
Polyrotaxanes Bearing Miscellaneous Ring Systems as the Wheel Components

2.4.1
Cucurbituril

Cucurbituril is a cyclic bisurea that is characterized by both hydrophobic cavity and its entrances fringed by highly Lewis basic carbonyl groups [166]. While cucurbituril is insoluble in organic solvents, it can be dissolved in strongly acidic water. α,ω-Alkanediamines with 4–6 methylene chains are incorporated into the cavity of cucurbituril in the acidic water by hydrophobic interaction and hydrogen-bonding between ammonium salt and carbonyl groups. Thus, various bifunctional pseudorotaxanes have been prepared from cucurbituril and functional diamines, and their polymerization has been examined, although limited polymerization methods were available because the polymerization must be carried out in strongly acidic aqueous media. The characterization of the polyrotaxane obtained from cucurbituril is rather difficult due to the low solubility of the polymer.

Coordination polymerization of pseudorotaxane having pyridine ligands at the termini has been extensively studied. The structure of the polymers depends on the structure of ligand, metal and counterion used, and the spacer of the pseudorotaxane. In a typical example, Kim et al. reported that the coordination polymerization of pseudorotaxane consisting of bisammonium salt, cucurbituril, and terminal 4-pyridyl groups with copper(II) nitrate afforded linear polyrotaxane [167]. Coordination polymerization of 3-pyridyl pseudorotaxane with silver nitrate also afforded linear polyrotaxane [168]. This polymer has a helical structure in the crystal state. However, when square-planar transition metal complexes with two *cis*- labile ligands were used, macrocyclic complex with many cucurbituril wheel were obtained. These catenanes were denoted by molecular necklaces. In a typical example, the complexation of square-planar Pt complex with *cis*- nitro groups and pseudorotaxane with 4-pyridyl termini was successfully carried out to prepare various molecular necklaces [169]. In the same manner, Kim et al. succeeded in synthesizing various two- and three-dimensional polyrotaxane network systems [170–172]. Typical polyrotaxanes and polycatenanes thus obtained are listed in Table 1 (see also Table 3).

The complex of cucurbituril and bisammonium salt itself can be used as a monomer. A complex prepared from 1,6-hexanediamine and cucurbituril was reacted with acid chloride in the presence of base (Scheme 26) [173].

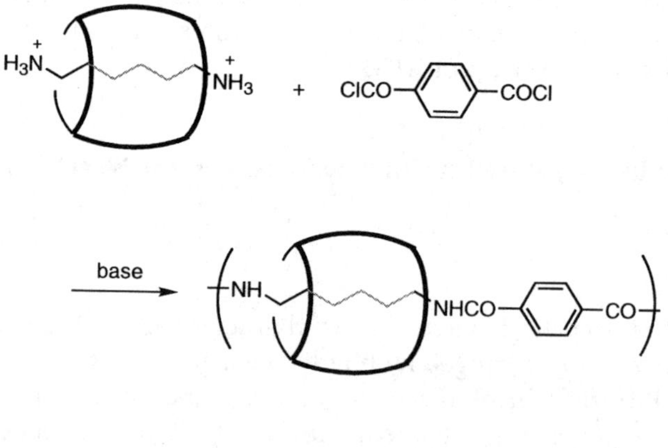

Scheme 26

Although Meschke et al. claimed that the pseudopolyrotaxane was obtained, the existence of pseudopolyrotaxane structure is still unclear because of the lack of the attractive interaction between polymer and cucurbituril.

Pseudopolyrotaxane was synthesized via threading of poly(4,4'-bipyridinium salt) into cucurbituril in water (Scheme 27) [174]. In the complex,

Table 1 Oligorotaxanes based on cucurbituril

Oligorotaxane	reference
	167
	169
	170

Table 1 (continued)

Oligorotaxane	reference
	171

cucurbituril beads are localized at the middle of the methylene spacer by hydrophobic and charge-dipole interactions.

1,3-Dipolar cycloaddition reaction of propargyl ammonium salt and 2-azidoethylammonium salt is greatly enhanced in the presence of cucurbituril because the hydrophobic cavity of cucurbituril is an excellent reaction field for the cyclocondensation [175]. Mock et al. obtained [2]rotaxane consisting

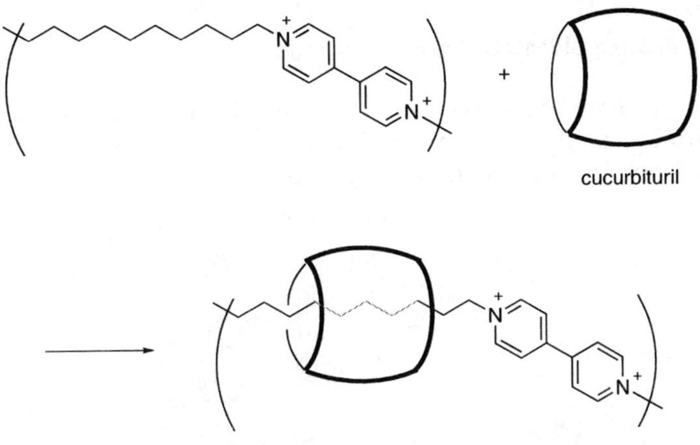

Scheme 27

of triazole bisammonium salt and cucurbituril, when monomers with bulky substituents were used [176]. This method was extended by Steinke et al. to the polyrotaxane synthesis as the bifunctional acetylene and azido compounds were used (Scheme 28) [177]. Meanwhile, linear polyamines such as poly(hexamethyleneimine) can be prepared by the reduction of nylon-66 forms pseudopolyrotaxane complex with cucurbituril in acidic aqueous solution [178].

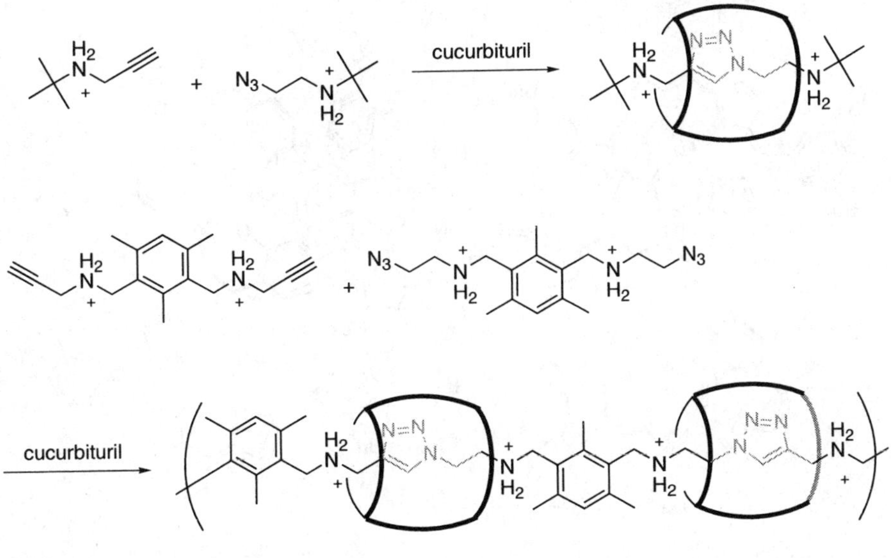

Scheme 28

2.4.2
Macrocycle Bearing Bidentate Nitrogen Ligand

Sauvage et al. have synthesized a variety of oligorotaxanes utilizing 1,10-phenanthroline-copper(I) complex [179]. Stable bis(1,10-phenanthroline) copper(I) complex has a tetrahedral coordination structure, and two

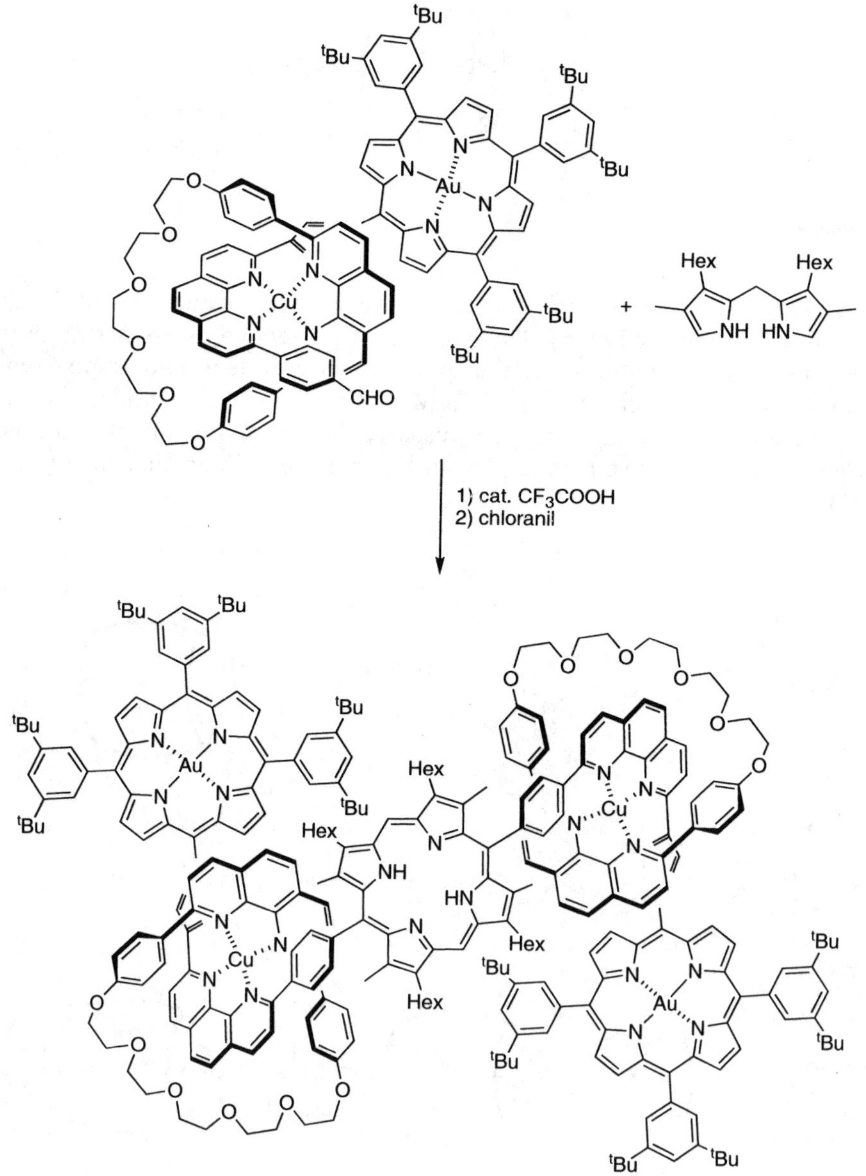

Scheme 29

phenanthroline ligands are perpendicular to each other. When 2,9-disubstituted 1,10-phenanthroline is used for the preparation of copper(I) complex, the cyclization at the substituents affords interlocked macrocycles. Based on this strategy, it has been proven that 2,9-disubstituted 1,10-phenanthroline-copper(I) complex is an excellent scaffold for the construction of interlocked compounds. A [3]rotaxane was obtained when a monoprotected pseudorotaxane consisting of formyl group was subjected to the porphyrin formation reaction [180] (Scheme 29). This technique can be applied to the synthesis of higher order rotaxanes. By the use of bisaldehyde instead of monoaldehyde and/or oligomeric phenanthroline as the axle, oligorotaxanes with polynuclear complex structure were prepared via the porphyrin formation reaction with bispyrrole [181, 182]. The demetallation of the products, the oligorotaxane-copper(I) complexes, with sodium cyanide yields free oligorotaxanes.

Swager et al. have shown that the electrochemical oxidation of metallorotaxanes consisting of bithienyl-substituted 2,2'-bipyridine and macrocyclic phenanthroline ligand leads to a film of polyrotaxane complex deposited on the electrode [183] (Scheme 30). These polyrotaxanes showed conductivity

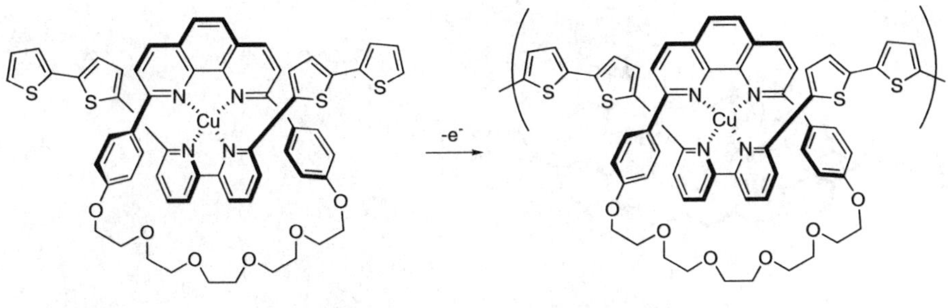

Scheme 30

due to the fully conjugated backbone. Treatment of the film with a chelating agent such as ethylenediamine resulted in the demetallation from the metallorotaxane unit. Retreatment with metal ions reversibly produced the original polyrotaxane. Similar result was confirmed by Sauvage et al. when bithienyl-substituted phenanthroline ligand was used [184, 185]. Further, Sauvage et al. prepared a ladder type of polyrotaxane from metallorotaxane monomer bearing two independently electropolymerizable thiophene groups that located both in the threading unit and the macrocyclic host [186].

2.4.3
Cyclophane

Cyclophanes are the class of synthetic receptors that have a hydrophobic cavity. Water-soluble cyclophane can make organic compounds soluble in water via the formation of inclusion complex. Anderson et al. synthesized rod-like organic compounds in water in the presence of water-soluble cyclophanes, and obtained corresponding rotaxanes. When oxidative coupling of an acetylenic derivative in the presence of a cyclophane, an oligorotaxane having an axle with a conjugated structure was dominantly obtained rather than non-complexed diacetylene.[187,188] (Scheme 31). This rotaxane is a prototype of "insulated molecular wire" that is expected as an electric wire in a molecular device.

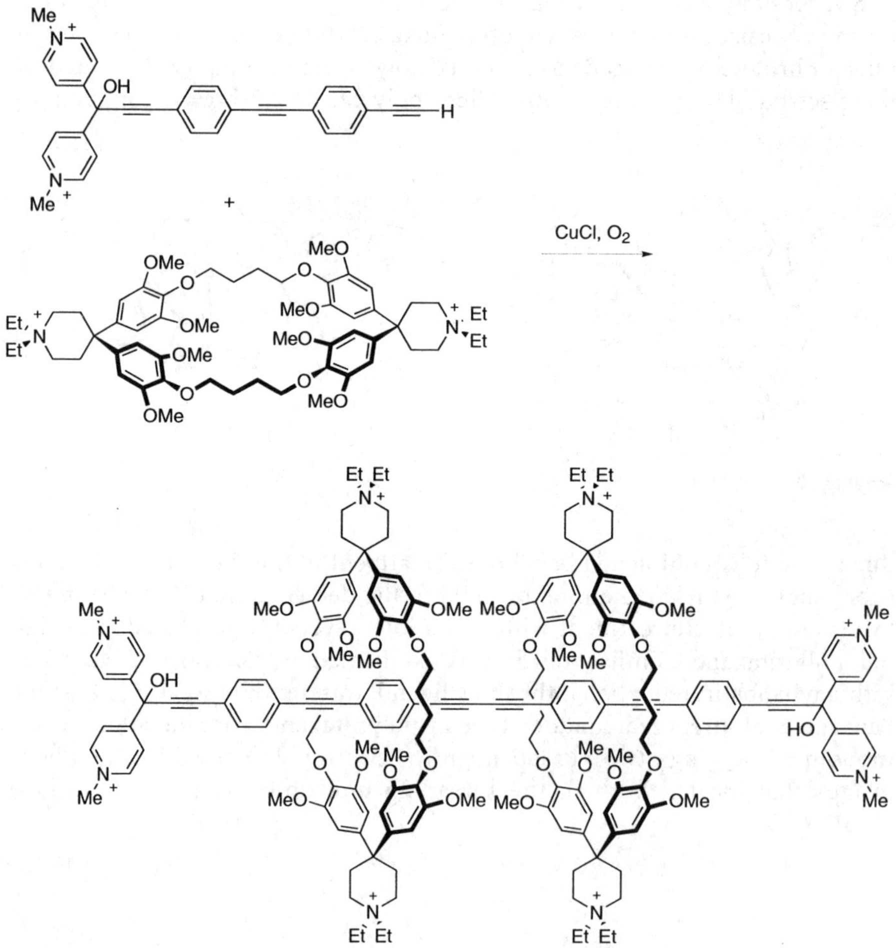

Scheme 31

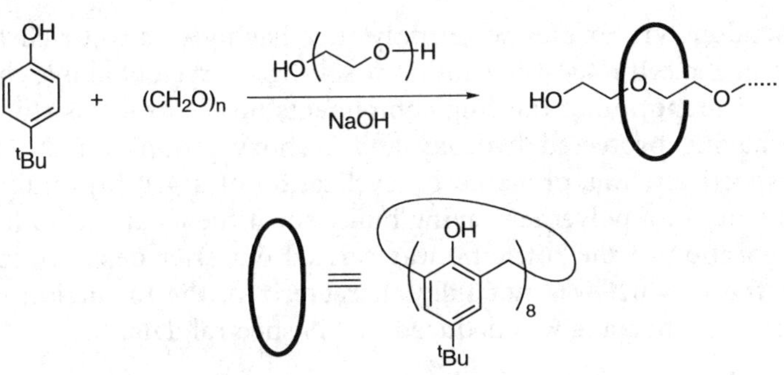

Scheme 32

Calixarene is one of the most easily accessible cyclophanes. Yamagishi et al. carried out the condensation of p-tert-butylphenol with paraformaldehyde in the presence of poly(ethylene glycol) (PEG), pseudopolyrotaxane consisting of PEG axle and calixarene wheel was obtained [189] (Scheme 32). The wheel components removed from pseudopolyrotaxane were found to be mainly calix[8]arene which was accompanied with a small amount of calix[4]arene. It was discussed that calix[8]arene threaded PEG most efficiently among a mixture of calixarenes.

2.4.4
Cyclophane Bearing Bis(4,4′-bipyridinium) Moiety

4,4′-Bipyridinium salt-based cyclophane forms stable inclusion complex with dialkoxy aromatic compounds via CT interaction. A few groups ob-

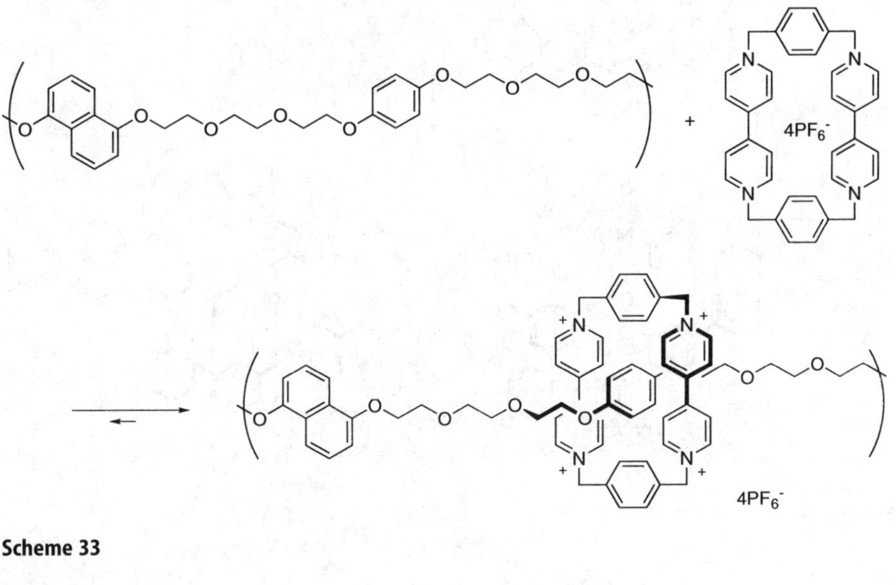

Scheme 33

tained pseudopolyrotaxanes when polyether having electron-rich aromatic rings was mixed with 4,4′-bipyridinium salt-based cyclophane (Scheme 33) [190, 191]. The motion of the ring components on the axle was studied. The rotaxane having protected hydroxy and carboxy groups on the axle and wheel, respectively, was prepared by cyclization of a 4,4′-bipyridinium salt in the presence of a polyether having bulky substituents at the both termini. Polycondensation of the rotaxane was carried out after deprotection [192]. Although the product was not fully characterized, the formation of daisy-chain-type oligorotaxane was deduced by MS spectral data.

2.4.5
Amide-Type Macrocycle

The hydrogen-bonding interaction between secondary amides is one of the most valuable interactions to be applied to the construction of interlocked compounds. There are two types of methods to construct interlocked com-

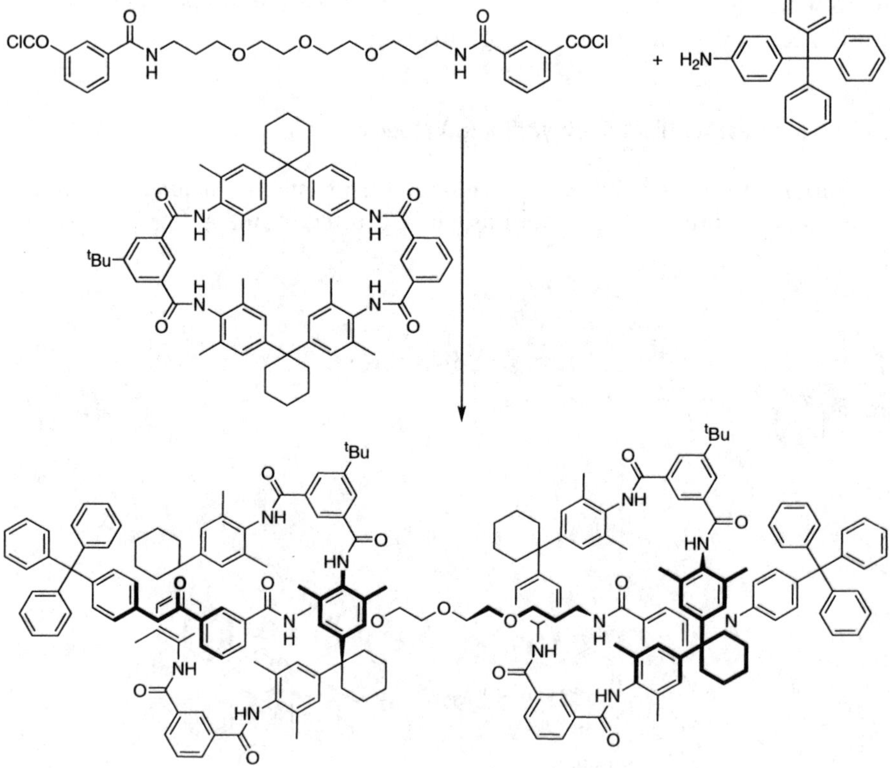

Scheme 34

pounds using the hydrogen-bonding interaction between secondary amides. One is macrocyclization to secondary amide-based macrocycles in the presence of rod-like secondary amide (clipping). The other is preparation of rod-like secondary amide in the presence of macrocyclic lactam (threading-end-capping). Vögtle et al. prepared oligorotaxanes by the threading-end-capping method, although the yield was very low (2% in the case of [3]rotaxane) (Scheme 34) [193].

When the cyclic component of the [3]rotaxane has low symmetry, planar chirality based on the direction of the rotation occurs (Scheme 35). Vuogtle

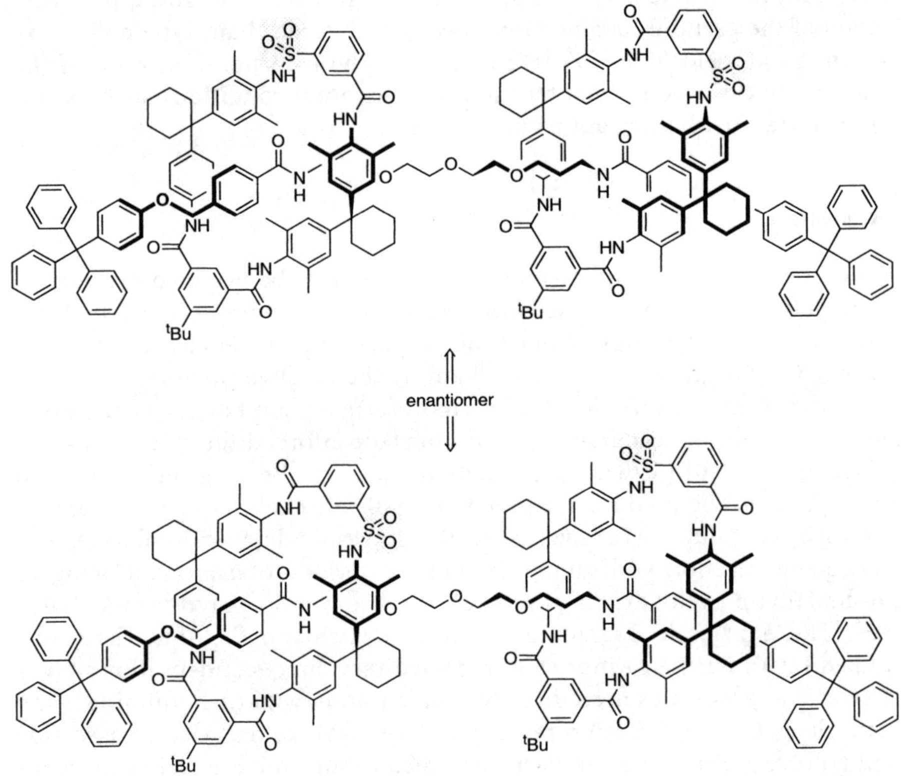

Scheme 35

et al. reported the isolation of both enantiomers and meso-type [3]rotaxanes by chiral HPLC technique [194, 195]. Since the selective alkylation of the sulfonamide group was achieved in the presence of carboxamide, various rotaxane assemblies were successfully prepared by the alkylation of secondary amide-based [2]rotaxane bearing sulfonamide moiety [196, 197].

3
Side Chain-Type Polyrotaxanes

3.1
Crown Ethers

Although very limited number of reactions are used for the end-capping of pseudorotaxane consisting of crown ether and secondary ammonium salt, polymerization of pseudorotaxane monomer is one of the most versatile methods to prepare polyrotaxane. Namely, Takata et al. demonstrated that a secondary ammonium salt having methacrylate group at the terminus formed the stable inclusion complex with DB24C8, and the radical polymerization of the pseudorotaxane monomer gave the side chain-type polyrotaxanes in good yield [198, 199] (Scheme 36). The incorporation ratio of the rotaxane unit was controlled by the polymerization conditions such as solvent and the copolymerization with styrene.

3.2
Cyclodextrin

One of the most promising methods to prepare side chain-type polyrotaxane is the polymer reaction of pseudorotaxane with a certain polymer that has reactive functional groups. Ritter et al. prepared a polymethacrylate bearing carboxylic acid group on the side chain as the reactive polymer. After the activation of the carboxylic acid by chloroformate, aminolysis of the polymer with an amine-terminated pseudorotaxane afforded the side chain-type polyrotaxane [200] (Scheme 37). Poly(ether sulfone) bearing carboxylic acid moiety can also be used for the preparation of side chain-type polyrotaxane. The carboxy group was converted to the corresponding acid chloride, and subsequent reaction with amine-terminated pseudorotaxane consisting of 2,6-dimethyl-β-cyclodextrin gave the side chain-type polyrotaxane [201–203]. The enzymatic degradation of cyclodextrin on the side chain was studied [204]. Further, Ritter et al. reported that direct condensation of carboxylic acid group was used directly, condensation with carbodiimide being available as the condensation reagent. Poly(methyl methacrylate)-based side chain polyrotaxane was similarly obtained from amine-terminated pseudorotaxane by the use of water-soluble carbodiimide [205].

Poly(benzimidazole) can be used as the reactive polymer for the preparation of side chain-type polyrotaxane via the alkylation on nitrogen. Osakada et al. showed that deprotonation of poly(benzimidazole) followed by N-alkylation with alkyl bromide-terminated pseudorotaxane consisting of 2,3,6-trimethyl-β-cyclodextrin gave a novel side chain-type polyrotaxane (Scheme 38) [206]. Poly(benzimidazole) having a long ω-hydroxyalkyl group as the side chain can form an inclusion complex with 2,3,6-trimethyl-β-cy-

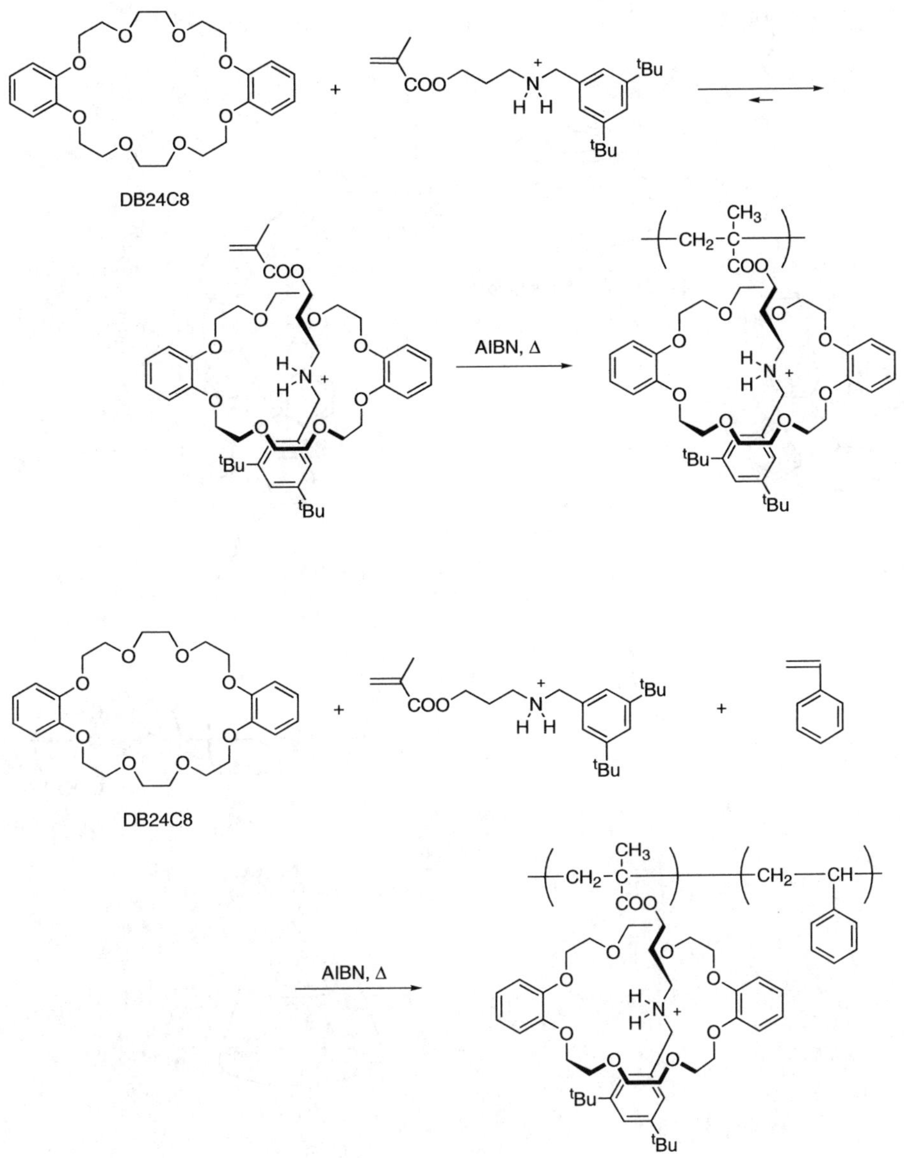

Scheme 36

clodextrin, which is eventually end-capped by the deprotonation of the terminal hydroxy group followed by complexation with sodium-crown ether complex as the bulky substituent [207].

A thiol with long alkyl chain having bulky substituent at the terminus was dissolved in water in the presence of cyclodextrin via the formation of

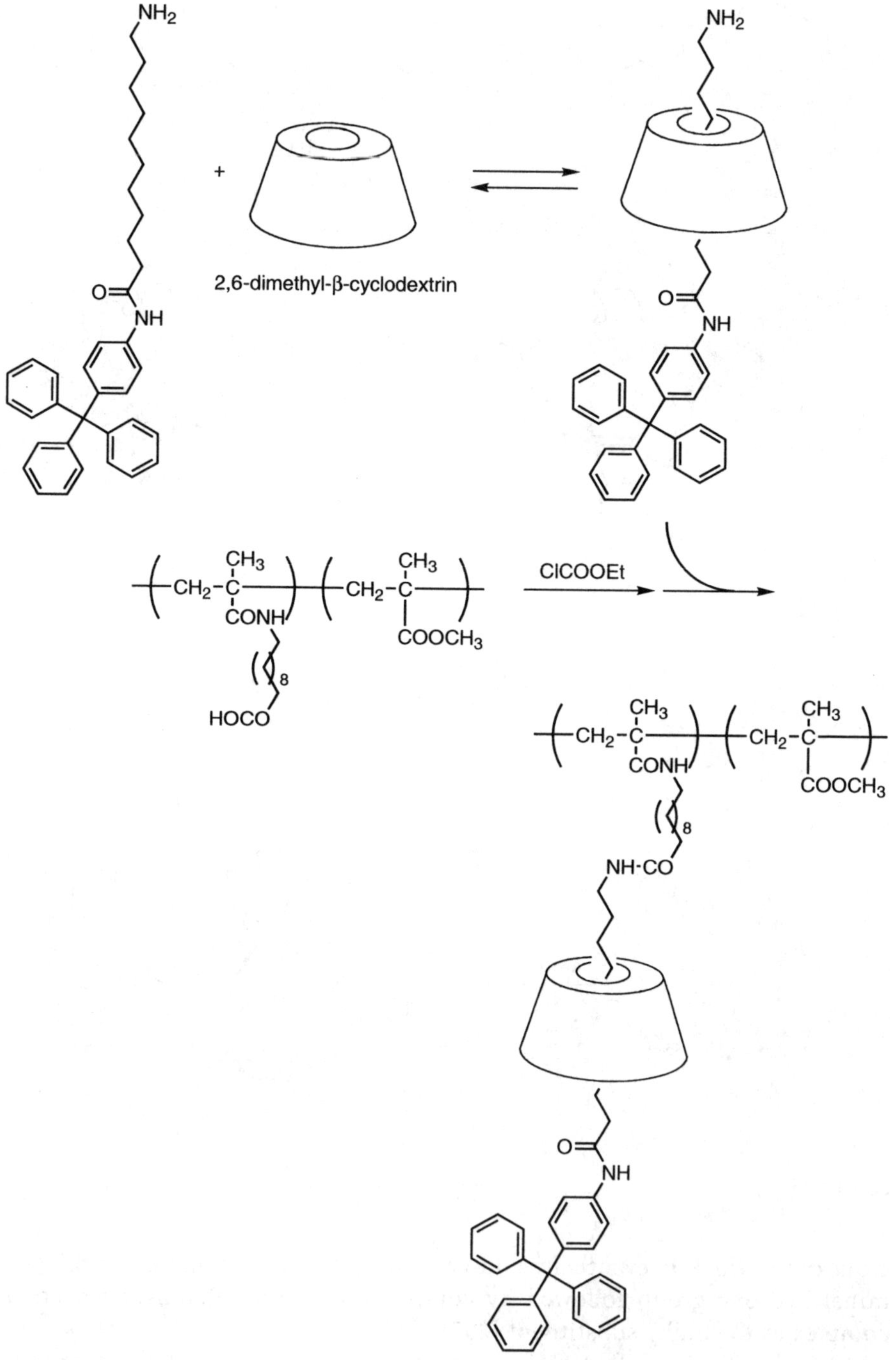
NH2
NH2
+
2,6-dimethyl-β-cyclodextrin
O
NH
O
NH
CH3
CH3
CH2-C
CH2-C
CONH
COOCH3
8
HOCO
ClCOOEt
CH3
CH3
CH2-C
CH2-C
CONH
COOCH3
8
NH-CO
O
NH

Scheme 37

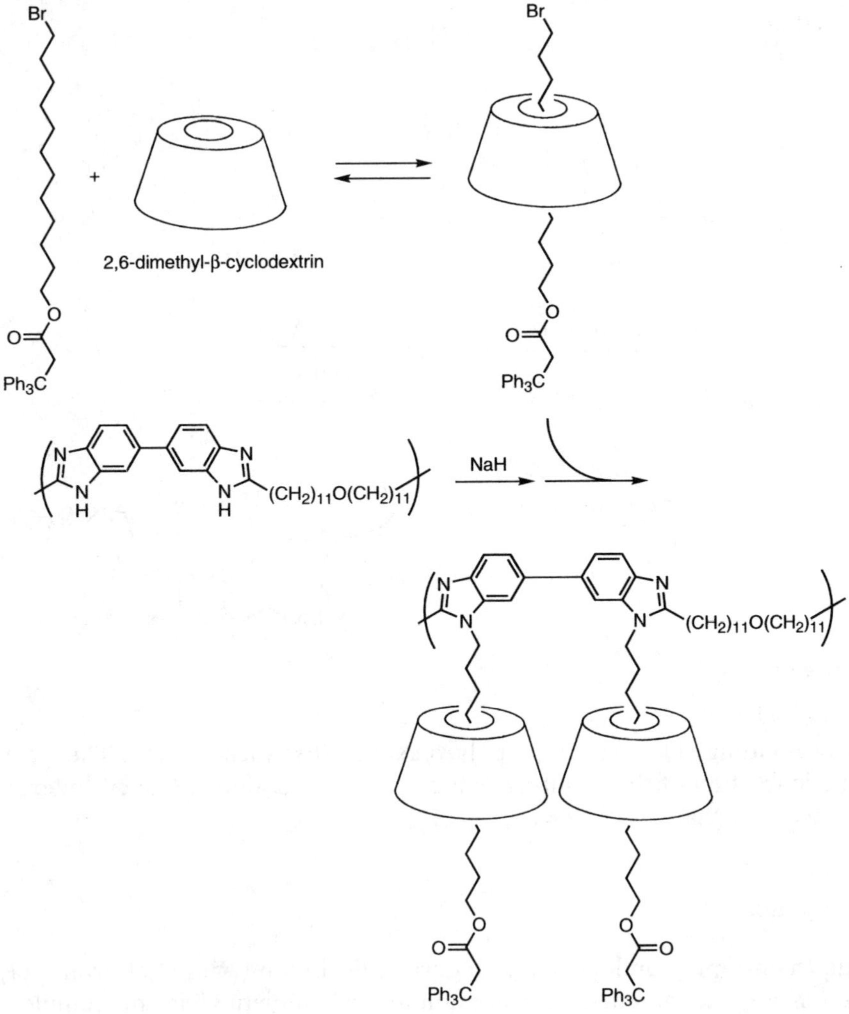

Scheme 38

pseudorotaxane. Kaifer et al. reported that the pseudorotaxane was bound on the surface of gold colloid to give rotaxane-functionalized gold particle [208]. When the terminal bulky group was the ferrocene group, the particle became electrochemically active.

Polymerization of pseudorotaxane is also a very effective method to prepare side chain-type polyrotaxanes in the case of cyclodextrin wheel. Ritter et al. claimed that the radical polymerization of pseudorotaxane consisting of 2,6-dimethyl-β-cyclodextrin and alkyl chain with the acrylamide group and the big polycycloalkane group (cholesterol) at the both termini gave the

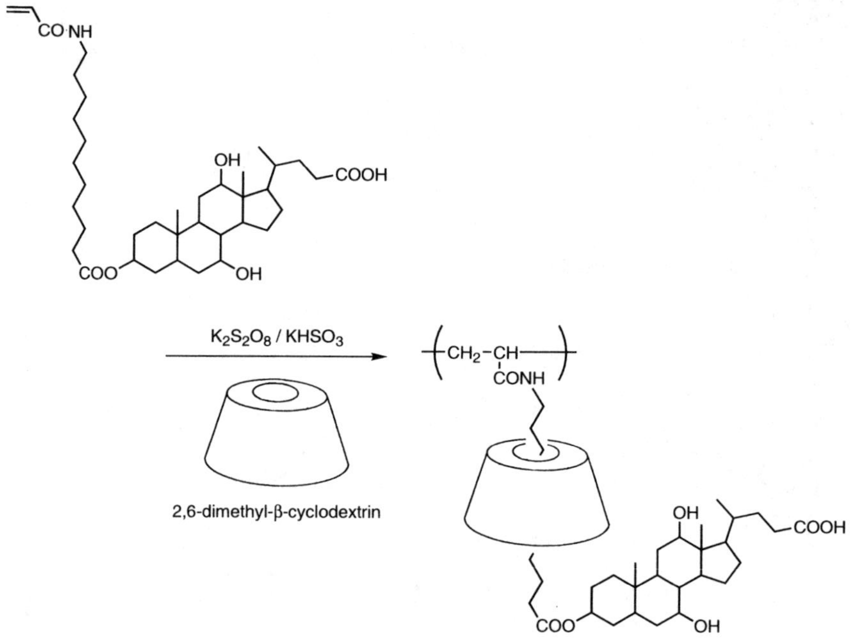

Scheme 39

corresponding side chain-type polyrotaxane [209] (Scheme 39). The water-soluble fraction of the resulting polymer contained almost one cyclodextrin on each side chain.

3.3
Cucurbituril

Side chain-type pseudopolyrotaxane was prepared by Kim et al. from polymers having tetramethylenebisammonium salt moieties via the complexa-

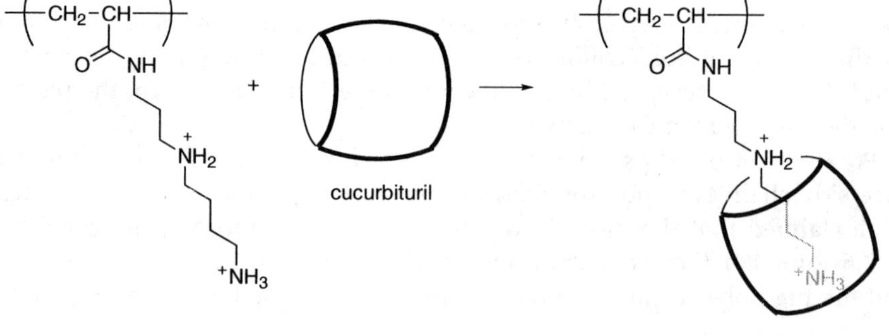

Scheme 40

tion with cucurbituril [210] (Scheme 40). End-capping of the resulting pseudopolyrotaxane has not been examined.

4
Synthesis and Application of Polymers Bearing Interlocked Structures Used for Monomer Linkage

4.1
"Topological" Polyrotaxanes

4.1.1
Poly[2]rotaxane

Stoddart and co-workers reported the attempt to construct a supramolecular poly[2]rotaxane (pseudopolyrotaxane) in 1998 [211]. They synthesized a self-complementary (plerotopic) monomer consisting of a 24-crown-8-ether moiety and a secondary ammonium salt group (Scheme 41). Upon recrystal-

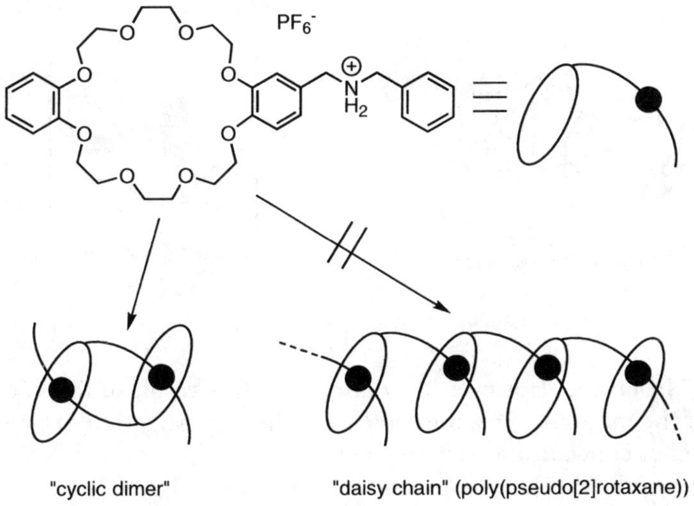

Scheme 41

lization, cyclic dimer was obtained instead of supramolecular poly[2]rotaxane. Further investigation revealed that the formation of the cyclic dimer dominated over that of other oligomer species even in solution. They also attempted to synthesize supramolecular [2]rotaxanes by taking advantage of a crown ether/bipyridinium salt system but this gave a similar result [212]. In this case, formation of at least a pentamer was confirmed by liquid secondary ion mass spectroscopy (LSI-MS) technique.

Their challenges went on. [2]Rotaxane, which has a formyl group attached to its ring component and a triphenylphosphonium salt moiety as a "surrogate" end-capping group, was designed with the aim of synthesizing poly[2]rotaxane through Wittig reaction-based polycondensation (Scheme 42) [213]. This attempt also resulted in predominant formation of

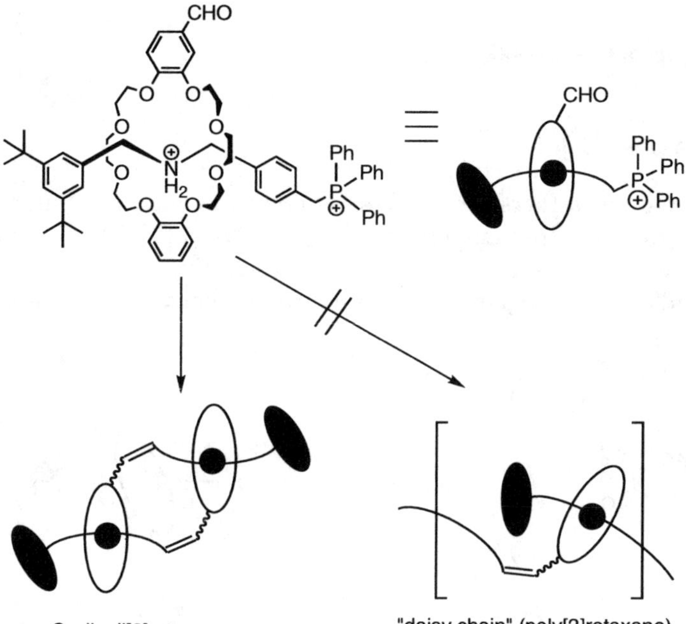

Cyclic di[2]rotaxane   "daisy chain" (poly[2]rotaxane)

Scheme 42

cyclic di[2]rotaxane. However, they attributed this result to the low concentration of the monomer (ca. 5 mmol/l). Further investigation of polymerization at higher concentrations is awaited.

Discrete cyclic di[2]rotaxanes and tri[2]rotaxanes have also been synthetic targets (Scheme 43). Kaneda's group reported the synthesis of cyclic di[2]rotaxane from a permethyl-α-cyclodextrin derivative having an azobenzene unit attached to the smaller rim [214]. They named it a "Janus" rotaxane. Easton's group also independently synthesized a cyclic di[2]rotaxane from an α-cyclodextrin derivative [215]. Harada and co-workers reported cyclic tri[2]rotaxane (bottom, Scheme 43) from an α-cyclodextrin having a cinnamoyl group in 38% yield [216]. Sauvage and co-workers built a cyclic di[2]rotaxane by their synthetic protocol using transition metal template (Scheme 44) [217]. The axle unit of each component has two different lig-

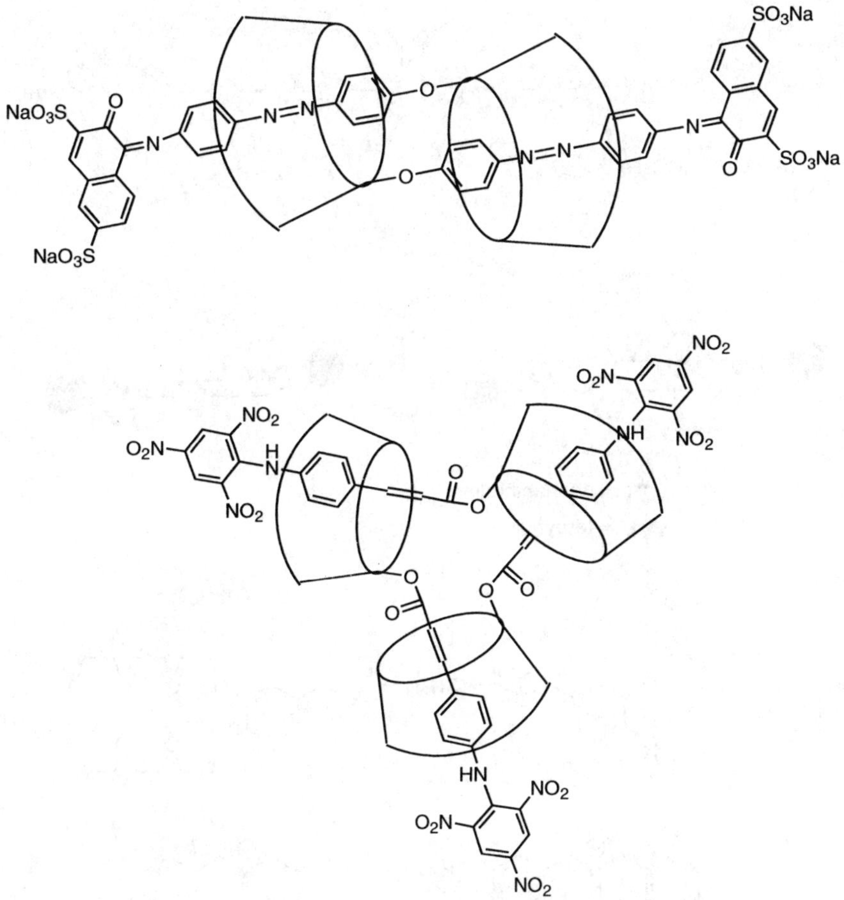

Scheme 43

ands, i.e., phenanthroline and terpyridine. The di[2]rotaxane can be stretched and contracted at will by changing metal cations. Upon contraction, the length of the di[2]rotaxane reduces from 83 Å to 65 Å. This change is roughly the same as that of the natural muscle and they called it "molecular muscle".

4.1.2
Poly[3]rotaxane

Since pseudorotaxane formation is an equilibrium process, it can essentially be controlled by the conditions such as concentration, temperature, solvent polarity, etc. In early 1999, Gibson's group demonstrated how it could be realized in the system of pseudopolyrotaxane (Scheme 45) [218]. They showed

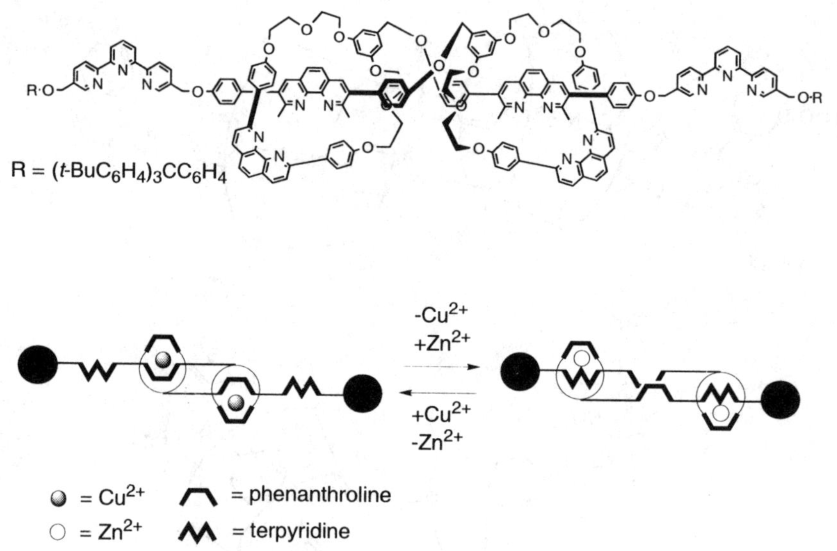

R = (*t*-BuC₆H₄)₃CC₆H₄

○ = Cu²⁺ ⋀ = phenanthroline
○ = Zn²⁺ ⋒ = terpyridine

Scheme 44

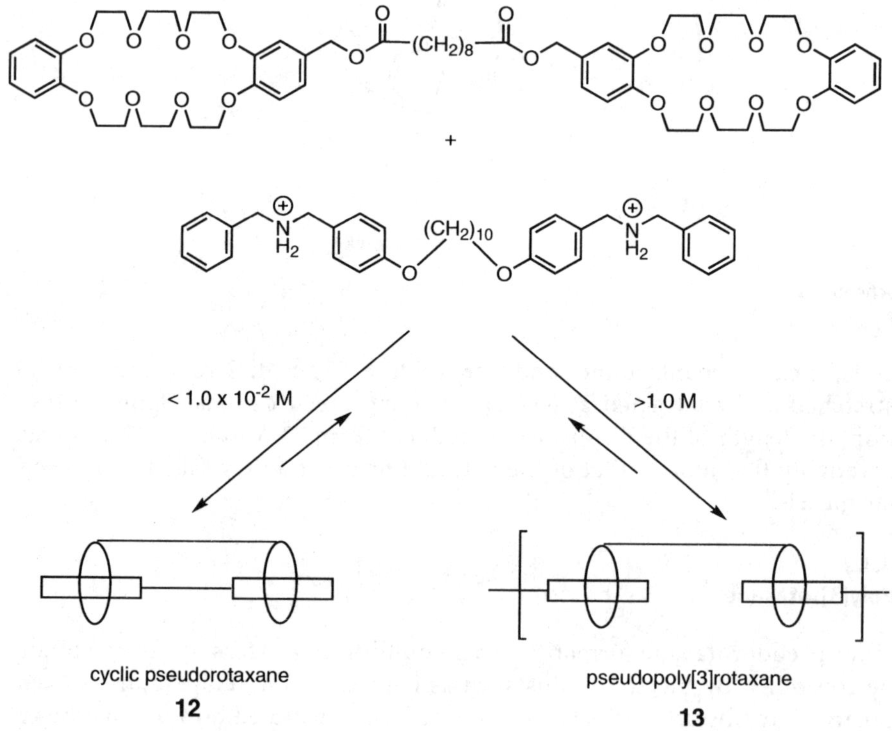

< 1.0 x 10⁻² M   >1.0 M

cyclic pseudorotaxane
12

pseudopoly[3]rotaxane
13

Scheme 45

that bis(24-crown-8 ether) and bis(secondary ammonium salt) formed cyclic dimer predominantly at low concentrations below 0.010 mol/l and resulted in the formation of linear pseudopoly[3]rotaxane (**13**) at high concentrations above 1.0 mol/l. They prepared flexible, creasible, amorphous, and transparent films by casting from 1:1 stoichiometric mixture of the two monomers. Thus, the formation of cyclic dimers can be suppressed by increasing feed concentrations. This encouraged chemists who wanted to synthesize poly[2]- or poly[3]rotaxanes.

Two years later, Takata and his co-workers reported a new protocol for the synthesis of polypseudorotaxanes (Scheme 46) [73]. They employed ho-

Scheme 46

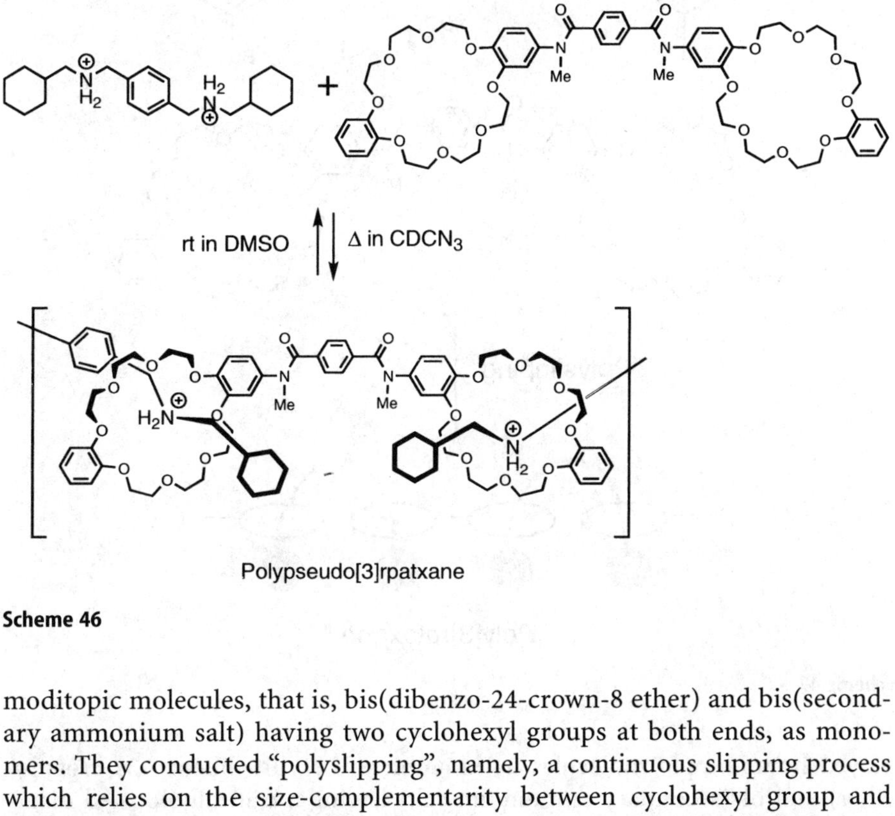

moditopic molecules, that is, bis(dibenzo-24-crown-8 ether) and bis(secondary ammonium salt) having two cyclohexyl groups at both ends, as monomers. They conducted "polyslipping", namely, a continuous slipping process which relies on the size-complementarity between cyclohexyl group and dibenzo-24-crown-8 ether. The polyrotaxane-like supramolecule (polypseudo[3]rotaxane) thus obtained behaves like a polyrotaxane under appropriate conditions, i.e., the polyrotaxane is kinetically stable at ambient temperature in less polar solvent. Upon immersion in a polar solvent such as DMSO at room temperature, however, the polyrotaxane was quantitatively disassembled into the monomers within 24 h. By definition, the polypseudo[3]rotaxane is not a polyrotaxane but a polypseudorotaxane, since the end-capping group is not bulky enough to prevent deslipping *under all circumstances*. Anyhow, the polypseudo[3]rotaxane has a more rotaxane-like character than other polypseudorotaxanes reported to date and can be categorized in a new class of polymeric compounds.

The synthesis of first poly[3]rotaxane, a real tour de force, was finally accomplished by Takata's group in early 2003 (Scheme 47) [66]. They em-

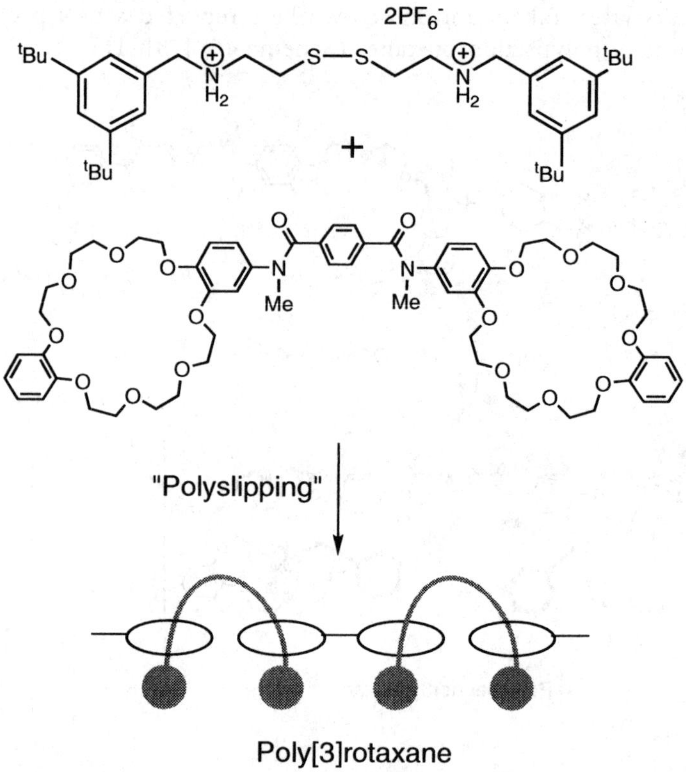

Scheme 47

ployed homoditopic molecules as monomers, i.e., bis(dibenzo-24-crown-8 ether) and bis(secondary ammonium salt) having a centrally-located disulfide linkage and two di-*tert*-butylphenyl groups at both termini. In this case, the di-*tert*-butylphenyl group is far too bulky to pass through the cavity of the dibenzo-24-crown-8 macroring. In fact, no reaction was observed even when a solution of a mixture of the two bifunctional monomers in CD3CN was heated up to 100 °C. Instead, they took advantage of the reversible nature of disulfide linkage. Upon addition of a catalytic amount of benzenethiol to a solution of an equimolar mixture of them in a mixed solvent system of CDCl3 and CD3CN, the thiol-disulfide interchange reaction started and the corresponding poly[3]rotaxane was formed through non-covalent step growth polymerization. Poly[3]rotaxane was purified by HPLC and its degree of polymerization (DP) was estimated to be 29, which corresponds a M_n of 28,000.

4.1.3
Crosslinked Polyrotaxane

Use of mechanical bonds as cross-linking points was first reported by Gibson's group [53, 219–222]. They obtained gel products in polycondensation with a 32-membered macrocyclic diol and with a 32-membered macrocyclic dicarboxylic acid [219–221]. The crosslinking through threading occurred to form three-dimensional polyrotaxane networks in the cases of polyamides [219], polyesters [220], and polyurethanes [221] (Scheme 48). According to their first report [219] direct polycondensation of the 32-membered crown ether and an aromatic diamine gave crosslinked amide polymer, which was insoluble in any solvents (Scheme 48). This remarkably low solubility was

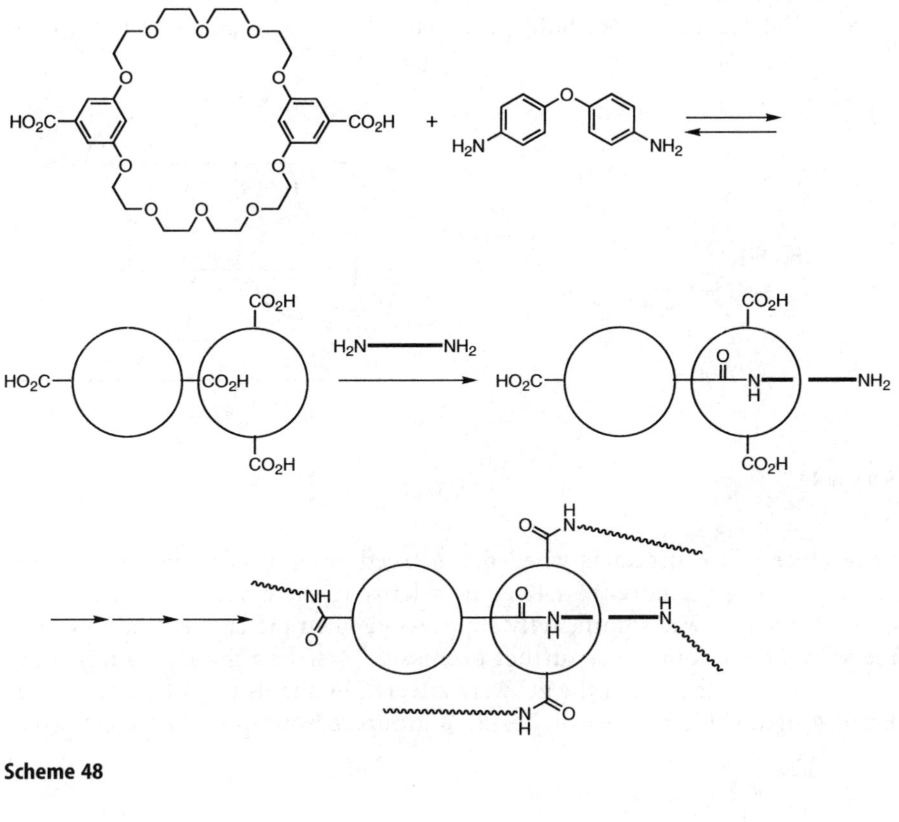

Scheme 48

the only evidence from which the rotaxane structures were deduced. Later on, Gibson et al. investigated cross-linked polyurethanes synthesized from a 32-membered macrocyclic diol and a diisocyanate in a similar manner to that employed for the crosslinked polymer by using a two-dimensional ^{1}H NMR technique, and they observed some NOEs which they attributed to the presence of the rotaxane structures [221]. They also obtained gel polymers from the reaction of poly(methacryloyl chloride) with a 32-membered crown ether having a hydroxymethyl group in pyridine [222]. Increasing the feed concentration to 0.71 mol/l yielded a material that was ca. 50% sol and 50% gel. The sol fraction was highly branched (MWD=214). They showed that in the polymers poly(methacrylate) chains were crosslinked through mechanical bonds by NMR spectroscopy and GPC. However, no specific properties derived from the mechanical cross-linkage were reported. They also obtained branched polyrotaxanes through the complexation of a 32-membered crown ether component attached to one polymer backbone with a dicationic bipyridinium component attached to another polymer backbone [53].

Zilkha's group also reported a series of papers describing the synthesis of mechanically crosslinked polymer through vinyl polymerization [223–225]. They employed the radical copolymerization of a 32-membered crown ether derivatives having one polymerizable double bond with styrene or methyl methacrylate (Scheme 49). Sufficient chain threading repeatedly took place

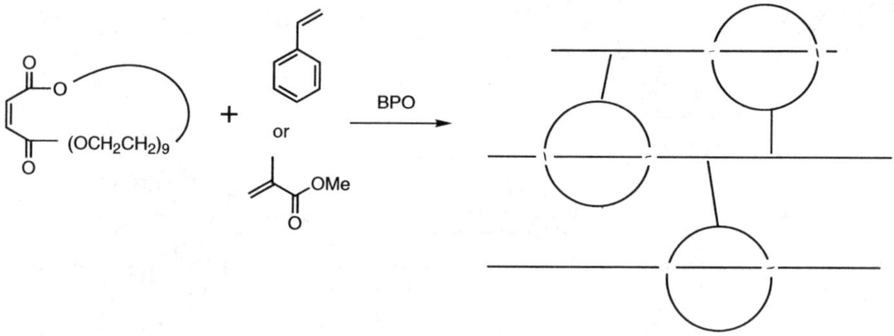

Scheme 49

to produce gelled products when the charged amount of vinyl comonomer remained to be no more than 40 equiv relative to the macrocyclic monomer. Chain threading was significantly suppressed when the concentration of the macrocyclic monomer was further decreased. Based on the DSC analyses, it was shown that the T_gs of the gel were affected by the degree of crosslinking. Employing a similar protocol, Tezuka's group recently prepared a gel product by radical copolymerization of a methacrylate-functionalized, cyclic poly(tetrahydrofuran) and methyl methacrylate [226].

Ito and co-worker first demonstrated in 2001 how nicely mechanical linkages worked as crosslinks of gel materials (Scheme 50) [227]. They prepared

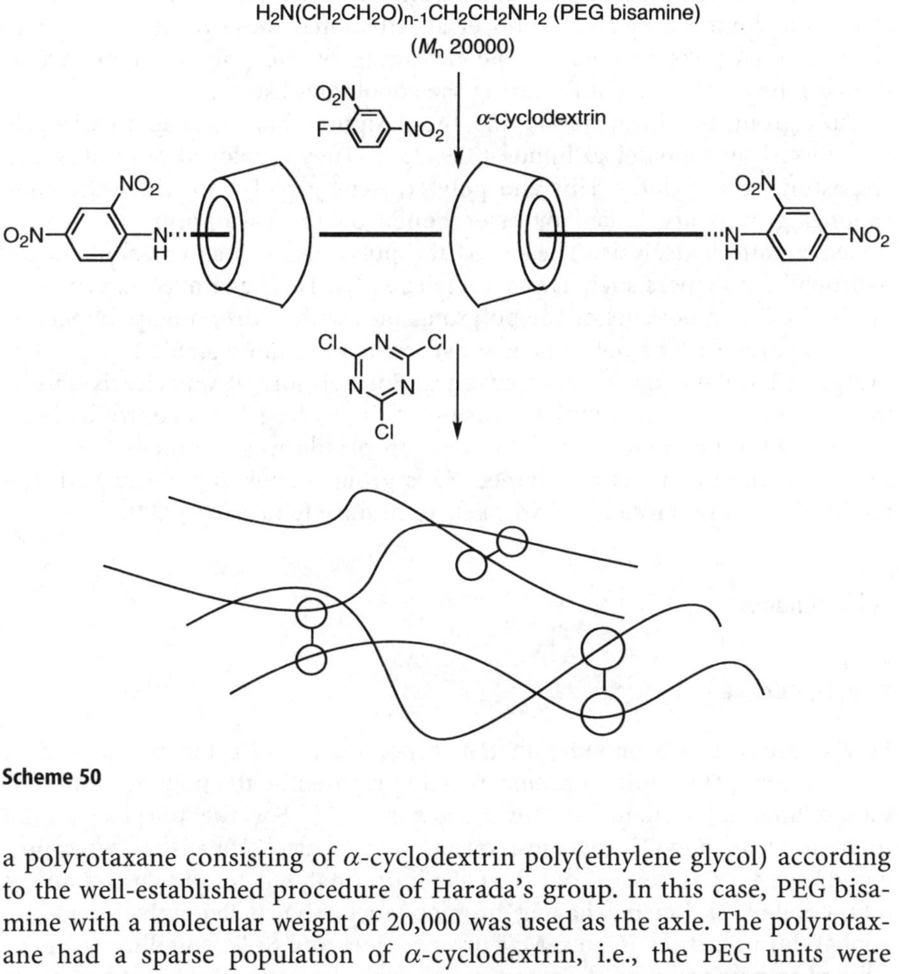

Scheme 50

a polyrotaxane consisting of α-cyclodextrin poly(ethylene glycol) according to the well-established procedure of Harada's group. In this case, PEG bisamine with a molecular weight of 20,000 was used as the axle. The polyrotaxane had a sparse population of α-cyclodextrin, i.e., the PEG units were "naked". Crosslinking was carried out by reaction with cyanuric chloride to yield the crosslinked polymer as a transparent gel. The gel polymer is super absorbent and can take up water to ca. 400 times the dry weight. The volume change is reversible without any pattern formations on the surface of the gel polymer unlike conventional chemical gels. The gel polymer is also flexible and tensile to twice its length. Ito et al. explained these properties based on the mechanical crosslinks and introduced the concept of "pulley effect".

Kubo et al. employed a similar protocol to those of Zilkha's group and Tezuka's group to prepare mechanically crosslinked polymers [229]. The copolymerization of a styrene-functionalized, cyclic polystyrene with *tert*-butyl acrylate in benzene at 70 °C yielded an insoluble polymer with a good swelling property. The insoluble material was converted to soluble polymers by cleavage of the macrocyclic units from the backbone chain of poly(styrene-co-*tert*-butyl acrylate). Kubo et al. attributed these results to its mechanical crosslinks formed by the threading of the polymer main chain through the macrocyclic unit during the copolymerization.

Yui's group has been developing the design of biodegradable hydrogels crosslinked by topological bonds [229–231]. They employed polyrotaxanes consisting of α-cyclodextrins and poly(ethylene glycol)s, in which the end-capping groups are linked by ester bonds on the assumption of in vivo nonenzymatic hydrolysis. They used the polyrotaxanes as crosslinkers for hydrophilic polymers such as poly(ethylene glycol). They linked intermolecularly α-CD components of the polyrotaxanes with hydrophilic polymers to obtain hydrogels. The polyrotaxane hydrogels were more stable towards hydrolysis than the original polyrotaxanes. Furthermore, it was clearly shown that the time to reach complete erosion of the hydrogel was controllable by the polyrotaxane contents and the erosion profile was dominated over by the M_n of the main chain polymer. Yui's group is pushing ahead with the applications of polyrotaxane hydrogels to tissue engineering [232].

4.2
Polycatenanes

4.2.1
Poly[2]catenane

Poly[2]catenane is a polymer that incorporates the [2]catenane subunit in the structure [233]. Poly[2]catenanes were prepared by the polymerization of bifunctional [2]catenanes or bis[2]catenanes [2]. For the purpose of the preparation of poly[2]catenanes, various functional [2]catenanes were prepared. Geerts et al. carried out the palladium-catalyzed coupling reactions of bifunctional [2]catenane having bromophenyl groups. Reductive coupling, Sonogashira reaction, Suzuki-Miyaura reaction, and Stille coupling were examined to prepare poly[2]catenanes, although the molecular weight was not so high in any case [234]. Polyester- and polyurethane-type poly[2]catenanes were prepared from [2]catenanes or bis[2]catenanes having hydroxy groups by Sauvage et al., Geerts et al., and Stoddart et al. [235–240]. Side chain-type poly[2]catenane could also be prepared in a similar manner [238]. Self-polycondensation of [2]catenane having both carboxylic acid and benzyl chloride moieties afforded polyester-type poly[2]catenane [241]. Shimada et al. reported the synthesis of polyamide-type poly[2]catenanes from [2]catenanes

Table 2 Poly[2]catenanes

Poly[2]catenane	reference
	234
	235, 239
	237, 241

having amino groups [242]. Polymerization of bis[2]catenanes having 2,2'-bipyridine units was carried out by Stoddart et al. via the coordination polymerization by the addition of silver ion [243]. Sauvage et al. also carried out the oxidative polymerization of [2]catenane having thiophene unit to yield conductive poly[2]catenane [244]. Although polymerization was not exam-

Table 2 (continued)

Poly[2]catenane	reference
	241, 243
	242
	244

ined, alkylation of bifunctional catenane having sulfonamide groups led to various oligomeric [2]catenanes [245]. Polycarbonate containing the [2]catenane subunit was prepared using bisphenol [2]catenane [246]. The typical structures of these poly[2]catenanes are listed in Table 2.

Physical properties of these poly[2]catenanes have been explored in expectation of unique properties based on the catenane structure [239, 246]. While various interesting physical properties were found in polyrotaxane, no characteristic property has been reported in [2]catenanes so far. Although poly[2]catenane has highly mobile moiety due to the mechanical bond, it has been suggested that the connection between [2]catenane subunit restricted the mobility in motion of [2]catenane. Further, intramolecular interaction in [2]catenane subunit may decrease its mobility. Secondary amide-based [2]catenanes can easily be prepared from commercially available compounds. Takata et al. found that the borane-reduction of the [2]catenanes afforded good yields of the amine-based [2]catenanes that can be useful for polymer synthesis [247, 148] (Scheme 51). Although the origi-

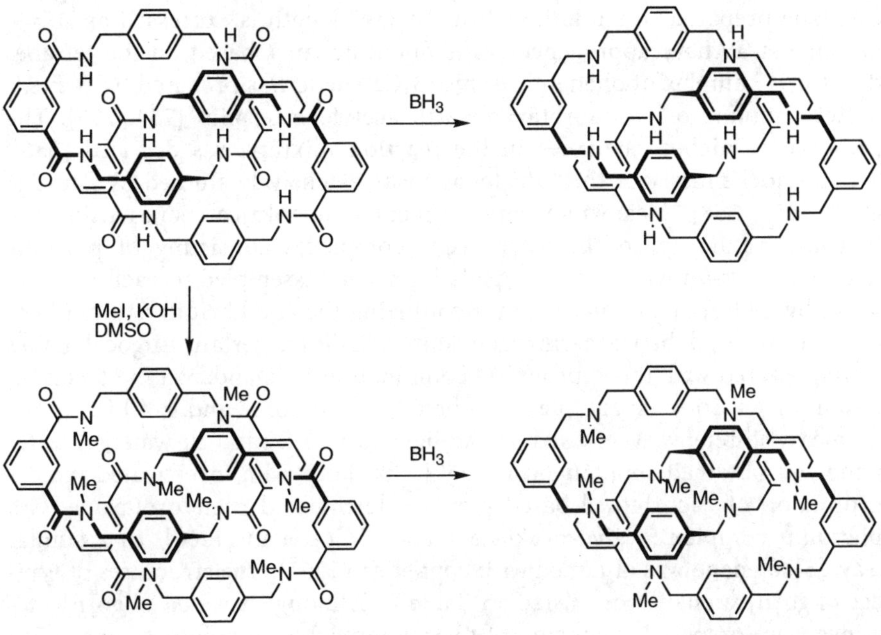

Scheme 51

nal [2]catenane has strong intramolecular hydrogen-bonding interaction between the two components that was necessary to construct interlocked structure, it disappeared after the reduction. Highly enhanced mobility of the resulting [2]catenane was confirmed by ^{1}H NMR spectra by Takata et al. The polymer derived from such an interlocked compound may show the physical properties that are characteristic of the poly[2]catenane [246].

4.2.2
Poly[n]catenane

Poly[n]catenane is a polymer that is composed only of cyclic components. Oligo[n]catenanes were successfully prepared by a stepwise manner. Stoddart has approached some oligocatenanes synthesized on the basis of the charge-transfer interaction between the electron-deficient cyclophane bearing 4,4'-bis(pyridinium salt) and the crown ether bearing electron-rich aromatics such as dialkoxybenzene and dialkoxynaphthalene. Various [3]catenanes were prepared by this approach [249–251]. Further, [5]catenane was prepared by a similar stepwise manner [252]. This catenane has been called olympiadane because its structure resembles the well-known symbol of the International Olympics. As the extension of this work, [6]catenane and [7]catenane were prepared [253]. The [7]catenane is the highest oligo[n]catenane prepared so far, although its "chain" length is expressed as DP=5. Sauvage et al. have approached some oligocatenanes based on the tetrahedral Cu(I) complex of phenanthroline. [3]Catenane was prepared by the oxidative coupling of pseudorotaxane with acetylene termini [254, 255]. The presence of higher catenanes in the reaction mixture was deduced [256]. The photochemistry of the [3]catenane was extensively studied by Sauvage et al. [257, 258]. Meanwhile, Fujita prepared oligo[n]catenanes using the thermodynamic approach. Macrocyclic complexes consisting of pyridine ligand and cis-(ethylenediamine)palladium were assembled to each other in water by hydrophobic interaction. By utilizing the equilibrium nature of coordination bond between nitrogen and palladium, certain oligocatenanes were prepared with the appropriate combination of ligands via the recombination of macrocyclic complexes. When 4,4'-bipyridine and 4,4'-bis(4-pyridylmethyl)biphenyl were used as the ligands in 1:2 ratio in water, [3]catenane was obtained spontaneously [259]. Similarly, Kim investigated the coordination of cucurbituril-based pyridine-terminated pseudorotaxane with platinum complex to give necklace-shaped [4]catenane [260]. In a similar way, [5]catenane was obtained with copper salt [261]. The structures of typical oligo[n]catenanes are listed in Table 3. Although various oligo[n]catenanes were prepared, extension of these approaches to poly[n]catenane has never been realized because the efficiency of the macrocyclization is far less than 100%.

Many researchers have tried to prepare poly[n]catenane. However, all attempts have so far been unsuccessful. The first proposal for the preparation of poly[n]catenane was illustrated in the early 1970s [262–265]. Hydrocarbon terminated by hydrophilic functional groups was spread on the organic solvent-water interface. It was claimed that connection of the resulting U-shaped molecules by appropriate cyclic compounds followed by ring-closure afforded poly[n]catenane. However, no direct evidence for the poly[n]catenane structure was reported. A more sophisticated proposal based on the

Table 3 Oligo[n]catenanes

Oligo[n]catenane	reference
	167
	169
	170

Table 3 (continued)

Oligo[n]catenane	reference
	254
	254, 255
	259

Table 3 (continued)

Oligo[n]catenane	reference
[structure depicting a [4]catenane assembly with three cucurbituril-like macrocycles interlocked via pyridyl-Pt(en) corners, 12NO$_3^-$ counterions, with legend showing the macrocycle equals a cucurbit[6]uril]	260

polycondensation of tetrafunctional monomer that has pre-interlocked structure was examined [179]. However, polymerization of the tetrafunctional monomer to form ladder polymer with 100% cyclization is required for the preparation of poly[n]catenane (Scheme 52). Since the complete cyclization is an extremely difficult task, preparation of [n]catenane via this methodology has still been unsuccessful.

Takata et al. have recently developed the directed ring-expansion method of catenane without destruction of the catenane structure [266]. In the modification of the catenane structure, the scission of one of the catenane rings generally causes the destruction of catenane structure. Takata et al. proposed that annulation followed by ring-scission is a safe way to modify the catenane structure with ring-expansion, while keeping the interlocked structure. A catenane having 1,3-diene-functionalized ring moiety underwent the Diels-Alder reaction followed by ozonolysis to afford the corresponding ring-expanded catenane in good yield (Scheme 53).

This new protocol for the modification of catenane may be logically used for the preparation of poly[n]catenane (Scheme 54). Namely, Diels-Alder

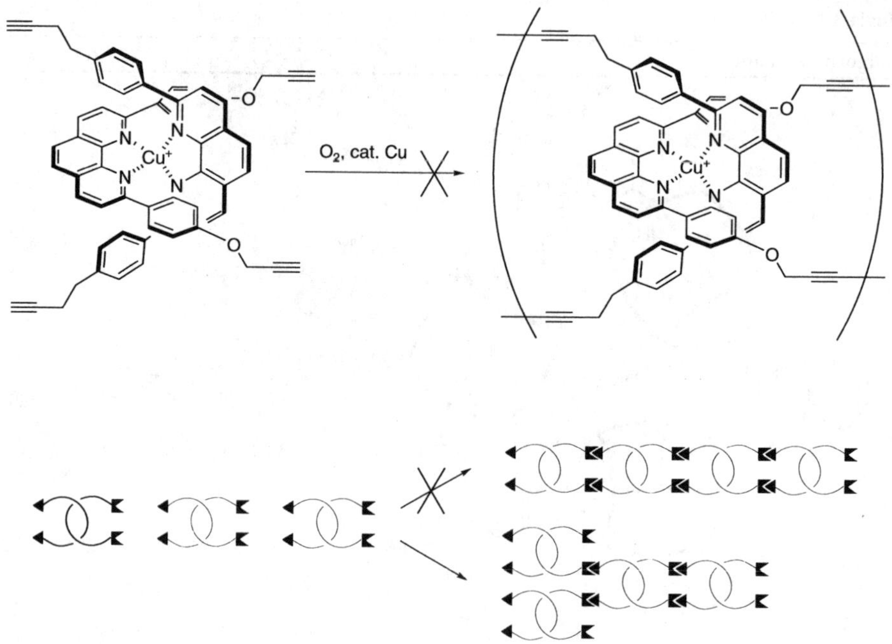

Scheme 52

polymerization of bis-1,3-diene-functionalized catenane with cyclic bis-dienophile may result in a formation of ladder polymer with a bridged catenane structure, as illustrated in Scheme 54. The scission of the formed double bonds by ozonolysis followed by the scission of x-y bonds yields the poly[n]catenane.

4.2.3
Polycatenane Network

Formation of polycatenane network has recently been suggested by Endo et al. during the thermal polymerization of cyclic disulfides such as 1,2-dithiane, where involvement of the cyclic polymers in the polydisulfide formed is proved by mass spectrometry [268] (Scheme 55). The elastic properties of the corresponding polydisulfide is believed to come from the polycatenane network structure.

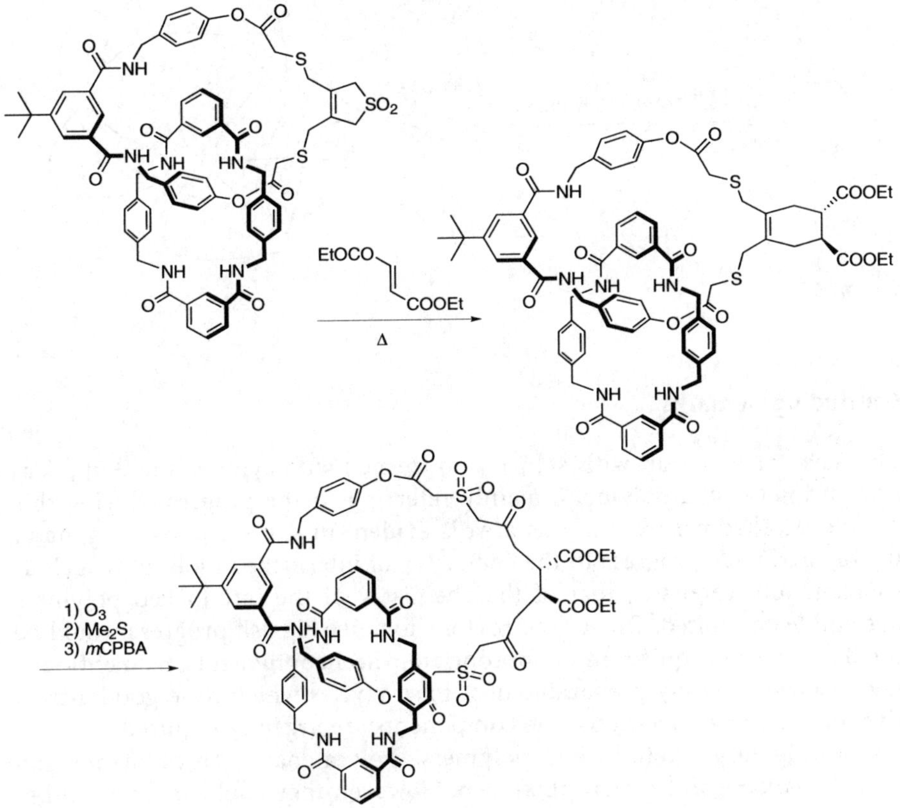

Scheme 53

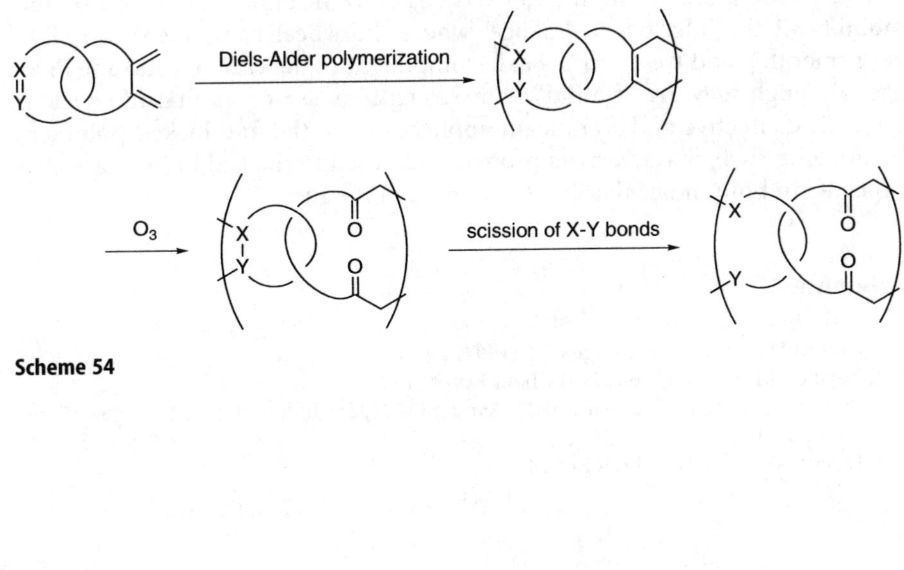

Scheme 54

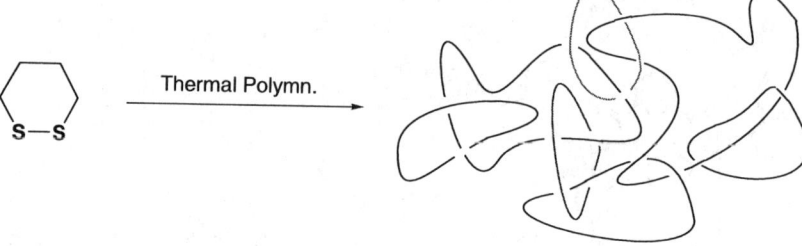

Scheme 55

5
Concluding Remarks

This chapter has dealt with subjects concerned with synthesis and application of interlocked polymers, mainly referring to the progress during this decade. Marked progress in this area is evident in some aspects. In spite of the big and wide progress in the chemistry of interlocked molecules such as rotaxane and catenane, that of the chemistry of the interlocked polymers has still been limited. One of the reasons for the sluggish progress would be the difficulty in acquisition of appropriate wheel components, as mentioned here. Cheap or easily preparable host macrocycles which have good attractive interaction with various axle components are urgently required.

Two big target interlocked polymers, polycatenane ([n]catenane) and poly[2]rotaxane still remain unknown. However, they could well be synthesized in the near future, since much effort has been directed toward these polymers so far. Applications of the interlocked polymers have been investigated from a variety of viewpoints of molecular level and bulk. The most characteristic point of the interlocked structure in bulk state would be the mobility of the interlocked linkage, where the wheel component can slide very smoothly and freely on the axle component in the case of rotaxane linkage, although they are "bound" with strength as strong as that of covalent bond. The effective and significant applications of the interlocked polymers by utilizing such characteristic properties of the interlocked linkage are also expected in both molecular and bulk levels in the future.

References

1. Gibson HW, Bheda MC, Engen PT (1994) 19:843
2. Raymo FM, Stoddart JF (1999) Chem Rev 99:1643
3. Rowan SJ, Cantrill SJ, Cousins GRL, Sanders JKM, Stodddart JF (2002) Angew Chem Int Ed 41:898
4. Hirada A (2001) Acc Chem Res 34:456

5. Furusho Y, Kihara N, Takata T (2001) Kobunshikako (Japanese) 50:114
6. Takata T, Kihara N, Furusho Y (2001) Kobunshi (Japanese) 50:770
7. Hubin TJ, Busch DH (2000) Coord Chem Rev 200/202:5
8. Sauvage JP, Dietrich-Buchecker CO (1999) Molecular catenanes, rotaxanes, and knots. VCH-Wiley, Weinheim
9. Harada A (1998) Acta Polym 49:3
10. Nepogodiev SA, Stoddart JF (1998) Chem Rev 98:1959
11. Harada A (1997) Kikansosetsu "Chemistry directed toward supramolecules" (Japanese). Gakkai Shuppan Center, p 206
12. Amabilino DB, Stoddart DJ (1995) Chem Rev 95:2725
13. Pease AR, Jeppeson JO, Stoddart JF, Lop YI, Collier CP, Heath JR (2001) Acc Chem Res 34:433
14. Ballardini R, Balzani V, Credi A, Gandolfi MT, Venturi M (2001) Acc Chem Res 34:445
15. Schalley CA, Beizai K, Vogtle F (2001) Acc Chem Res 34:465
16. Collin JP, Buchecker CD, Gabina P, Molero MCJ, Sauvage JP (2001) Acc Chem Res 32:477
17. Kihara N, Takata T (2001) Yukigosei Kagaku Kyukaishi (J Synth Org Chem Jpn) (Japanese) 59:206
18. Harada A (2000) Kagaku to Kogyo (Japanese) 74:470
19. Furusho Y, Takata T (2000) Kagaku to Kogyo (Japanese) 74:477
20. Balzani V, Credi A, Raymo FM, Stoddart JF (2000) Angew Chem Int Ed 39:3348
21. Takata T, Kihara N (2000) Rev Heteroatom Chem 22:197
22. Fyfe MCT, Stoddart JF (1999) Adv Supramol 5:1
23. Blanco MJ, Jimenez MC, Chambron JC, Heitz V, Linke M, Sauvage JP (1999) Chem Soc Rev 28:293
24. Vogtle F, Safarowsky O, Heim C, Affeld A, Braun O, Morhry A (1999) Pure Appl Chem 71:247
25. Nakashima N (1998) Kobunshi (Japanese) 47:546
26. Balzani V, Lopez MG, Stoddart JF (1998) Acc Chem Res 31:405
27. Sauvage JP (1998) Acc Chem Res 31:611
28. Vögtle F, Jäger R, Händel M, Ottens-Hildebrandt S (1996) Pure Appl Chem 68:225
29. Kubo M, Ito T (2002) Mirai Zairyo (Japanese) 2(12):18
30. Okumura Y, Ito K (2002) Mirai Zairyo (Japanese) 2(12):8
31. Okumura Y, Ito K (2002) Nippon Gomu Kyokaishi (Japanese) 75:36
32. Okumura Y, Ito K (2002) Kobunshi (Japanese) 51:252
33. Yui N (2001) Kobunshi (Japanese) 50:312
34. Shimomura T, Ito K (2000) Kobunshi Kako (Japanese) 49:361
35. Ito K, Shimomura T (1999) Kagaku to Kogyo (Japanese) 52:1170
36. Ooya T, Fujita H, Yui N (1999) Hyoumen (Japanese) 37:232
37. Ooya T, Yui N (1998) Kino Zairyou (Japanese) 18:56
38. Yui N (1998) Kino Zairyou (Japanese) 18:52
39. Gibson HW, Bheda M, Engen PT, Shen TX, Sze J, Wu C, Joardar S, Ward TC, Lecavalier PR (1991) Makromol Chem Macromol Symp 42/43:395
40. Shen YX, Xie D, Gibson HW (1994) J Am Chem Soc 116:537
41. Shen YX, Gibson HW (1992) Macromolecules 25:2058
42. Nagapudi K, Hunt J, Shepherd C, Baker J, Beckham HW (1999) Macromol Chem Phys 200:2541
43. Gibson HW, Liu S, Gong C, Ji Q, Joseph E (1997) Macromolecules 30:3711
44. Gong C, Ji Q, Subramaniam C, Gibson HW (1998) Macromolecules 31:1814

45. Gong C, Gibson HW (1996) Macromolecules 29:7029
46. Lee SH, Engen PT, Gibson HW (1997) Macromolecules 30:337
47. Houk KN, Menzer S, Newton SP, Raymo FM, Stoddart JF, Williams DJ (1999) J Am Chem Soc 121:1479
48. Loveday D, Wilkes GL, Bheda MC, Shen YX, Gibson HW (1995) J M S Pure Appl Chem A32:1
49. Ashton PR, Ballardini R, Balzani V, Belohradsky M, Golfi MT, Philp D, Prodi L, Raymo FM, Reddington MV, Spencer N, Stoddart JF, Venturi M, Williams DJ (1996) J Am Chem Soc 118:4931
50. Amabilino DB, Ashton PR, Balzani V, Brown CL, Credi A, Fréchet JMJ, Leon JW, Raymo FM, Spencer N, Stoddart JF, Venturi M (1996) J Am Chem Soc 118:12012
51. Ambilino DB, Ashton PR, Belohradsky M, Raymo FM, Stoddart JF (1995) J Chem Soc Chem Commun 747
52. Gong C, Gibson HW (1998) Angew Chem Int Ed 37:310
53. Gong C, Gibson HW (1998) Macromol Chem Phys 199:1801
54. Asakawa M, Ashton PR, Ballardini R, Balzani V, Belohradsky M, Gadolfi MT, Kocian O, Prodi L, Raymo FM, Stoddart JF, Venturi M (1997) J Am Chem Soc 119:302
55. Amabilino DB, Ashton PR, Belohradsky M, Raymo FM, Stoddart JF (1995) J Chem Soc Chem Commun 751
56. Amabilino DB, Asakawa M, Ashton PR, Ballardini R, Balzani V, Belohradsky M, Credi A, Higuchi M, Raymo FM, Shimizu T, Stoddart JF, Venturi M, Yase K (1998) New J Chem 959
57. Ashton PR, Glink PT, Stoddart JF, Menzer S, Tasker PA, White AJP, Williams DJ (1996) Tetrahedron Lett 37:6217
58. Ashton PR, Glink PT, Stoddart JF, Tasker PA, White AJP, Williams DJ (1996) Chem Eur J 2:729
59. Ashton PR, Glink PT, Stoddart JF, Menzer S, Tasker PA, White AJP, Williams DJ (1996) Tetrahedron Lett 37:6217
60. Kawasaki H, Kihara N, Takata T (1999) Chem Lett 1015
61. Kihara N, Shin JI, Ohga Y, Takata T (2001) Chem Lett 592
62. Watanabe N, Yagi T, Kihara N, Takata T (2002) Chem Commun 2720
63. Kihara N, Nakakoji N, Takta T (2002) Chem Lett 924
64. Kolchinski AG, Alcock NW, Roesner RA, Busch DH (1998) Chem Commun 1437
65. Furusho Y, Oku T, Hasegawa T, Tsuboi A, Kihara N, Takata T (2003) Chem Eur J 9:2895
66. Oku T, Furusho Y, Takata T (2003) J Polym Sci Part A Polym Sci 41:119
67. Rowan SJ, Cantrill SJ, Stoddart JF (1999) Org Lett 1:129
68. Rowan SJ, Stoddart JF (2000) J Am Chem Soc 122:164
69. Chiu SH, Elizarov AM, Glink PT, Stoddart JF (2002) Org Lett 4:3561
70. Elizarov AM, Chiu SH, Glink PT, Stoddart JF (2002) Org Lett 4:679
71. Ashton PR, Fyfe MCT, Schiavo C, Stoddart JF, White AJP, Williams DJ (1998) Tetrahedron Lett 39:5455
72. Ashton PR, Baxter I, Fyfe MCT, Raymo FM, Spencer N, Stoddart JF, White AJP, Williams DJ (1998) J Am Chem Soc 120:2297
73. Sohgawa Y, Fujimori H, Shoji J, Furusho Y, Kihara N, Takata T (2001) Chem Lett 774
74. Kihara N, Tachibana Y, Kawasaki H, Takata T (2000) Chem Lett 506
75. Loeb SJ, Wisner JA (1998) Angew Chem Int Ed 37:2838
76. Loeb SJ, Wisner JA (2000) Chem Commun 845
77. Ihata O, Kato T (1999) Polym Bull 42
78. Ogata N, Sanui K, Wada J (1976) J Polym Sic Polym Lett Ed 14:459

79. Maciejewski M, Panasiewicz M, Jarminska D (1978) J Macromol Sci Chem A12:7018
80. Maciejewski M (1979) J Macromol Sci Chem A13:77
81. Maciejewski M, Gwizdowski A, Peczak P, Pietrzak A (1979) J Macromol Sci Chem A13:87
82. Maciejewski M (1979) J Macromol Sci Chem A13:1175
83. Maciejewski M, Durski Z (1981) J Macromol Sci Chem A16:441
84. Steinbrunn MB, Wenz G (1996) Angew Chem Int Ed Engl 35:2139
85. Steinbrunn MB, Landfester K, Wenz G (1997) Tetrahedron 53:15575
86. Yamaguchi I, Takenaka Y, Osakada K, Yamamoto T (1999) Macromolecules 32:2051
87. Yamaguchi I, Osakada K, Yamamoto T (2000) Chem Commun 1335
88. Harada A, Kamachi M (1990) Macromolecules 23:2821
89. Harada A, Kamachi M (1990) J Chem Soc Chem Commun 1322
90. Harada A, Li J, Kamachi M (1993) Macromolecules 26:5698
91. Harada A, Li J, Kamachi M (1994) Macromolecules 27:4538
92. Harada A, Okada M, Li J, Kamachi M (1995) Macromolecules 28:8406
93. It has reported recently that β-cyclodextrin forms a pseudorotaxane with PEG in a heterogeneous mixture of PEG-water (1/1) and β-cyclodextrin. Its structure was established by X-ray crystallography. Udachin KA, Wilson LD, Ripmeester (2000) J Am Chem Soc 122:12375
94. Harada A, Li J, Kamachi M (1993) Chem Lett 237
95. Harada A, Li J, Suzuki S, Kamachi M (1993) Macromolecules 26:5267
96. Li J, Harada A, Kamachi M (1994) Bull Chem Soc Jpn 67:2808
97. Harada A, Suzuki S, Okada M, Kamachi M (1996) Macromolecules 29:5611
98. Harada A, Adachi H, Kawguchi Y, Okada M, Kamachi M (1996) Polym J 28:159
99. Harada A, Nishiyama T, Kawaguchi Y, Okada M, Kamachi M (1997) Macromolecules 30:7115
100. Harada A, Kawaguchi Y, Nishiyama T, Kamachi M (1997) Macromol Rapid Commun 18:535
101. Harada A, Okada M, Kamachi M (1998) Bull Chem Soc Jpn 71:535
102. Harada A, Okada M (1999) Polymer J 31:1095
103. Okada M, Kamachi M, Harada A (1999) J Phys Chem B 103:2607
104. Okada M, Kamachi M, Harada A (1999) Macromolecules 32:7202
105. Harada A, Okumura H, Okada M, Suzuki S, Kamachi M (2000) Chem Lett 548
106. Kamitori S, Matsuzaka O, Kondo S, Muraoka S, Okuyama K, Noguchi K, Okada M, Harada A (2000) Macromolecules 33:1500
107. Kawaguchi Y, Nishiyama T, Okada M, Kamachi M, Harada A (2000) Macromolecules 33:4472
108. Wenz G, Keller B (1992) Angew Chem Int Ed Engl 31:197
109. Herrman W, Keller B, Wenz G (1997) Macromolecules 30:4966
110. Weickenmeier M, Wenz G (1997) Macromol Rapid Commun 18:1109
111. Okumura H, Okada M, Kawaguchi Y, Harada A (2000) Macromolecules 33:4297
112. Okumura H, Kawaguchi Y, Harada A (2001) Macromolecules 34:6338
113. Okumura H, Kawaguchi Y, Harada A (2002) Macromol Rapid Commun 23:781
114. Harada A, Li J, Kamachi M (1992) Nature 356:325
115. Harada A, Li J, Nakamitsu T, Kamachi M (1993) J Org Chem 58:7524
116. Harada A, Li J, Kamachi M (1994) J Am Chem Soc 114:3192
117. Herrmann W, Schneider M, Wenz G (1997) Angew Chem Int Ed Engl 36:2511
118. Yamaguchi I, Osakada K, Yamamoto T (1996) J Am Chem Soc 118:1811
119. Harada A, Li J, Kamachi M (1994) Nature 370:126

120. Shigekawa H, Miyake K, Sumaoka J, Harada A, Komiyama M (2000) J Am Chem Soc 122:5411
121. Anderson S, Anderson HL (1996) Angew Chem Int Ed Engl 35:1956
122. Anderson S, Alpin RT, Goodson T III, Maciel AC, Rumbles G, Ryan JF, Anderson HL (1998) J Chem Soc Perkin Trans 1 2383
123. Yoshida K, Shimomura T, Ito K, Hayakawa R (1999) Langmuir 15:910
124. Shimomura T, Yoshida K, Ito K, Hayakawa R (2000) Polym Adv Technol 11:837
125. Yamaguchi I, Nurulla I, Yamamoto T (2000) Kobunshi Ronbunshu 57:472
126. Taylor PN, O'Connell MJ, McNeill LA, Hall MJ, Aplin RT, Anderson HL (2000) Angew Chem Int Ed Engl 39:3456
127. Cacialli F, Wilson JS, Michels JJ, Daniel C, Silva C, Friend RH, Severin N, Samorì P, Rabe J, O'Connell MJ, Taylor PN, Anderson HL (2002) Nature Mater 1:160
128. Ooya Y, Mori H, Terano L, Yui N (1995) Macromol Rapid Commun 16:259
129. Ooya T, Yui N (1997) J Biomater Sci Polym Ed 8:437
130. Yui N, Ooya T, Kumeno T (1998) Bioconj Chem 9:118
131. Ooya T, Kumeno T, Yui N (1998) J Biomater Sci Polym Edn 9:313
132. Ooya T, Yui N (1998) Macromol Chem Phys 199:2311
133. Watanabe J, Ooya T, Yui N (1998) Chem Lett 1031
134. Ooya T, Yui N (1999) J Control Release 58:251
135. Watanabe J, Ooya T, Yui N (1999) J Biomater Sci Polym Edn 10:603
136. Ooya T, Arizono K, Yui N (2000) Polym Adv Technol 11:642
137. Watanabe J, Ooya T, Yui N (2000) J Artif Organs 3:136
138. Huh KM, Ooya T, Sasaki S, Yui N (2001) Macromolecules 34:2402
139. Yui N, Ooya T, Kawshima T, Saito Y, Tamai I, Sai Y, Tsuji A (2002) Bioconj Chem 13:582
140. Fujita H, Ooya T, Kurisawa M, Mori H, Terano M, Yui N (1996) Macromol Rapid Commun 42:237
141. Fujita H, Ooya T, Yui N (1999) Macromol Chem Phys 200:706
142. Fujita H, Ooya T, Yui N (1999) Macromolecules 32:2534
143. Ikeda T, Ooya T, Yui N (1999) Polym J 31:658
144. Fujita H, Ooya T, Yui N (1999) Polym J 31:1099
145. Ikeda T, Ooya T, Yui N (2000) Polym Adv Technol 11:830
146. Ikeda T, Watabe N, Ooya T, Yui N (2001) Macromol Chem Phys 202:1338
147. Ikeda T, Ooya T, Yui N (2000) Macromol Rapid Commun 21:1257
148. Ikeda T, Hirota E, Ooya T, Yui N (2001) Langmuir 17:234
149. Ichi T, Watanabe J, Ooya T, Yui N (2001) Biomacromolecules 2:204
150. Huh KM, Tomita H, Ooya T, Lee KW, Sasaki S, Yui N (2002) Macromolecules 35:3775
151. Huh KM, Ooya T, Sasaki S, Yui N (2001) Macromolecules 34:2402
152. Huh KM, Tomita H, Ooya T, Lee WK, Sasaki S, Yui N (2002) Macromolecules 35:3775
153. Ooya T, Yui N (2002) J Control Release 80:219
154. Park HD, Lee WK, Ooya T, Park KD, Kim YH, Yui N (2002) J Biomed Mater Res 60:186
155. Ooya T, Eguchi M, Ozaki A, Yui N (2002) Int J Pharm 242:47
156. Harada A, Li J, Kamachi M (1993) Nature 364:516
157. Ikeda T, Lee WK, Ooya T, Yui N (2003) J Phys Chem B 107:14
158. Ooya T, Kobayashi N, Ichi T, Sasaki S, Yui N (2004) Sci Technol Adv Mater (in press)
159. Liu Y, You CC, Zhang HY, Kang SZ, Zhu CF, Wang C (2001) Nano Lett 1:613
160. Ikeda E, Okumura Y, Shimomura T, Ito K, Hayakawa R (2000) J Chem Phys 112:4321
161. Okumura Y, Ito K, Hayakawa Y (1998) Phys Rev Lett 80:5003
162. Saito M, Shimomura T, Okumura Y, Ito K, Hayakawa R (2001) J Chem Phys 114:1

163. Okumura Y, Ito K, Hayakawa R, Nishi T (2000) Langmuir 16:10278
164. Shimomura T, Akai T, Abe T, Ito K (2002) J Chem Phys 116:1753
165. Akai T, Abe T, Shimomura T, Ito K (2001) Jpn J Appl Phys 40:L1327
166. Mock WL, Irra TA, Wepsiec JP, Adhya M (1989) J Org Chem 54:5302
167. Whang D, Jeon YM, Heo J, Kim K (1996) J Am Chem Soc 118:11333
168. Whang D, Heo J, Kim CA, Kim K (1997) Chem Commun 2361
169. Park KM, Kim SY, Heo J, Whang D, Sakamoto S, Yamaguchi K, Kim K (2002) J Am Chem Soc 124:2140
170. Whang D, Kim K (1997) J Am Chem Soc 119:451
171. Lee E, Heo J, Kim K (2000) Angew Chem Int Ed 39:2699
172. Lee E, Kim J, Heo J, Whang D, Kim K (2001) Angew Chem Int Ed 40:399
173. Meschke C, Buschmann HJ, Schollmeyer E (1998) Macromol Rapid Commun 19:59
174. Choi SW, Lee JW, Ko YH, Kim K (2002) Macromolecules 35:3526
175. Mock WL (1995) Top Curr Chem 175:1
176. Mock WL, Irra TA, Websiec JP, Adhya M (1989) J Org Chem 54:5302
177. Tuncel D, Steinke JHG (1999) Chem Commun 1509
178. Tuncel D, Steinke JHG (2001) Chem Commun 253
179. Sauvage JP (1990) Acc Chem Res 23:319
180. Chambron JC, Heitz B, Sauvage JP (1993) J Am Chem Soc 115:12378
181. Solladié N, Chambron JC, Sauvage JP (1999) J Am Chem Soc 121:3684
182. Solladié N, Chambron JC, Dietrich-Buchecker CO, Sauvage JP (1996) Angew Chem Int Ed Engl 35:906
183. Zhu S, Carroll PJ, Swager TM (1996) J Am Chem Soc 118:8713
184. Vidal PL, Billon M, Divisia-Blohorn B, Bidan G, Kern JM, Sauvage JP (1998) Chem Commun 629
185. Vidal PL, Divisia-Blohorn B, Bidan G, Kern JM, Sauvage JP, Hazemann JL (1999) Inorg Chem 38:4203
186. Buey J, Swager TM (2000) Angew Chem Int Ed 39:608
187. Anderson S, Anderson H (1996) Angew Chem Int Ed Engl 35:1956
188. Anderson S, Aplin RT, Claridge TDW, Goodson T III, Maciel AC, Rumbles G, Ryan JF, Anderson HL (1998) J Chem Soc Perkin Trans I 2383
189. Yamagishi T, Kawahara A, Kita J, Hoshima M, Umehara A, Ishida S, Nakamoto Y (2001) Macromolecules 34:6565
190. Mason PE, Parsons IW, Tolley MS (1996) Angew Chem Int Ed Engl 35:2238
191. Hodge P, Monvisade P, Owen GJ, Heatley F, Pang Y (2000) New J Chem 24:703
192. Werts MPL, van den Boogaard M, Hadziioannou G, Tsivgoulis GM (1999) Chem Commun 623
193. Vögtle F, Dünnwald T, Händel M, Jäger R, Meier S, Harder G (1996) Chem Eur J 2:640
194. Schmieder R, Hübner G, Seel C, Vögtle F (1999) Angew Chem Int Ed 38:3528
195. Reuter C, Mohry A, Sobanski A, Vögtle F (2000) Chem Eur J 6:1674
196. Jäger R, Händel M, Harren J, Rissanen K, Vögtle F (1996) Liebigs Ann 1201
197. Dünnwald T, Jäger R, Vögtle F (1997) Chem Eur J 3:2043
198. Takata T, Kawasaki H, Asai S, Kihara N, Furusho Y (1999) Chem Lett 111
199. Takata T, Kawasaki H, Kihara N, Furusho Y (2001) Macromolecules 34:5449
200. Born M, Ritter H (1991) Makromol Chem Rapid Commun 12:471
201. Born M, Koch T, Ritter H (1995) Macromol Chem Phys 196:1761
202. Born M, Ritter H (1996) Macromol Rapid Commun 17:197
203. Born M, Koch T, Ritter H (1994) Acta Polym 45:68
204. Born M, Koch T, Ritter H (1995) Macromol Chem Phys 196:1761

205. Born M, Ritter H (1995) Angew Chem Int Ed Engl 34:309
206. Yamaguchi I, Osakada K, Yamamoto T (1997) Macromolecules 30:4288
207. Yamaguchi I, Osakada K, Yamamoto T (2000) Macromolecules 33:2315
208. Liu J, Xu R, Kaifer AE (1998) Langmuir 14:7337
209. Noll O, Ritter H (1998) Macromol Chem Phys 199:791
210. Tan Y, Choi SW, Lee JW, Ko YH, Kim K (2002) Macromolecules 35:7161
211. Ashton P, Baxter I, Cantrill SJ, Fyfe MCT, Glink PT, Stoddart JF, White AJP, Williams DJ (1998) Angew Chem Int Ed Engl 37:1294
212. Ashton PR, Parsons IW, Raymo FM, Stoddart JF, White AJP, Williams DJ, Wolf R (1998) Angew Chem Int Ed Engl 37:1913
213. Rowan SJ, Cantrill SJ, Stoddart JF, White AJP, Williams DJ (2000) Org Lett 2:759
214. Fujimoto T, Sakata Y, Kaneda T (2000) Chem Commun 2143
215. Onagi H, Easton CJ, Lincoln SF (2001) Org Lett 3:1041
216. Hoshino T, Miyauchi M, Kawaguchi Y, Yamaguchi H, Harada A (2000) J Am Chem Soc 122:9876
217. Jímenez MC, Dietrich-Buchecker C, Sauvage JP (2000) Angew Chem Int Ed Engl 39:3284
218. Yamaguchi N, Gibson HW (1999) Angew Chem Int Ed Engl 38:143
219. Delaviz Y, Gibson HW (1992) Macromolecules 25:4859
220. Gibson HW, Nagveker DS, Powell J, Gong C, Bryant WS (1997) Tetrahedron 53:15197
221. Gong C, Gibson HW (1997) J Am Chem Soc 119:8585
222. Gong C, Gibson HW (1997) J Am Chem Soc 119:5862
223. Zada A, Avny Y, Zilkha A (1999) Eur Polym J 35:1159
224. Zada A, Avny Y, Zilkha A (2000) Eur Polym J 36:351
225. Zada A, Avny Y, Zilkha A (2000) Eur Polym J 36:359
226. Oike H, Mouri T, Tezuka Y (2001) Macromolecules 34:6229
227. Okumura Y, Ito K (2001) Adv Mater 13:485
228. Kubo M, Hibino T, Tamura M, Uno T, Itoh T (2002) Macromolecules 35:5816
229. Watanabe J, Ooya T, Yui N (1999) J Biomater Sci Polym Edn 10:1275
230. Watanabe J, Ooya T, Park KD, Kim YH, Yui N (2000) J Biomater Sci Polym Edn 11:1333
231. Ichi T, Watanabe J, Ooya T, Yui N (2001) Biomacromolecules 2:204
232. Lee WK, Ichi T, Ooya T, Yamamoto M, Kato M, Yui N (2004) J Biomed Mater Res (in press)
233. Geerts Y (1999) Polycatenanes, poly[2]catenanes, and polymeric catenanes. In: Sauvage JP, Dietrich-Buchecker C (eds) Molecular catenanes, rotaxanes, and knots. VCH, Weinheim, p 247
234. Geerts Y, Muscat D, Müllen K (1995) Macromol Chem Phys 196:3425
235. Weidmann JL, Kern JM, Sauvage JP, Geerts Y, Muscat D, Müllen K (1996) Chem Commun 1243
236. Muscat D, Witte A, Köhler W, Müllen K, Geerts Y (1997) Macromol Rapid Commun 18:233
237. Menzer S, White AJP, Williams DJ, Belohradsky M, Hamers C, Raymo FM, Shipway AN, Stoddart JF (1998) Macromolecules 31:295
238. Hamers C, Raymo FM, Stoddart JF (1998) Eur J Org Chem 2109
239. Weidmann JL, Kern JM, Sauvage JP, Muscat D, Mullins S, Köhler W, Rosenauer C, Räder HJ, Martin K, Geerts Y (1999) Chem Eur J 5:1841
240. Muscat D, Köhler W, Räder HJ, Martin K, Mullins S, Müller B, Müllen K, Geerts Y (1999) Macromolecules 32:1737
241. Raymo FM, Stoddart JF (1999) Polym Mater Sci Eng 80:33

242. Shimada S, Ishikawa K, Tamaoki N (1998) Acta Chem Scand 52:374
243. Hamers C, Kocian O, Raymo FM, Stoddart JF (1998) Adv Mater 10:1366
244. Simone DL, Swager TM (2000) J Am Chem Soc 122:9300
245. Jäger R, Schmidt T, Karbach D, Vögtle F (1996) Synlett 723
246. Fustin CA, Bailly C, Clarkson GJ, De Groote P, Galow TH, Leigh DA, Robertson D, Slawin AMZ, Wong JKY (2003) J Am Chem Soc 125:2200
247. Takata T, Shoji J, Furusho Y (1997) Chem Lett 881
248. Furusho Y, Shoji J, Watanabe N, Kihara N, Adachi T, Takata T (2001) Bull Chem Soc Jpn 74:139
249. Ashton PR, Brown CL, Chrystal EJT, Goodnow TT, Kaifer AE, Parry KP, Slawin AMZ, Spencer N, Stoddart JF, Williams DJ (1991) Angew Chem Int Ed Engl 30:1039
250. Ashton PR, Brown CL, Chrystal EJT, Parry KP, Pietraszkiewicz M, Spencer N, Stoddart JF (1991) Angew Chem Int Ed Engl 30:1042
251. Asakawa M, Ashton PR, Menzer S, Raymo FM, Stroddart JF, White AJP, Williams DJ (1996) Chem Eur J 2:877
252. Ambilino DB, Ashton PR, Reder AS, Spencer N, Stoddart JF (1994) Angew Chem Int Ed Engl 33:1286
253. Amabilino DB, Ashton PR, Balzani V, Boyd SE, Credi A, Lee JY, Menzer S, Stoddart JF, Venturi M, Williams DJ (1998) J Am Chem Soc 120:4295
254. Dietrich-Buchecker CO, Guilhem J, Khémiss AK, Kintzinger JP, Pascard C, Sauvage JP (1987) Angew Chem Int Ed Engl 26:661
255. Dietrich-Buchecker CO, Hemmert C, Khémiss AK, Sauvage JP (1990) J Am Chem Soc 112:8002
256. Bitsch F, Dietrich-Buchecker CO, Khémiss AK, Sauvage JP, van Dorsselaer A (1991) J Am Chem Soc 113:4023
257. Armaroli N, Balzani V, Barigelletti F, De Cola L, Sauvage JP, Hemmert C (1991) J Am Chem Soc 113:4033
258. Armaroli N, Balzani V, Barigelletti F, De Cola L, Flamigni L, Sauvage JP, Hemmert C (1994) J Am Chem Soc 116:5211
259. Hori A, Kumazawa K, Kusukawa T, Chand DK, Fujita M, Sakamoto S, Yamaguchi K (2001) Chem Eur J 7:4142
260. Whang D, Park KM, Heo J, Ashton P, Kim K (1998) J Am Chem Soc 120:4899
261. Roh SG, Park KM, Park GJ, Sakamoto S, Yamaguchi K, Kim K (1999) Angew Chem Int Ed 38:638
262. Koragoumis G, Pandi-Agathokli I (1970) Prakt Akad Athenon 45:118
263. Karagoumis G, Kontaraki E (1973) Prakt Akad Athenon 48:197
264. Karagoumis G, Pandi-Agathokli I, Kontaraki E, Nikolelis D (1975) Prakt Akad Athenon 49:501
265. Karagoumis G, Pandi-Agathokli I (1972) J Pract Panellion Chem Synedriou 2:213
266. Watanabe N, Kihara N, Takata T (2001) Org Lett 3:3519
267. Endo K, Shiori T, Murata N (2000) Nippon Gomu Kyokaishi 73:392

Transition Metal-Mediated Polymerization of Isocyanides

Michinori Suginome[1] (✉) · Yoshihiko Ito[2]

[1] Department of Synthetic Chemistry and Biological Chemistry,
Graduate School of Engineering, Kyoto University, 615-8510 Nishikyo-ku, Kyoto, Japan
suginome@sbchem.kyoto-u.ac.jp
[2] Department of Molecular Science and Technology, Faculty of Engineering,
Doshisha University, 610-0394 Kyotanabe, Kyoto, Japan

1	Introduction	78
2	Polymerization of Monoisocyanide	81
2.1	Synthesis of Racemic Poly(isocyanide)s	82
2.1.1	Ni-Catalyzed Systems	82
2.1.2	Pd-Catalyzed Systems	92
2.1.3	Co- and Rh-catalyzed Systems	96
2.2	Synthesis of Chiral Non-Racemic Poly(isocyanide)s	97
2.2.1	Screw-Sense Induction by Incorporation of Chiral Non-racemic Isocyanide	99
2.2.2	Screw-Sense Induction by Selective Inhibition of the Growth of a One Screw-Sense (Chiral Poisoning)	106
2.2.3	Screw-sense Induction by a Chiral Chain-end Group at the Non-Propagating Terminus	107
2.2.4	Screw-sense Induction by a Chiral Ligand at the Propagating Terminus	110
2.3	Polymerization of Isocyanides Having Functional Groups	111
3	Polymerization of Diisocyanobenzenes	118
3.1	Synthesis of Racemic Poly(quinoxaline-2,3-diyl)s	119
3.2	Synthesis of Non-Racemic Poly(quinoxaline-2,3-diyl)s	122
3.2.1	Asymmetric Polymerization via Optical Resolution of Living Oligomers	123
3.2.2	Asymmetric Polymerization by Chiral Initiators	126
3.2.3	Asymmetric Block Copolymerization	131
4	Concluding Remarks	133
References		134

Abstract Much interest has focused on the polymerization of isocyanide, which can be regarded as a stable *N*-substituted iminocarbene. Transition metal complexes have served as the most efficient and versatile initiators for the polymerization of isocyanides. This review covers a variety of studies carried out on transition metal mediated polymerization of isocyanides, from mechanistic studies to functionalized polymer synthesis. Emphasis is placed on asymmetric polymerization, which leads to the formation of optically active, rigid rod helical poly(isocyanide)s. Polymerization of 1,2-diisocyanobenzenes is also dealt with in this review. The living aromatizing polymerization provides for otherwise inaccessible poly(quinoxaline-2,3-diyl)s, which also adopt non-racemizable rigid rod helical structures. Asymmetric synthesis of poly(quinoxaline-2,3-diyl)s is achieved by screw-sense selective polymerization of 1,2-diisocyanobenzenes using chiral organopalladium complexes as initiators.

Keywords Transition metal complexes · Living polymerization · Rigid rod helical structure · Optically active polymer · Asymmetric polymerization

1
Introduction

Isocyanides have attracted much attention in synthetic organic chemistry, and have led to the development of a variety of useful synthetic transformations [1]. In particular, carbon–carbon bond forming reactions using isocyanides as key reagents have been extensively studied during the past 50 years, as they allow the realization of useful multi-component assembly reactions, such as the Ugi and Passerini reactions [2]. These have once again been highlighted in recent years with the development of combinatorial chemistry. The characteristic reactions of isocyanides are mostly ascribed to the unique reactivity of the isocyano carbon atom, which can be represented by a divalent, carbene-like electronic structure (Scheme 1). In this respect,

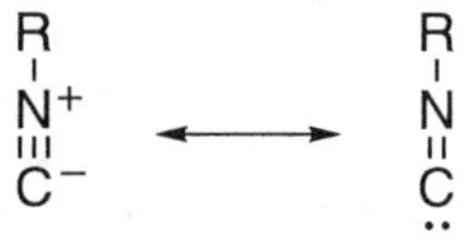

$$R-N^+\equiv C^- \longleftrightarrow R-N=C:$$

Scheme 1

isocyanide can be characterized as being a "stable carbene" that bears an N-arylimino group.

Inspired by the contribution of the carbene-like resonance structure, the homopolymerization of isocyanide giving rise to the formation of poly(isocyanide) has attracted much attention [3, 4]. On storage, or distillation, isocyanides that lack bulky N-substituents tend to form solid materials, which had been supposed to be poly(isocyanide)s. However, this "polymerization", (or resinification), largely depended upon the nature of the glass surface of the apparatus used for storage or distillation and, therefore, was poorly reproducible. Moreover, no structural information was provided for these materials, making the evaluation of the polymerization systems difficult. The historical background has already been overviewed by Millich in two reviews published in 1972 and 1980 [3, 4].

Much effort has been expended to realize the efficient polymerization of isocyanides using promoters or catalysts. Such effort can be classified using three criteria: those where the polymerization occurs via radical intermediates (Scheme 2), via anionic intermediates (Scheme 3), and via carbocationic intermediates (Scheme 4). Only the third criterion has been fruitful, and has led to the findings of the remarkable catalytic activity of homogenous and

class 1: radical-mediated polymerization (not successful)

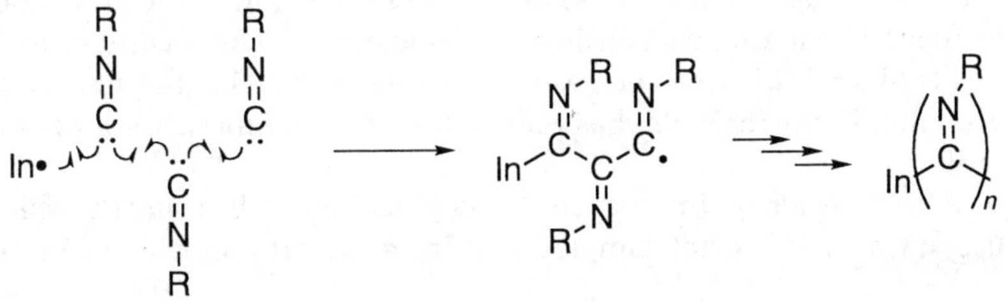

Scheme 2

heterogeneous acid derivatives. In 1966, Yamamoto and Hagihara reported a homogeneous system using $BF_3 \cdot OEt_2$ or $SnCl_4$ in the polymerization of phenyl isocyanide and cyclohexyl isocyanide [5–7]. The properties of the polymers were carefully investigated by polymerization degree measurements,

class 2: nucleophile-mediated polymerization (not successful)

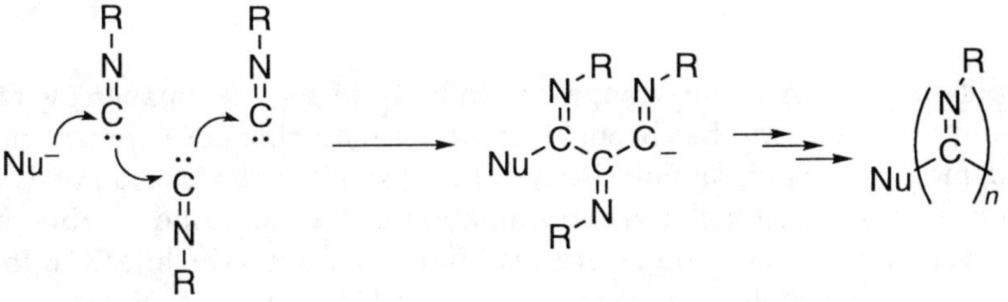

Scheme 3

IR spectroscopy, elemental analysis, and X-ray diffraction studies. Millich et al. demonstrated that acid-treated ground glass showed remarkable catalytic activity in the polymerization of isocyanides [8, 9]. Their heterogeneous

class 3: acid-mediated polymerization

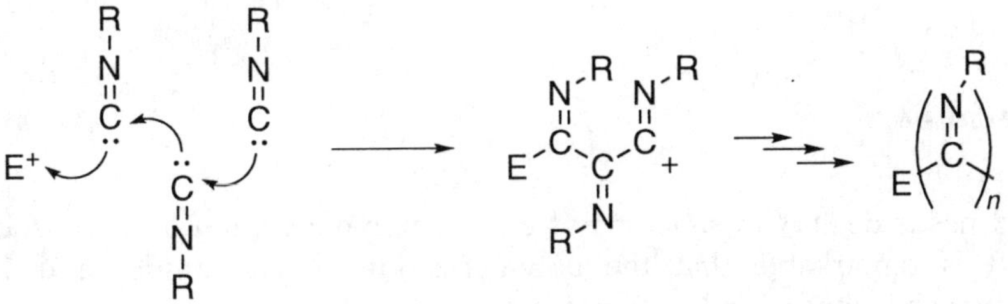

Scheme 4

acid-catalyzed system was carefully designed to avoid any undesirable side reactions of isocyanide and to avoid the destruction of the poly(isocyanide)s formed, which are acid sensitive to some extent. The second criterion using nucleophilic initiators, such as Grignard reagents, alkyllithium reagents, and alkylaluminum reagents, has only led to the oligomerization of isocyanides [10].

Recent progress in isocyanide polymerization has largely relied on the use of transition metal complexes acting as initiators (Scheme 5). The fourth

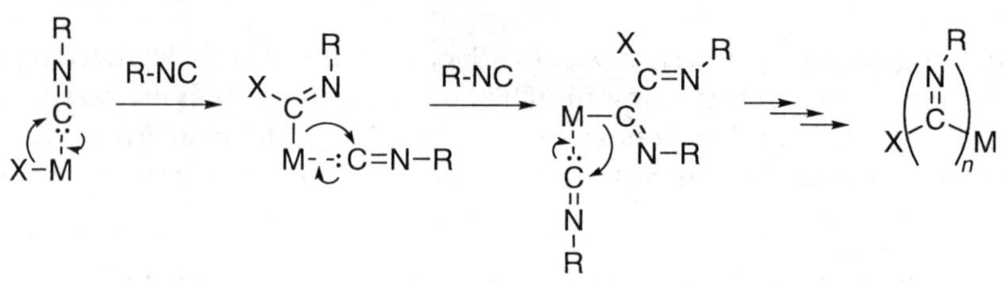

class 4: transition metal-mediated polymerization

Scheme 5

polymerization strategy appeared initially in the 1966 paper by Yamamoto and Hagihara, and has brought about a remarkable development in the fundamental research on poly(isocyanide)s [5]. Here, the isocyanide is activated on the transition metal via coordination at the isocyanide carbon, then undergoing successive migratory insertion reactions, which lead to the formation of high molecular weight polymers [11, 12]. Various transition metal complexes are now used in the polymerization of isocyanide. An important application of transition metal mediated systems is in the polymerization of 1,2-diisocyanobenzenes, leading to the synthesis of poly(quinoxaline-2,3-diyl)s 1 (Scheme 6). In this review, the polymerization of 1,2-diisocyanoben-

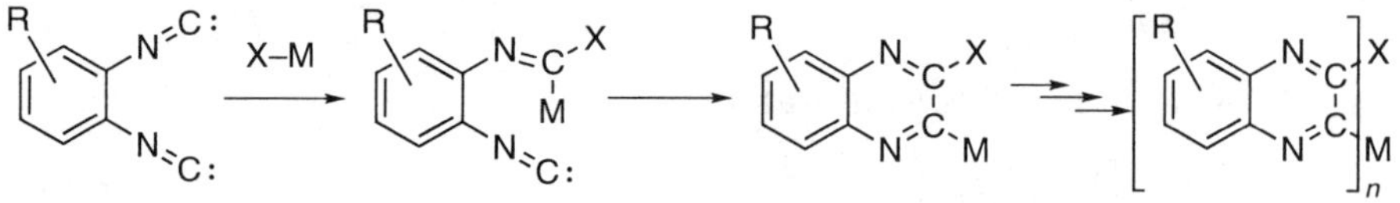

Scheme 6

zenes is dealt with separately from the polymerization of monoisocyanides. It is remarkable that the polymerization of isocyanides and 1,2-diisocyanobenzenes can be extended to asymmetric polymerization systems using optically active transition metal complexes as chiral initiators. Using this strategy, rigid helical structures of poly(isocyanide)s and poly(quinoxaline-2,3-diyl)s can be successfully constructed with a preferred right- or left-handed helical sense.

This review focuses on the transition metal mediated polymerization of isocyanides and 1,2-diisocyanobenzenes. Other approaches, including acid-catalyzed polymerization and main-group metal mediated polymerization, are also included, but only if they are closely related to the above topic.

In this review, monoisocyanide polymers are denoted as "poly(isocyanide)s" rather than using the systematic name, "poly(iminomethylene)s", because the former name is more familiar to non-experts in this area. For example, poly(N-tert-butyliminomethylene) is denoted as poly(tert-butyl isocyanide). On the other hand, the systematic name "poly(quinoxaline-2,3-diyl)" is used for polymers of 1,2-diisocyanobenzene to differentiate the fact that the polymerization process is accompanied by the formation of quinoxaline rings.

2
Polymerization of Monoisocyanide

In the late 1960s, the chemistry of isocyanide polymers arrived at a turning point, with several important findings on transition metal catalyst systems for polymerization emerging. Important contributions from some independent research groups are exemplified here to provide an overview of the initial stages of transition metal-catalyzed polymerization of isocyanides. In 1966, Yamamoto and Hagihara reported that some cobalt and nickel complexes effectively catalyze the polymerization of cyclohexyl isocyanide in benzene [5]. They reported that cobalt complexes, such as $Co_2(CO)_8$, Cp_2Co, and $CpCo(CO)_2$, exhibit higher catalytic activity than nickel complexes, such as $Ni(CO)_4$, Cp_2Ni, and $[CpNi(CO)]_2$. It was noted that the polymerization of cyclohexyl isocyanide in the presence of cobaltocene catalysts proceeds even at room temperature, yielding poly(N-cyclohexyliminomethylene) in almost quantitative yields. In the following year, Saegusa, Ito, and Kobayashi discovered Cu-catalyzed polymerization of cyclohexyl isocyanide during their investigation into copper-catalyzed reaction of isocyanide with alcohols [13]. The catalytic activity of the chlorides of nickel(II), cobalt(II), and palladium(II) in the polymerization of cyclohexyl isocyanide was also mentioned in the report, although no experimental details were given. In 1969, Otsuka reported that cyclohexyl isocyanide polymerized in the presence of a cyclic nickel complex, which was isolated in the reaction of the tetrakis(tert-butyl isocyanide)nickel(0) complex with methyl iodide in a high yield [14]. This cyclic nickel complex can be regarded as being a living isocyanide oligomer, which, for the first time, suggested a mechanism for the polymerization.

Following these reports, Nolte and Drenth's group began to make an important contribution. In a series of studies originating in 1973, the nickel-catalyzed polymerization of isocyanides were investigated in detail, to establish a general procedure, in which nickel complexes, such as $NiCl_2$ and $Ni(acac)_2$, were used as catalysts in ethanol [15]. A significant effect of molecular oxygen on the polymerization rate was later indicated by Deming and Novak [16]. Nickel systems are still the most common synthetic access to poly(isocyanide)s.

The significant contribution by the Nolte and Drenth group in this area was that, for the first time, they demonstrated the existence of a non-racemic helical structure for poly(isocyanide)s. Optical resolution using a chiral HPLC technique or asymmetric polymerization led to the isolation of optically active polymers, whose chirality was supposed to be solely due to the main chain helicity. Their effort, in conjunction with that of Novak's and Takahashi's significant contributions to asymmetric polymerization, will be discussed in the next section. Non-asymmetric and asymmetric polymerizations will be described separately in the following sections.

2.1
Synthesis of Racemic Poly(isocyanide)s

2.1.1
Ni-Catalyzed Systems

As has already been described, the nickel-catalyzed-system is currently the most general protocol for the polymerization of isocyanides. An initial report [5] described that $Ni(CO)_4$, $Ni(CO)_3(PPh_3)$, Cp_2Ni, and $CpNi(CO)_2$ show high catalytic activity in the polymerization of cyclohexyl isocyanide in benzene, yielding poly(isocyanide) 2 as a white powder, although the nickel catalysts are a little less active than the corresponding cobalt catalysts (Scheme 7). In a typical experiment, the polymerization of cyclohexyl isocy-

CyNC $\xrightarrow[\substack{\text{benzene, 100 °C, 5 h} \\ \text{(in sealed tubes)}}]{\substack{Ni(CO)_4, Cp_2Ni \\ \text{or } [CpNi(CO)]_2}}$ $\left(\begin{array}{c} N^{\text{-}Cy} \\ \parallel \\ C \end{array}\right)_n$ 2

$Ni(CO)_4$ (4 mol%): 32% conv.
Cp_2Ni (2 mol%): 58% yield
$[CpNi(CO)]_2$ (2 mol%): 50% yield

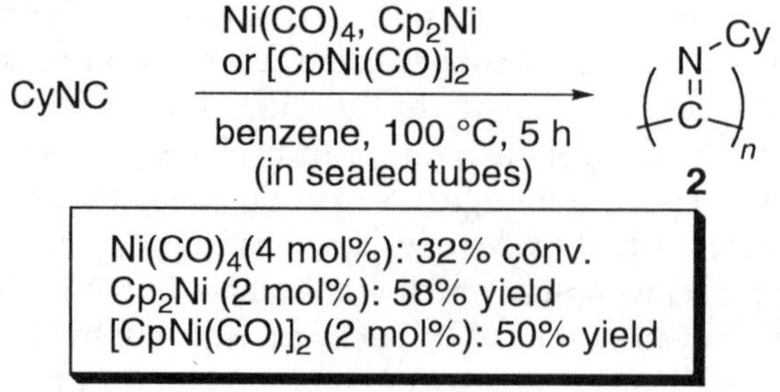

Scheme 7

anide (1.9 g) was carried out at 100 °C in benzene (10 mL) in a sealed tube in the presence of Cp_2Ni (1.8 mol%) for 5 h to afford poly(cyclohexyl isocyanide) **2** in a 58% yield.

The initiation step of polymerization was studied using the stoichiometric reactions of the $(PPh_3)CpRNi(II)$ complexes **3** with isocyanides (Scheme 8) [17, 18]. Insertion of the isocyanides into the carbon–nickel

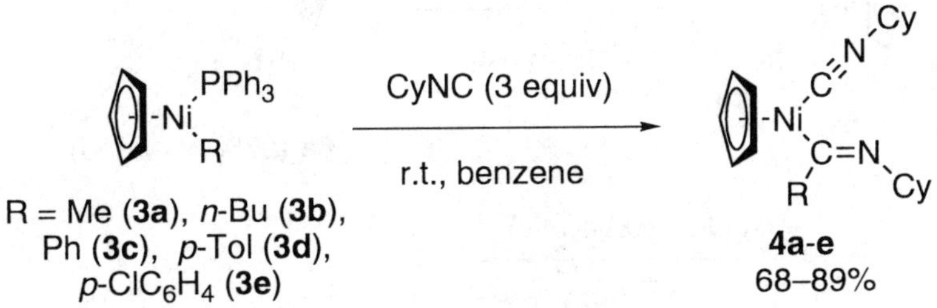

Scheme 8

bonds of **3** afforded the corresponding iminomethylnickel(II) complexes **4** in high yields, with a second molecule of the isocyanide coordinated onto nickel. The migrating organic group (R) on the nickel could be a methyl, butyl, phenyl, or p-substituted phenyl group. Bulky aryl groups on the nickel, such as an o-tolyl or a mesityl group, failed to migrate. Moreover, the reaction depended upon the nature of the organic group on the isocyanides. Thus, tert-butyl isocyanide and 2,6-dimetylphenyl isocyanide afforded the corresponding insertion products in only poor yields. No further polymerization of cyclohexyl isocyanide was observed in the nickel system.

Recently, the CpNi system was reinvestigated to reveal the system's catalytic activity toward aryl isocyanides [19]. The $CpNiR(PPh_3)$ complexes **3a** and **5a–c**, where R=methyl or alkynyl group, polymerized p- or m-substituted aryl isocyanides in THF at room temperature (Scheme 9). Alkyl iso-

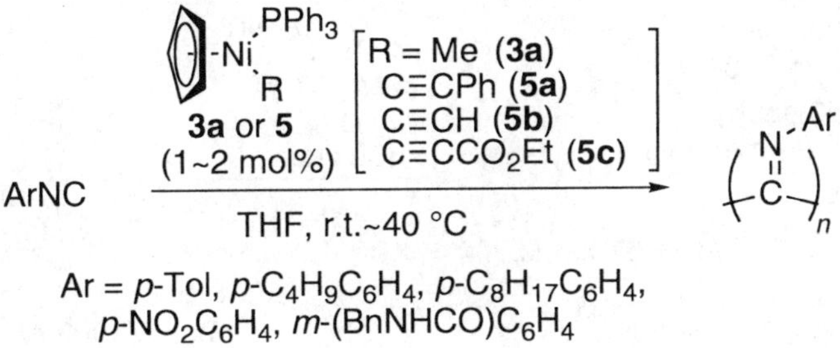

Scheme 9

cyanides and sterically demanding aryl isocyanides, such as 2,6-dimethylphenyl isocyanide, did not undergo polymerization, which is in good agreement with previous studies. The stoichiometric reaction of **5a** with three equivalents of aryl isocyanide yielded the bis(imino) complex **6a** (Scheme 10), the formation of which is in sharp contrast to the formation of

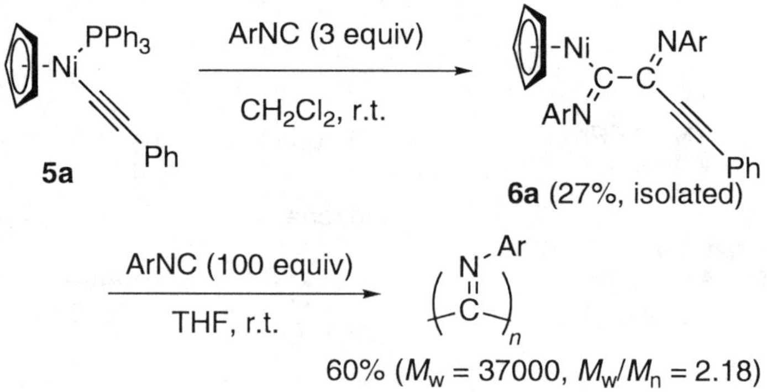

Scheme 10

the monoimino complex **4** in the reaction of alkyl isocyanide (Scheme 8). The double insertion complex **6a** served as an initiator for further polymerization of the aryl isocyanide. From these results, the double insertion complex **6** is likely to be the intermediate in the polymerization process, which proceeds through successive insertions of isocyanides into the Ni–C bond of the poly(imino)nickel intermediates. Although this mechanism suggests a living character for the polymerization, the given polydispersity index was only 2.18 for the polymerization of *m*-(benzylaminocarbonyl)phenyl isocyanide with CpNi(CCPh)(PPh$_3$), indicating that chain-termination and/or chain-transfer may occur during the polymerization process. Nevertheless,

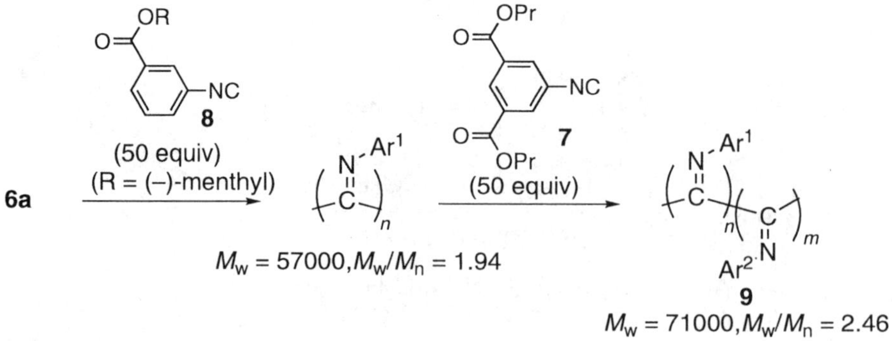

Scheme 11

the polymerization process displayed some living properties, as exemplified by the successful block copolymerization by the addition of a second monomer **7** (50 equiv) after complete consumption of the first monomer **8** (50 equiv), yielding the diblock copolymer **9** (Scheme 11).

A more product-like intermediate for the nickel-catalyzed polymerization was isolated in the reaction of the tetrakis(*tert*-butyl isocyanide)nickel(0) complex **10** with organic halides [14, 20] (Scheme 12). In the reaction of **10**

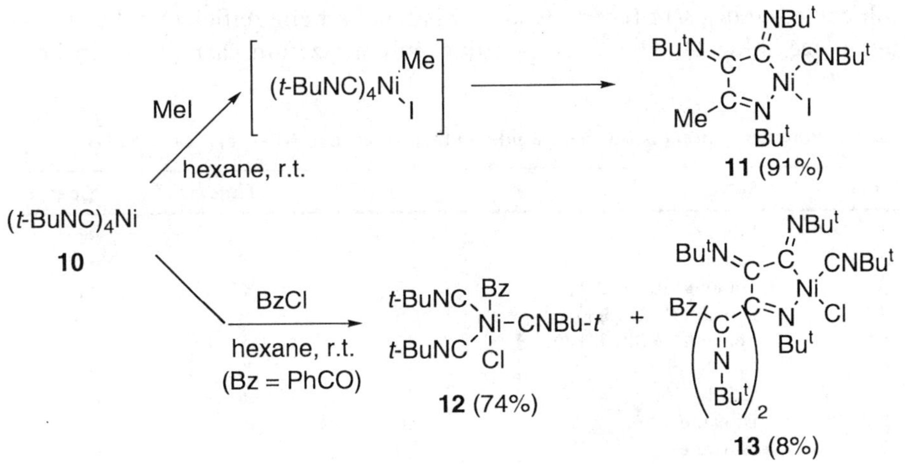

Scheme 12

with methyl iodide, a cyclic nickel(II) complex **11** was isolated in a high yield. Complex **11** may be formed through the triple insertion of *tert*-butyl isocyanide into the methyl–nickel bond, which is formed by oxidative addition of MeI to the nickel(0) species. An analogous reaction of the nickel(0) complex with benzoyl chloride yielded a pentacoordinated tris(isocyanide)benzoylnickel(II) chloride **12**, along with a pentaimino nickel complex **13** in a low yield. Both the cyclic polyimino nickel complexes **11** and **13** involve intramolecular coordination of the imino nitrogen to nickel, which may stabilize the cyclic structure. Complex **13** was found to serve as a catalyst for the polymerization of cyclohexyl isocyanide, yielding poly(cyclohexyl isocyanide) at temperatures between 30 and 60 °C.

Following these reports, a more general Ni-catalyzed system was reported by Nolte, Drenth, and co-workers, who used $NiCl_2$, $Ni(acac)_2$, and $NiCl_2$(EtOH)(*t*-BuNC) as catalysts (Scheme 13) [15, 21]. Among these three nickel catalysts, $NiCl_2$ appears to be the least effective, as the low solubility of nickel chloride makes the polymerization system a heterogeneous system. The remarkable acceleration effect of protic solvents on the polymerization rate was also discovered. Ethanol (entry 1), dioxane-water (3/1) (entry 4), and

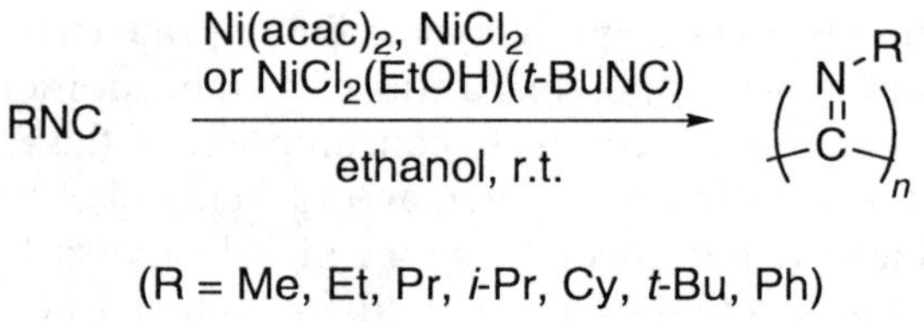

Scheme 13

$$\text{RNC} \xrightarrow[\text{ethanol, r.t.}]{\text{Ni(acac)}_2,\ \text{NiCl}_2 \text{ or NiCl}_2(\text{EtOH})(t\text{-BuNC})} \left(\begin{array}{c} \text{N}^{\text{R}} \\ \parallel \\ \text{C} \end{array}\right)_n$$

(R = Me, Et, Pr, i-Pr, Cy, t-Bu, Ph)

toluene-ethanol (3/1) (entry 3) are classified as being efficient solvent systems (see Table 1). The corresponding polymerization carried out in ben-

Table 1 Polymerization of ethyl isocyanide in the presence of Ni(acac)$_2$ (ca. 2 mol%)

Entry	Solvent	Temp °C	Time h	Yield %
1	Ethanol	25	5	96
2	Aniline	25	5	69
3	Toluene/EtOH (3/1)	25	5	82
4	Dioxane/H$_2$O (3/1)	25	5	99
5	Benzene (with CCl$_3$CO$_2$H)	25	5	66
6	(Me$_2$N)$_3$PO	25	5	9
7	CHCl$_3$	40	5	69
8	Dioxane	40	5	6
9	Benzene	40	5	2
10	Benzene	40	79	22

zene, toluene, or dioxane is much slower than polymerization carried out in ethanol. Nevertheless, the addition of a protic compound to an aprotic solvent system makes polymerization possible. Thus, polymerization of ethyl isocyanide in benzene in the presence of Ni(acac)$_2$ (2 mol%) with a catalytic quantity of trichloroacetic acid (1 mol%) afforded poly(ethyl isocyanide) in a 66% yield at room temperature after 5 h (entry 5), whereas the corresponding reaction in the absence of added acid afforded the same polymer in only a 2% yield, even at 40 °C (entry 9).

The superior catalytic activity of Ni catalysts over other transition metal complexes was proven by a comparison of a series of metal acetylacetonates as catalysts for the polymerization of ethyl isocyanide in chloroform (Scheme 14) [15]. Fe(acac)$_3$, Mn(acac)$_2$, Zn(acac)$_2$, and Cd(acac)$_2$ showed either no, or an extremely low, catalytic activity, whereas Cr(acac)$_3$ and Co(acac)$_3$ afforded almost 10% yield of poly(isocyanide). Co(acac)$_2$, Pd(acac)$_2$, and Cu(acac)$_2$ exhibited moderate activity, resulting in the formation of the polymer in 29–61% yields. Ni(acac)$_2$ showed almost the same catalytic activity as Co$_2$(CO)$_8$, whose remarkable catalytic activity had already been established by Yamamoto et al. [5].

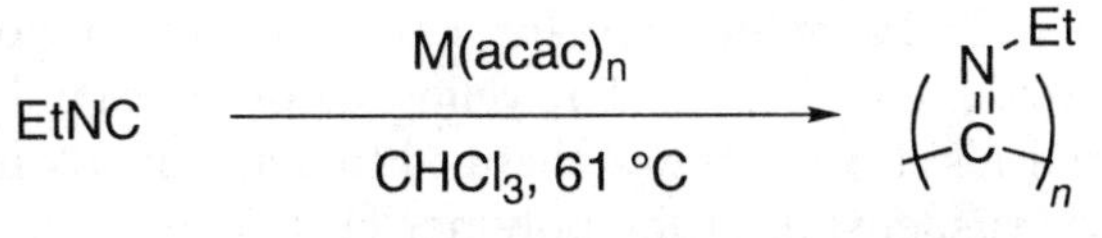

EtNC $\xrightarrow[\text{CHCl}_3, 61\,°\text{C}]{M(acac)_n}$ $\left(\begin{array}{c}N^{-Et}\\ \|\\ C\end{array}\right)_n$

Fe(acac)$_3$ 0%	Ni(acac)$_2$ 100%	Zn(acac)$_2$ 0%
Cr(acac)$_3$ 8%	Pd(acac)$_2$ 53%	Cd(acac)$_2$ 0%
Co(acac)$_3$ 9%	Cu(acac)$_2$ 29%	
Co(acac)$_2$ 61%	Mn(acac)$_2$ 1%	Co$_2$(CO)$_8$ 100%

Scheme 14

The physical properties of poly(isocyanide)s prepared by nickel catalysis have been studied [15]. All the polymers show strong IR absorption bands between 1625 and 1635 cm^{-1}, which are in good agreement with the poly(isocyanide)s obtained previously by acid-catalyzed polymerization. Polymers derived from ethyl, 1-phenylethyl, and *tert*-butyl isocyanides are soluble in benzene, whereas those derived from methyl, butyl, isopropyl, cyclohexyl, and phenyl isocyanides are insoluble. Poly(*tert*-butyl isocyanide) shows a considerably low degree of polymerization (ca. 15), which is presumably due to steric hindrance. More recently, the synthesis and physical properties of poly(isocyanide)s have been systematically studied [22]. In this study, the polymerization of a series of isocyanides was carried out under common reaction conditions (an initial isocyanide concentration of ca. 1 mol L^{-1}, 0.5 mol% NiCl$_2$, ethanol, room temperature, two days). The results indicated that the solubility of poly(aryl isocyanide)s was strongly dependent upon the substituents on the aromatic rings of the isocyanides. In general, *o*-substituted aryl isocyanides are more prone to yield CHCl$_3$-insoluble polymers. Among the alkyl isocyanides, the primary alkyl isocyanides mostly yield CHCl$_3$-soluble polymers. Exceptions are for methyl and neopentyl isocyanides, which afford insoluble polymers as the major fraction. The formation of soluble polymers from butyl isocyanide is in sharp contrast to earlier reports [15], in which an insoluble polymer was obtained in the Ni(acac)$_2$ mediated polymerization, indicating that the choice of catalyst may have some impact on the nature of the resulting polymer.

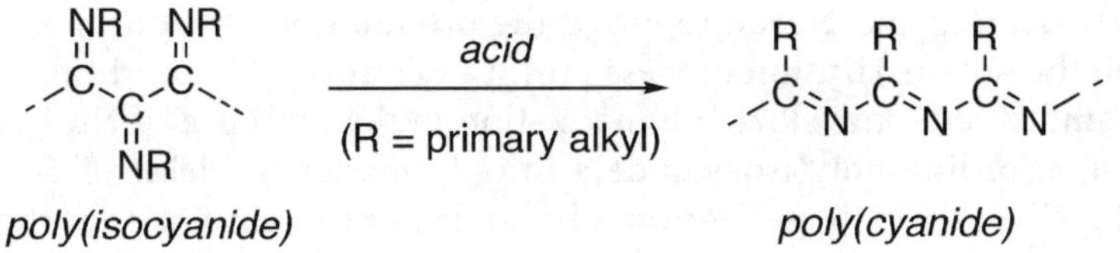

poly(isocyanide) $\xrightarrow[\text{(R = primary alkyl)}]{\text{acid}}$ poly(cyanide)

Scheme 15

All the polymers had a yellowish color when prepared in polar solvents at temperatures between 0 and 25 °C. The yellow color of the poly(isocyanide)s derived from *prim*-alkyl isocyanides changed to black on addition of an acid to the solution or suspension of the polymer [15]. This change in color was not observed for polymers derived from *sec*- and *tert*-alkyl isocyanides. The structure of the black polymer was assigned to poly(ethyl cyanide) from spectroscopic and conductivity measurements (Scheme 15).

A proposed mechanism for Ni-mediated polymerization was given by Drenth and Nolte (Scheme 16) [23]. They presumed the involvement of a cationic tetrakis(isocyanide)nickel(II) complex **A** before the initiation step.

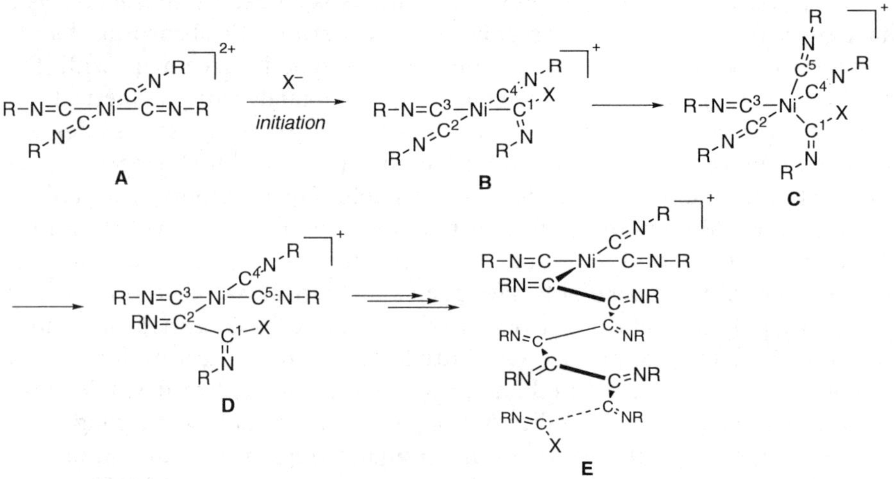

Scheme 16

A nucleophile (X⁻) attacks one of the isocyano carbon atoms, generating the α-iminomethylnickel species **B**. Then, coordination of the fifth isocyanide to the nickel center (to form **C**) is followed by the migration of the iminomethyl group onto the neighboring isocyano carbon atom (C^2), forming a dimeric intermediate **D**. This associative migration step is repeated to form the poly(isocyanide) **E**. Nolte and Drenth called this possible mechanism a "merry-go-round" mechanism, and it was also proposed as the key mechanism for the screw-sense selective polymerizations discussed later.

In 1991, Deming and Novak reported the pronounced effect of molecular oxygen on the polymerization process [16]. Under anaerobic conditions, isocyanides underwent only slow polymerization in the presence of $NiCl_2$, even in ethanol, affording poly(isocyanide)s in only moderate yields of 50–60% after 20 h. When the polymerization was carried out in air at 1 atm, the polymerization was completed within 1 h. Air at 1 atm was sufficient to maintain the active Ni(II) species in the system. If oxygen was continuously bubbled into the reaction system, then no polymerization took place, but the oxidation of isocyanide occurred, leading to the formation of isocyanate. These results, in conjunction with the additional data from ESR spectroscopy, cyclic voltammetry, and bulk magnetic susceptibility measurements, provided evidence that a Ni(II)–Ni(I) redox reaction may be involved in the polymerization mechanism. An excess of isocyanide in the polymerization system can lead to a reduction in the Ni(II) intermediate to form the Ni(I) species, which is only moderately active in the propagation stage.

Deming and Novak then began to develop a modified nickel-initiated system for a highly efficient polymerization of isocyanides [16]. They used the η^3-allylnickel trifluoroacetate dimer **14**, whose allyl group may serve as an initial migrating group, thus facilitating the initiation step (Scheme 17).

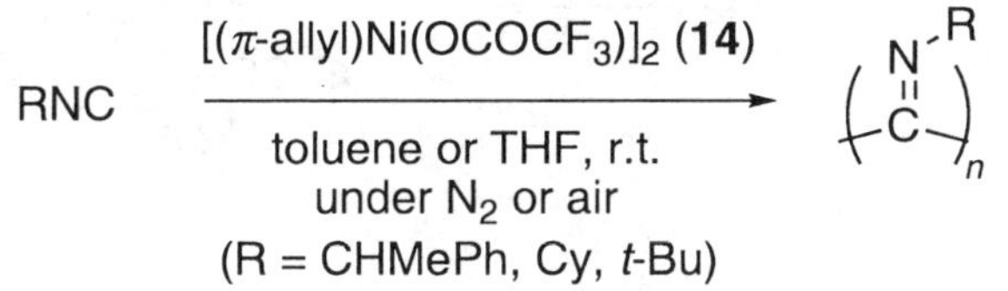

$$\text{RNC} \xrightarrow[\substack{\text{toluene or THF, r.t.} \\ \text{under N}_2 \text{ or air} \\ (R = \text{CHMePh, Cy, }t\text{-Bu})}]{[(\pi\text{-allyl})\text{Ni}(\text{OCOCF}_3)]_2 \ (\mathbf{14})} \left(\underset{\text{C}}{\overset{\text{N}^{\,R}}{\|}} \right)_n$$

Scheme 17

With the nickel complex in an aprotic solvent, such as THF, cyclohexyl isocyanide was polymerized instantaneously, affording the corresponding polymer in a quantitative manner. Furthermore, the successful application of this new catalyst system was demonstrated by the quantitative polymerization of *tert*-butyl isocyanide (DP=ca. 27), which is one of the least reactive isocyanides.

The new nickel system using **14** still showed non-living chain growth *in THF*, as evidenced by the formation of a high molecular weight polymer, even with the low conversion of the monomer (M_n=57,000 for a 5% conversion), in the polymerization of 1-phenylethyl isocyanide. However, an attempted polymerization *in toluene* was found to proceed via a living mechanism with an almost tenfold increase in catalyst activity [24]. The molecular weight of the polymer increased linearly with increasing ratio of monomer/catalyst. The molecular weight distribution found in the living polymerization system ranged from 1.1 to 1.6. Once again, an accelerant effect of oxygen was observed. Moreover, the polymerization of isocyanide under oxygen was a zero order reaction for the monomer, whereas polymerization under nitrogen was a first order reaction for the monomer. This change in the rate-determining step was attributed to the Ni(I)–Ni(II) redox reaction, which was also observed in the NiX$_2$ catalyzed polymerization.

The mechanism for the initiation step of the polymerization was investigated using the stoichiometric reaction of η^3-allylnickel trifluoroacetate (14) with 1-phenylethyl isocyanide labeled with ^{13}C at the isocyano carbon atom (Scheme 18) [24, 25]. ^{13}C NMR measurements revealed that addition of three

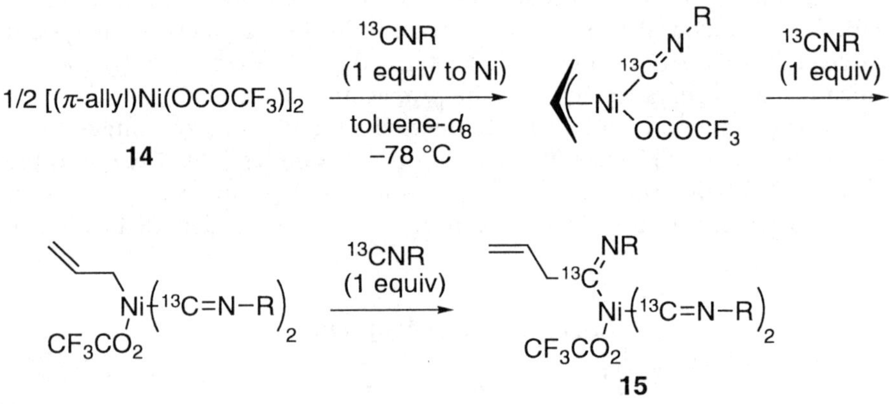

Scheme 18

equivalents of isocyanide to a toluene-d_8 solution of 14 at a temperature of −60 °C resulted in the formation of 1-imino-3-buten-1-ylbis(isocyanide) nickel trifluoroacetate 15.

As for the propagation mechanism in the presence of oxygen, Deming and Novak initially proposed that molecular oxygen may serve as an "oxidant", which oxidizes the inactive Ni(I) species to the more catalytic active Ni(II) species [16]. Soon after, they revised this idea based on the results of additional experiments [25, 26]. Their revised proposal for the effect of oxygen was based upon the "spin trap" mechanism, as depicted in Scheme 19. The coordination of isocyanide to the bis(isocyanide)(CF$_3$CO$_2$)nickel(II) complex F bearing an allyl or a propagating poly(iminomethylene) group induces metal–ligand charge transfer to afford the Ni(I) complex G bearing a radical cationic isocyanide ligand. The Ni(I) complex G may be involved in polymerization under an inert atmosphere. On the other hand, in the presence of oxygen, the radical cationic isocyanide ligand is trapped by oxygen to form the Ni(I) complex H. Coordination of isocyanide followed by migratory insertion may constitute the propagation sequences.

The nickel(II) complex 14 is also known as a catalyst for the polymerization of 1,3-butadiene. The versatility of this nickel catalyst was used in the synthesis of butadiene-isocyanide block copolymers [27]. In the so-called "change of mechanism" block copolymerization, 1,3-butadiene was initially polymerized with 14, yielding a butadiene polymer bearing living η^3-allyl-nickel termini (Scheme 20). Addition of *tert*-butyl isocyanide to this living

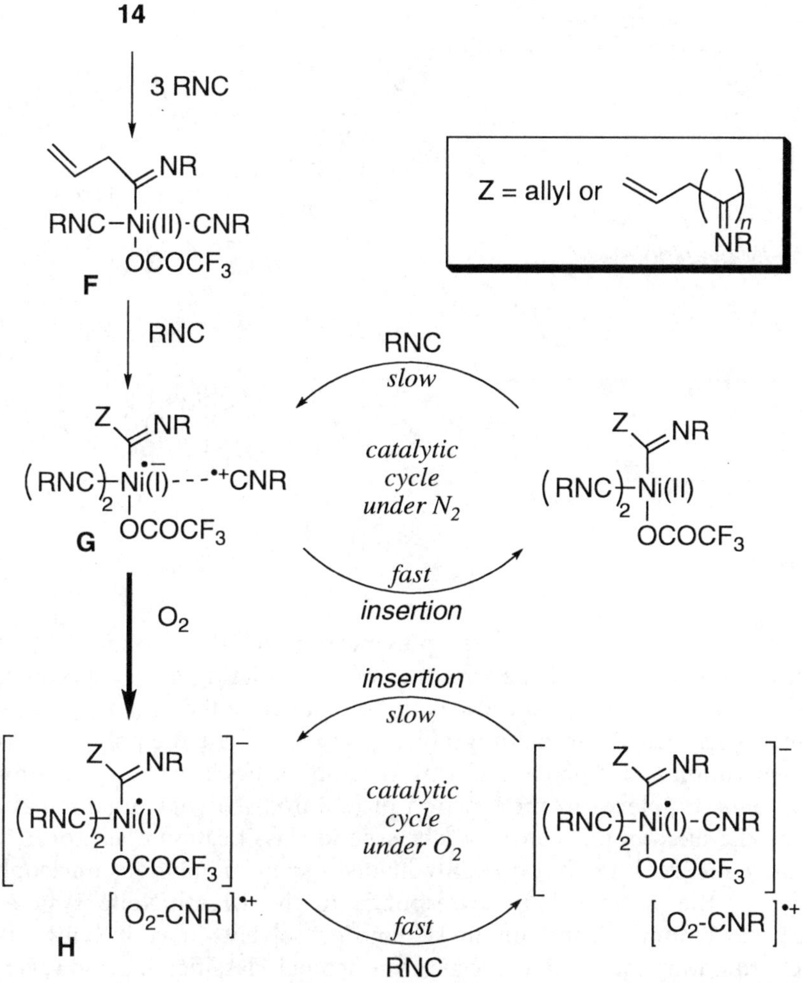

Scheme 19

system allowed for the polymerization of isocyanide via the aforementioned mechanism, yielding the butadiene-isocyanide copolymer **16a**. This copolymerization provides a way for the synthesis of copolymers possessing elastomeric butadiene segments and rigid rod helical isocyanide segments in their main chain, and was extended to the synthesis of isocyanide-butadiene-isocyanide triblock copolymers using bifunctional initiators bearing two η^3-allylnickel moieties [28]. Furthermore, the "change of mechanism" copolymerization of isocyanide with terminal allenes, yielding the block copolymer **16b**, has been reported using catalyst **14** with PPh$_3$ [29].

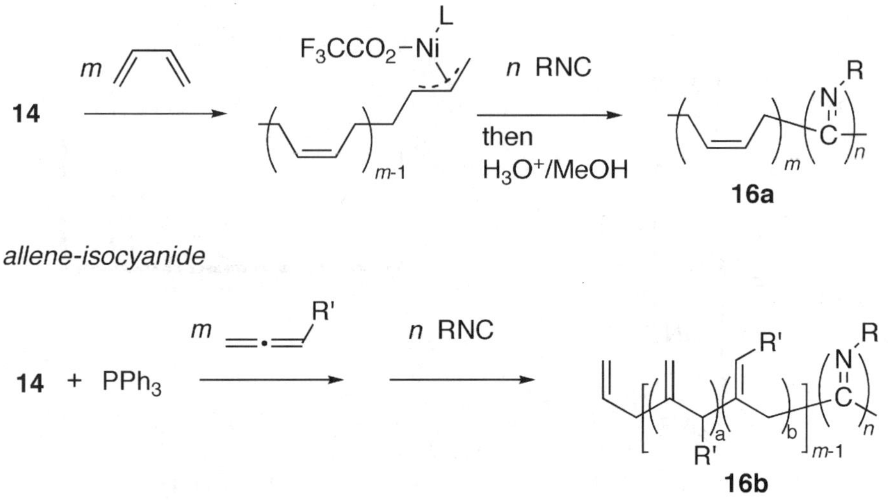

Scheme 20

To summarize the Ni-mediated polymerizations, there seems to be two classifications in terms of the structure of the initiating nickel species. One involves the organonickel species as initiators, where the organic groups on the nickel serve as the initial migrating group to initiate the polymerization. Polymerization with CpNi and allylnickel complexes, as well as formation of oligoimino complexes in the reaction of $(t\text{-BuNC})_4$Ni with organic halides, may be classified in this category. The second classification does not involve definitive nucleophiles, but certainly involves some anonymous nucleophilic species in the system. This corresponds to the Drenth/Nolte system, in which the anionic ligand on nickel and/or solvents may initially attack the coordinating isocyanide. Despite the formal classification, however, all Ni-mediated polymerizations may be mechanistically related, being characterized regarding their propagation steps involving successive migratory insertions. The Drenth/Nolte system has been extended to asymmetric polymerization, where optically active amines are used as the definitive nucleophilic species.

2.1.2
Pd-Catalyzed Systems

The first multiple insertion of isocyanide into the Pd–C bond of organopalladium complexes was reported by Yamamoto in 1970 (Scheme 21) [30]. The reaction of one, two, and three equivalents of cyclohexyl isocyanide with the *trans*-iodobis(phosphine)methylpalladium(II) complexes **17** yielded the monoimino (**18**), bisimino (**19**), and trisimino (**20**) palladium complexes, re-

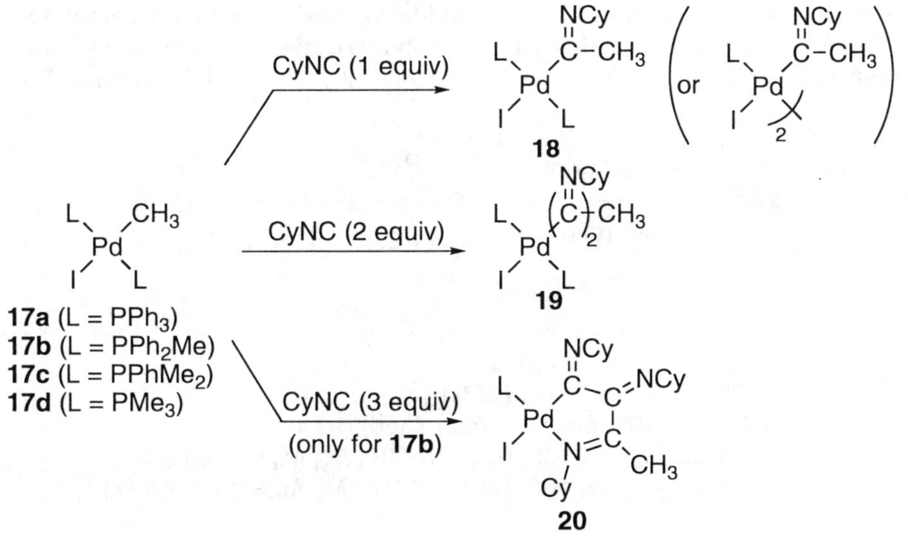

Scheme 21

spectively. The double insertion reaction, yielding **19**, is applicable for a wide variety of phosphine ligands, including PPh$_3$, PPh$_2$Me, PPhMe$_2$, PMe$_3$, and PBu$_3$, although the corresponding palladium complex **17e** bearing bulky PPh$_2$Cy groups fails to undergo the double insertion reaction, but selectively gives rise to a mono insertion product. The triple insertion reaction has a strict preference for the phosphine ligand. Thus, only PPh$_2$Me-coordinated palladium complex **17b** undergoes the triple insertion reaction with cyclohexyl isocyanide. It was noted that no further insertion of cyclohexyl isocyanide to the tris(imino) complex took place, presumably because of stabilization by the intramolecular coordination of the imino nitrogen to palladium [31].

No appreciable development had been noted in the palladium-catalyzed polymerization of isocyanide until Takashi's group reported a new palladium initiator for the polymerization of aryl isocyanides [32–34]. They tested a series of alkynylpalladium complexes, including mononuclear palladium acetylides **21**, (µ-ethynediyl)dipalladium complexes **22**, and Pd–Pt heteronuclear µ-ethynediyl complexes **23**, as initiators for the polymerization of isocyanides (Fig. 1) [34]. Although the mononuclear acetylides **21** only under-

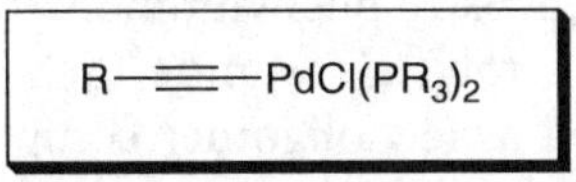

R≡≡PdCl(PR$_3$)$_2$

R = H (**21a**), Ph (**21b**), PdCl(PR$_3$)$_2$ (**22**), PtCl(PR$_3$)$_2$ (**23**)

Fig. 1

went a single insertion of aryl isocyanide to yield monoimino complexes, complexes **22** and **23** initiated the polymerization of aryl isocyanides (100 equiv) under reflux in THF (Scheme 22). The Pt–Pd heteronuclear

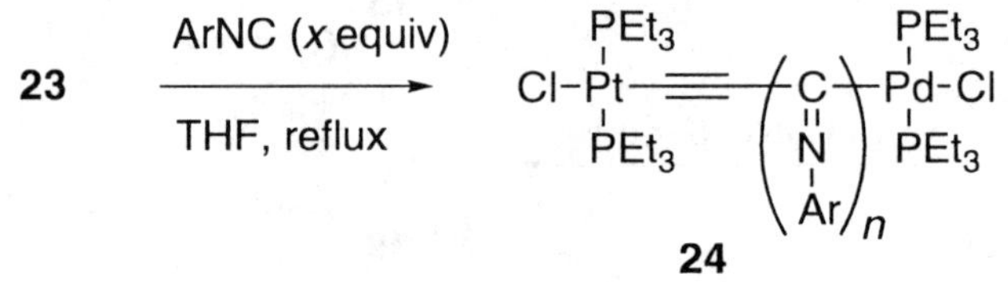

23 $\xrightarrow{\text{ArNC }(x\text{ equiv})}_{\text{THF, reflux}}$
$$\text{Cl-Pt(PEt}_3\text{)}_2\text{-[C(=NAr)]}_n\text{-Pd(PEt}_3\text{)}_2\text{-Cl}$$
24

Ar = Ph, x = 2: (n = 2) (94%)
Ar = Ph, x = 10: (n ~ 10) (47%)
Ar = Ph, x = 100: (M_n not determined) (92%)
Ar = p-BuC$_6$H$_4$, x = 100: (M_n = 13000, M_w/M_n = 1.02) (87%)
Ar = p-OcC$_6$H$_4$, x = 100: (M_n = 17000, M_w/M_n = 1.01) (88%)

Scheme 22

complexes **23** were finally selected as being the most effective catalysts, based on oligomerization experiments in which an improved oligomer distribution was achieved by the heteronuclear initiators **23**. The oligomerization study also revealed that the insertion of isocyanide took place only at the Pd–C bond of the heteronuclear complex, leaving the Pt–C bond intact. Even in the reaction of the dipalladium complexes **22**, successive insertion took place exclusivly at one of the two Pd–C bonds. No explanation has been provided for this remarkable non-reacting palladium or platinum moiety in the dinuclear initiators.

The polymers prepared from 100 equiv of phenyl isocyanide exhibited only low solubility, and this led to precipitation during the polymerization process. This difficulty was overcome by the use of aryl isocyanide having alkyl chains (e.g., butyl and octyl), which yielded polymers that were soluble in benzene and THF, but insoluble in methanol [34]. The polymers prepared from the alkyl-substituted aryl isocyanides exhibited quite narrow polydispersity indices that were quite close to 1. For example, p-butylphenyl and p-octylphenyl isocyanides exhibited polydispersity indices of 1.02 (M_n=13,000) and 1.01 (M_n=17,000), respectively, suggesting a living mechanism for this polymerization system. The living nature of the polymerization was further demonstrated by the achievement of block copolymerization. Thus, an oligomer complex **24** constituted of N-phenyliminomethylene (n~10) units was isolated, and reacted with p-tolyl isocyanide (Scheme 23). Oligomerization was re-initiated, and yielded a new oligomer complex **25a** having N-phenyl- and N-p-tolyliminomethylene units. Block polymers **25b** with higher molecular weight were also prepared successfully using this protocol (Scheme 24).

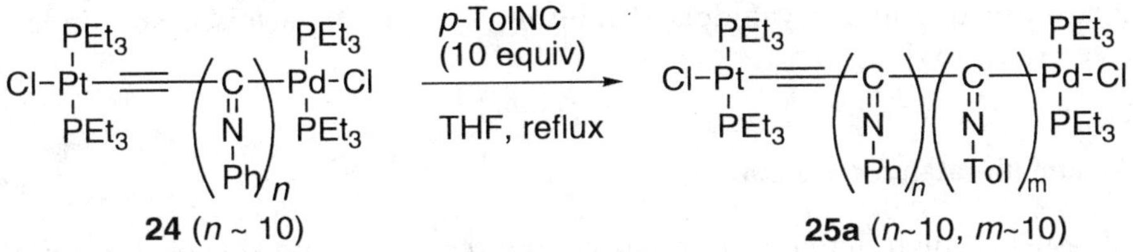

Scheme 23

$$24\ (n \sim 10) \xrightarrow[\text{THF, reflux}]{p\text{-TolNC (10 equiv)}} 25a\ (n{\sim}10,\ m{\sim}10)$$

Cl–Pt(PEt$_3$)$_2$≡(C(=NPh))$_n$–Pd(PEt$_3$)$_2$–Cl → Cl–Pt(PEt$_3$)$_2$≡(C(=NPh))$_n$(C(=NTol))$_m$–Pd(PEt$_3$)$_2$–Cl

These results suggest that the living palladium termini were quite stable, and were able to maintain catalytic activity even on isolation.

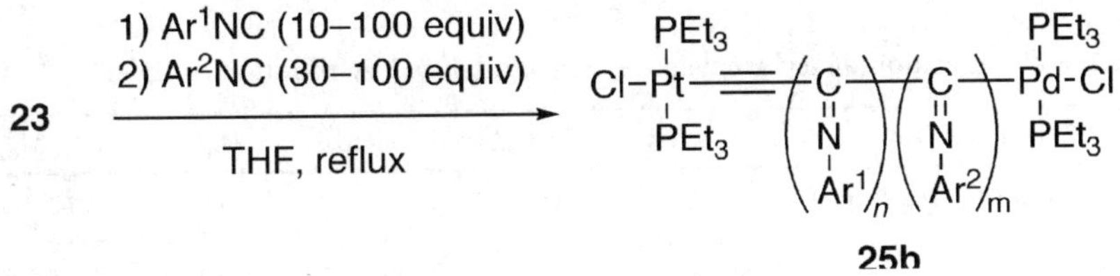

Scheme 24

$$23 \xrightarrow[\text{THF, reflux}]{\substack{1)\ \text{Ar}^1\text{NC (10–100 equiv)} \\ 2)\ \text{Ar}^2\text{NC (30–100 equiv)}}} 25b$$

Cl–Pt(PEt$_3$)$_2$≡(C(=NAr^1))$_n$(C(=NAr^2))$_m$–Pd(PEt$_3$)$_2$–Cl

Multifunctional initiators bearing two (**26a**) and three (**26b**) Pd-Pt μ-ethynediyl units have been prepared and successfully used for the synthesis of

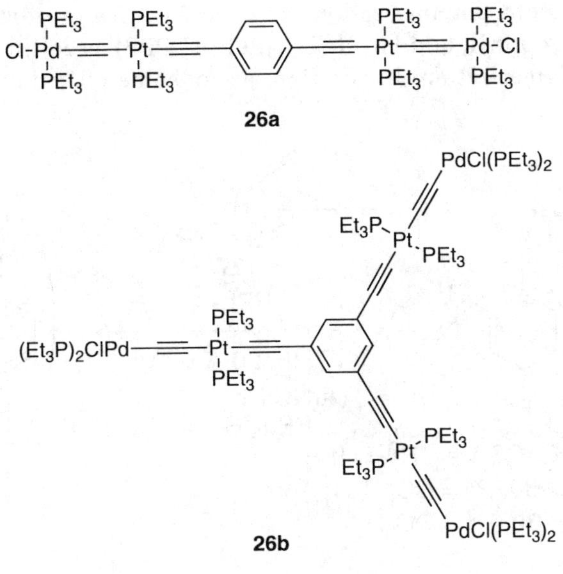

26a: Cl–Pd(PEt$_3$)$_2$≡Pt(PEt$_3$)$_2$≡–C$_6$H$_4$–≡Pt(PEt$_3$)$_2$≡Pd(PEt$_3$)$_2$–Cl

26b: tris-substituted benzene core with three Pt(PEt$_3$)$_2$ and three PdCl(PEt$_3$)$_2$ arms connected via ethynediyl linkages; one arm shown as (Et$_3$P)$_2$ClPd≡Pt(PEt$_3$)$_2$≡–

Fig. 2

multi-armed poly(isocyanide)s exhibiting a narrow polydispersity index (see Fig. 2) [35].

2.1.3
Co- and Rh-catalyzed Systems

Although a high activity of cobalt complexes was found during the early stages of research in this area, Group 9 metal complexes have rarely been examined in detail as catalysts for the polymerization of isocyanides. Recently, organorhodium complexes have been found to catalyze the polymerization of sterically encumbered aryl isocyanides [36]. This polymerization system consisted of arylrhodium complexes 27, bearing o-substituents on the aryl group (Table 2), and PPh$_3$, which had to be used in excess (10 equiv based

Table 2 Polymerization of aryl isocyanides 28a–c catalyzed by rhodium complexes 27

Entry	Initiator (Ar)	ArNC	ArNC/Rh	Yield %	$M_n/10^3$	M_w/M_n
1	27a (2,6-Me$_2$C$_6$H$_3$)	28a	25	83	4.7	1.12
2	27a	28a	50	80	9.3	1.19
3	27a	28a	75	99	14.1	1.26
4	27a	28a	100	92	17.5	1.47
5	27a	28b	50	83	22.6	1.34
6	27b (2,4,6-i-Pr$_3$C$_6$H$_2$)	28a	50	82	8.9	1.29
7	27b	28c	50	98	16.6	1.25
8	27c (2-PhC$_6$H$_4$)	28a	50	90	9.5	1.26

on Rh) for smooth polymerization (Scheme 25). Use of other phosphines, such as PCy$_3$, P(OPh)$_3$, and Ph$_2$P(CH$_2$)$_3$PPh$_2$ (DPPP) as additives did not result in any polymerization. In addition, use of some related organorhodium

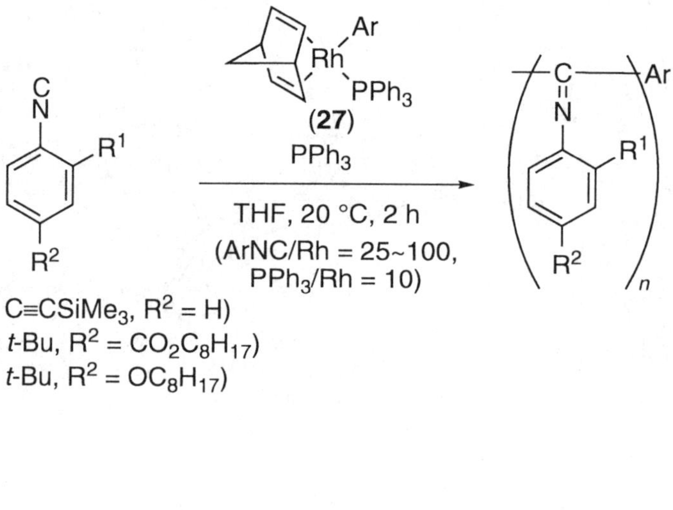

28a (R^1 = C≡CSiMe$_3$, R^2 = H)
28b (R^1 = t-Bu, R^2 = CO$_2$C$_8$H$_{17}$)
28c (R^1 = t-Bu, R^2 = OC$_8$H$_{17}$)

Scheme 25

complexes, such as Rh(C≡CPh)(nbd)(PPh$_3$)$_2$, Rh(Me)(nbd)(PPh$_3$)$_2$, and Rh(Ph)(nbd)(PPh$_3$) yielded a mixture of oligomers and polymers, suggesting that the organic group on the rhodium plays a crucial roll in the polymerization process. This new catalyst system is particularly suitable for the polymerization of o-substituted aryl isocyanides 28a–c (Table 2). For example, 2-(trimethylsilylethynyl)phenyl isocyanide 28a affords a yellow-brown polymer in the presence of complexes 27a–c with PPh$_3$ in THF at 20 °C (Entries 1–4, 6, and 8). The resulting polymer was soluble in common organic solvents. Aryl isocyanides 28b and 28c bearing o-tert-butyl groups were similarly polymerized under the same reaction conditions (Entries 5 and 8), although those with no o-substituent or with a primary alkyl o-substituent failed to afford high-molecular-weight polymers. The conventional NiCl$_2$-catalyzed system did not catalyze the polymerization of 28a–c.

A mechanism for Rh-catalyzed polymerization was proposed to involve successive insertions of isocyanides into the Rh–C bond [36]. This was proven from a terminal analysis of the resulting polymer using ^{1}H NMR spectroscopy, which suggested the presence of a 2,6-dimethylphenyl group at the end of the polymerized polymer using 27a. The M_n value of the polymer was in good agreement with the value calculated from the initial monomer/catalyst ratio. The polydispersity index ranged from 1.1 to 1.5, and increased almost linearly with increasing molecular weight. This tendency, along with the failure of block copolymerization by the addition of a second monomer after the complete consumption of the first monomer, suggested that the propagating Rh terminus was not robust enough to grant a flawless living nature to this polymerization process.

2.2
Synthesis of Chiral Non-Racemic Poly(isocyanide)s

The existence of a non-racemic helical conformation in poly(isocyanide)s was first suggested by Millich et al. in 1969 [37]. They carried out polymerizations of optically pure d- and l-1-phenylethyl isocyanide 29 in the presence of an acidic ground glass catalyst (Scheme 26). The observation of a

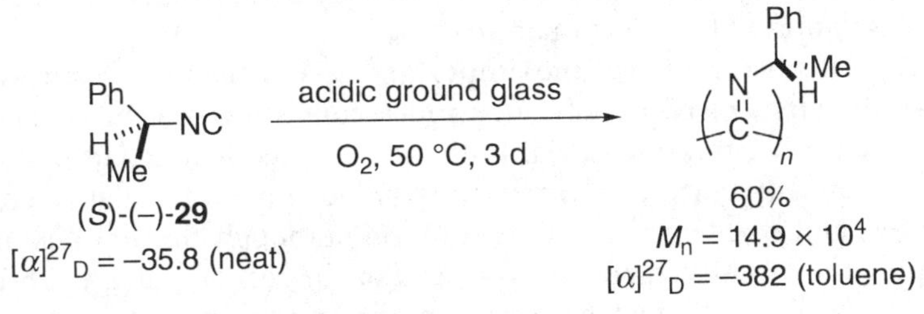

(S)-(–)-29
$[\alpha]^{27}_D = -35.8$ (neat)

acidic ground glass
O_2, 50 °C, 3 d

60%
$M_n = 14.9 \times 10^4$
$[\alpha]^{27}_D = -382$ (toluene)

Scheme 26

significant increase in the optical rotation through polymerization led them to the assumption that non-racemic helical conformation was induced at the main chain. Related polymerizations of chiral non-racemic isocyanides using transition metal catalysts are described in the following sections. Some review articles on the synthesis of helical polymers have been published [38–43].

The first optically active poly(isocyanide)s devoid of chiral groups on the nitrogen atom were isolated by optical resolution via column chromatography on the chiral stationary phase [44, 45]. Right- and left-handed helices of poly(*tert*-butyl isocyanide) ((±)-**30**) (degree of polymerization ca. 20), which was prepared as a racemate by nickel-catalyzed polymerization, were successfully resolved on poly((S)- or (R)-sec-butyl isocyanide), which served as a chiral support for column chromatography (Scheme 27). Using poly((S)-

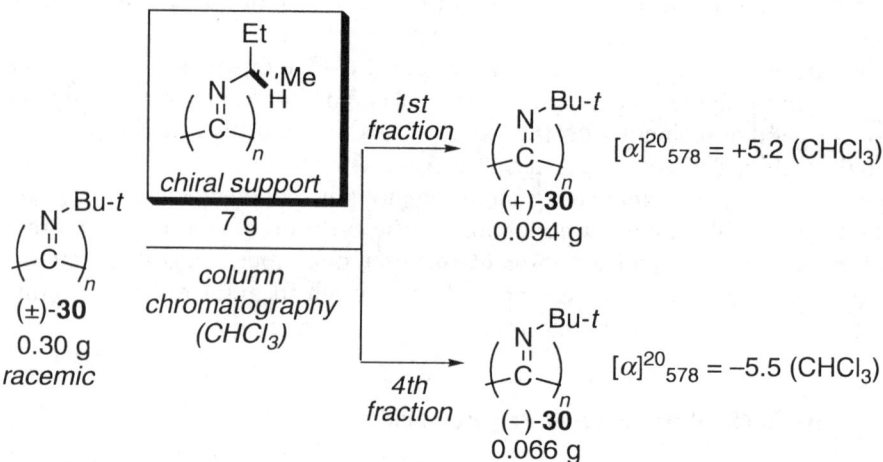

Scheme 27

sec-butyl isocyanide) as a chiral support, dextrorotatory polyisocyanide (+)-**30** was initially eluted, followed by the elution of the levorotatory polymer (−)-**30** when chloroform was used as the eluent. Further fractionation allowed the isolation of (−)-**30** with an $[\alpha]^{20}_{578}$ value up to −16.

These experiments revealed the important stereochemical features of polyisocyanide. First, racemic polyisocyanides consisting of achiral repeating units exist as a 1:1 mixture of right- and left-handed helices. Second, the right- or left-handed helical conformation is stable enough to exist as either enantiomer at ambient temperature in solution, although the stability may depend upon the N-substituent as well as the degree of polymerization. Based on these ideas, efficient systems for asymmetric polymerization of isocyanides have been further pursued.

The following subsections describe the asymmetric polymerization of isocyanides using four classifications based upon the mechanism of the asymmetric induction. The first two subsections deal with homo- and copolymerization of *chiral, non-racemic* isocyanides, and asymmetric polymerization of *achiral* isocyanides by chiral enantiopure nickel complexes are described in the final two subsections.

2.2.1
Screw-Sense Induction by Incorporation of Chiral Non-racemic Isocyanide

Nickel-catalyzed polymerization of optically active isocyanides have been studied with particular attention being paid to the screw-sense induction to the polymer main chain (Scheme 28) [46–48]. Representative homopoly-

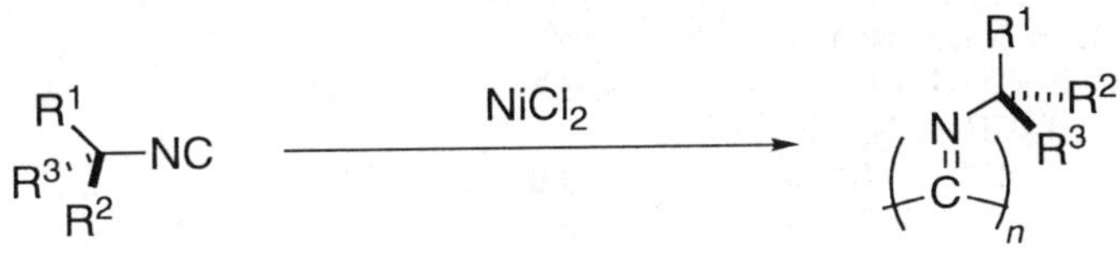

Scheme 28

mers derived from optically active isocyanides are shown in Table 3, with the specific rotations of the starting isocyanides and the resulting polymers.

It is interesting to compare the optical rotations of poly((S)-1-phenylethyl isocyanide) obtained from the Ni-catalyzed reaction (Table 3, Entries 1 and 2) with the optical rotation of the corresponding polymer obtained from acid-catalyzed polymerization (Scheme 26). Both polymerizations yield levorotating poly(isocyanide)s, whose specific rotations are almost ten times larger than those of the starting isocyanide 29. These results suggest that the sense, as well as the degree of asymmetric induction to the polymer main chain, may be independent of the polymerization mechanism. This assumption is also supported by the polymerization of optically pure 1-phenylethyl isocyanide 29 using a [(π-allyl)Ni(O$_2$CCF$_3$)]$_2$ catalyst (14) under O$_2$ [49]. For all polymerizations listed in Table 3, the preferential formation of one helical sense is suggested by a comparison of the specific rotations of the starting isocyanide with the resulting polymer. More precisely, the CD spectra can be used to estimate the sense and degree of asymmetric induction. In general, it has been suggested that a positive couplet (a negative-to-positive transition with an increase in wavelength) corresponds to M-helical conformation [48]. The helical sense indicated in Table 3 have been assumed using this criterion. The quantitative determination of the screw-sense excess seems to involve considerable difficulty. Nolte and Drenth used CD spectroscopy for

Table 3 Polymerization of optically active isocyanide by nickel chloride

Entry	Isocyanide	Rotation of monomer/deg	Rotation of polymers/deg	Screw-sense	Reference
1	(S)-PhCHMe(NC)	−55.0[a]	−458[a]	M	46, 48
2		−40.9[b]	−350[b]		56
3	(S)-PhCH$_2$CH$_2$CHMe(NC)	+70.6[a]	+22[a]		46, 48
4	(S)-EtO$_2$CCHMe(NC)	+21.2[a]	−355[a]		46, 48
5		+16.7[b]	−280[b]		56
6	(S)-MeO$_2$CCH(i-Pr)(NC)	+5.5[b]	−110[b]	M	56
7	(S)-i-Pr O$_2$CCH(i-Pr)(NC)	+32.7[b]	−24[b]	M	56
8	(S)-t-Bu O$_2$CCH(i-Pr)(NC)	+29.5[b]	+32.5[b]	M	56
9	(S)-CH$_3$CO$_2$CH$_2$CHMe(NC)	+74.0[a]	−190[a]		46, 48
10	(S)-CH$_3$CO$_2$CH$_2$CH(i-Pr)(NC)	+72.5[a]	−172[a]		48
11	(R)-CH$_3$CO$_2$CH$_2$CHPh(NC)	−129[a]	−156[a]		48
12	(S)-HC≡CCHEt(NC)	−11.9[a]	−102[a]		48
13	(S)-H$_2$C=CCHEt(NC)	+90.2[a]	+27.6[a]		48
14	(S)-Ph$_2$P(O)CHMe(NC)	+35.3[a]	−910[a]		48
15	(R)-C$_6$H$_{13}$CHMe(NC)	−66.3[c]	+47.3[c]	M	47
16	(R)-Me$_2$CHCHMe(NC)	−23.5[c]	−16.2[c]	M	47
17	(R)-Me$_2$CHCH$_2$CHMe(NC)	−68.8[c]	+68.9[c]	M	47

[a] $[\alpha]^{20}_{578}$
[b] $[\alpha]^{20}_{D}$
[c] $[\alpha]^{22}_{578}$

their estimation of the screw-sense excess of isocyanide polymers (Entries 15–17 in Table 3), which did not possess chromophores other than the polymer main chain [47]. They assumed that the maximum $\Delta\epsilon$ values at 320–340 nm, which may arise solely from the helicity of the main chain, were a measure of screw-sense excess of the polymer. Screw-sense excesses of 20% (Entry 15), 56% (Entry 16), and 62% (Entry 17) were extrapolated from a maximum $\Delta\epsilon$ value for poly(tert-butyl isocyanide) that was obtained by complete optical resolution. However, it was later suggested that such extrapolation may not be warranted generally [50]. A question was raised as to the long-accepted assignment of a stereoregular 4/1 helical rod structure to poly(isocyanide)s.

An empirical rule for the prediction of screw-sense induction in the polymerization of optically active isocyanides was proposed [46]. Chiral isocyanides ((S)(M)(L)C-NC, where S, M, and L stand for the smallest, medium, and the largest organic groups), which have no coordinating substituents, polymerize to P-screw (right-handed helix) if the sequence S→M→L, as viewed from nitrogen atom to the stereogenic carbon center, is clockwise (typically corresponding to (R)-isocyanide). For those bearing a coordinating group (Y), the P-screw polymerization is supposed to be pre-

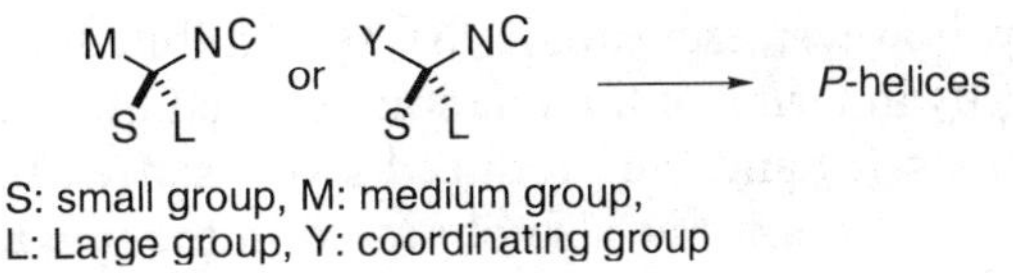

Fig. 3

S: small group, M: medium group,
L: Large group, Y: coordinating group

ferred, if the sequence S→Y→L, as viewed from the nitrogen atom to the stereogenic carbon center, is clockwise (Fig. 3).

In the nickel-catalyzed reaction system, a screw-sense induction with a chiral aryl isocyanide bearing stereogenic carbon centers far distant from the isocyano functionalities has been reported (Scheme 29) [51]. Polymer-

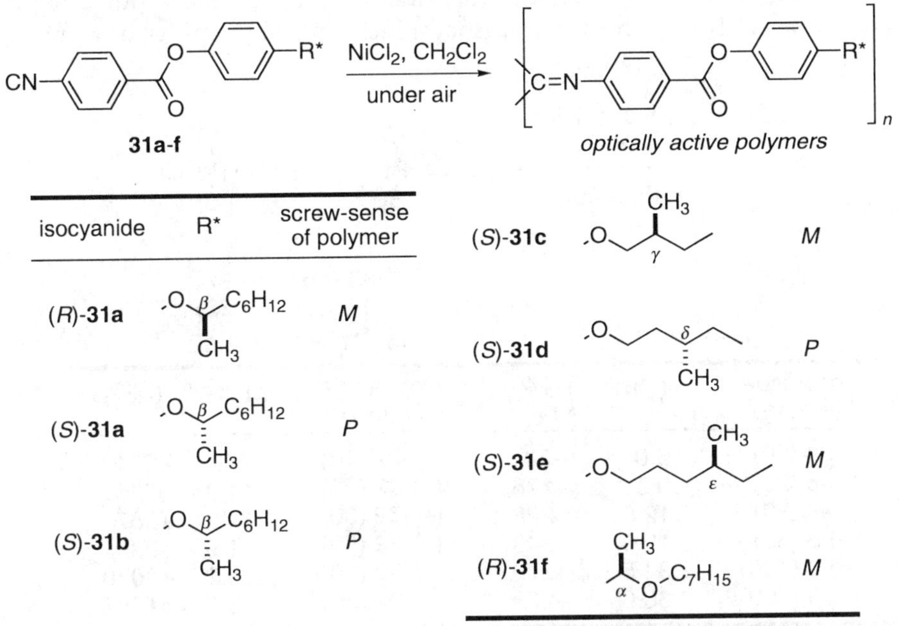

Scheme 29

isocyanide	R*	screw-sense of polymer
(R)-31a	O-β-C₆H₁₂, CH₃	M
(S)-31a	O-β-C₆H₁₂, CH₃	P
(S)-31b	O-β-C₆H₁₂, CH₃	P
(S)-31c	O-γ-CH₃	M
(S)-31d	O-δ-CH₃	P
(S)-31e	O-ε-CH₃	M
(R)-31f	α-CH₃, O-C₇H₁₅	M

ization of the enantioenriched isocyanide (R)-31a in the presence of a NiCl$_2$ catalyst under aerobic conditions yielded a poly(isocyanide) that exhibited a large positive optical rotation and a strong CD absorption signal. The helical sense of the polymer was assigned to M, based on its CD spectrum. An enantiomeric P-helical polymer was obtained using (S)-31a as the monomer. Similar screw-sense induction was also observed in the polymerization of related isocyanides 31b–f. Isocyanide (R)-31b and (S)-31b afforded M and P-helices, respectively, as expected from the results on the polymerization of

31a. Interestingly, however, isocyanide (S)-**31c** exhibited an opposite screw-sense induction (M) in terms of the relation between the absolute configuration at the side chain (S) and the produced screw-sense (M). An alternation of the screw-sense induction was also observed in the polymerization of **31d** and **31e**. This phenomenon may be related to the alternance rule for cholesteric liquid crystals, in which the odd or even position of the stereogenic centers with regard to the phenyl ring alters the sense of the helical arrangement of the molecules. The only exception, however, is the polymerization of **31f** bearing an (R)-stereogenic center α to the phenyl ring, which yielded an M helical sense polymer in the same manner as the polymerization of (R)-**31a**, which produces an M screw-sense polymer.

The palladium-catalyzed living polymerization involving the Pt–Pd heteronuclear complex **23** was found to be suitable for screw-sense induction using optically active isocyanides (Scheme 30) [52, 53]. Since this catalyst system was only effective for aryl isocyanides, optically pure p- or m-men-

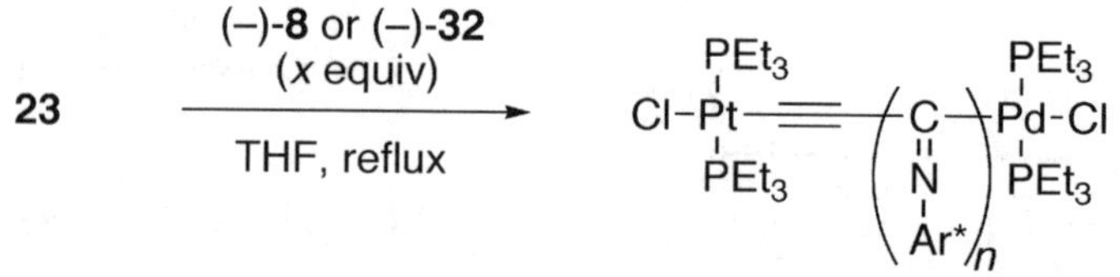

isocyanide (equiv)	$M_w/10^3$	$[\alpha]^{20}_D$	isocyanide (equiv)	$M_w/10^3$	$[\alpha]^{20}_D$
(−)-**8** (10)	6.0	+22	(−)-**32** (10)	7.7	+354
(−)-**8** (20)	9.5	+176	(−)-**32** (20)	11.0	+786
(−)-**8** (30)	12.0	+228	(−)-**32** (30)	14.2	+867
(−)-**8** (50)	19.0	+243	(−)-**32** (50)	21.9	+998
(−)-**8** (100)	34.5	+270	(−)-**32** (100)	38.2	+1070
(−)-**8** (200)	66.0	+262	(−)-**32** (200)	73.2	+1057

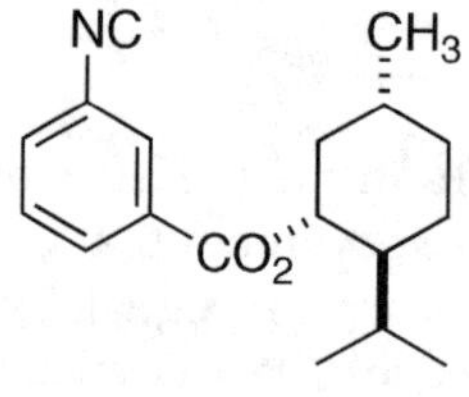

(−)-**8** ($[\alpha]^{20}_D = -83$)

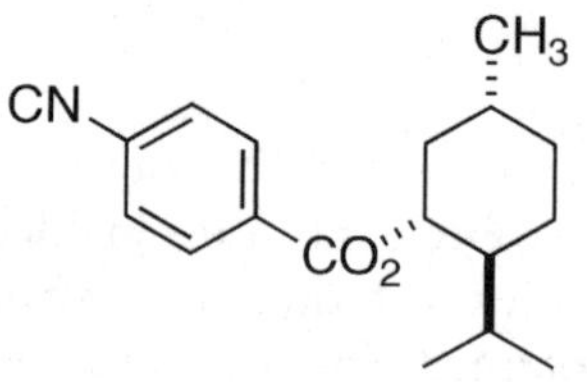

(−)-**32** ($[\alpha]^{20}_D = -75$)

Scheme 30

thyloxycarbonylphenyl isocyanides 32 and 8 were used. Although the stereogenic centers in the aryl isocyanides are located distal from the polymer main chain, the polymerization successfully yielded one helix preferentially. For a degree of polymerization $(P_n)<50$, the specific rotation as well as the $\Delta\epsilon$ values in the CD spectra increased with increasing P_n. The values then became constant, suggesting that polymers with $P_n>50$ have predominantly single-handed helical structures.

One of the important features of this polymerization process is its perfect living nature, which permits block copolymerization [53]. The isolated living oligomers 33 (P_n=10, 20, and 30), derived from (−)-8, were used as initiators for the polymerization of achiral aryl isocyanides 7 and 35 (Scheme 31, Table 4). The intensity of the specific rotation largely depended

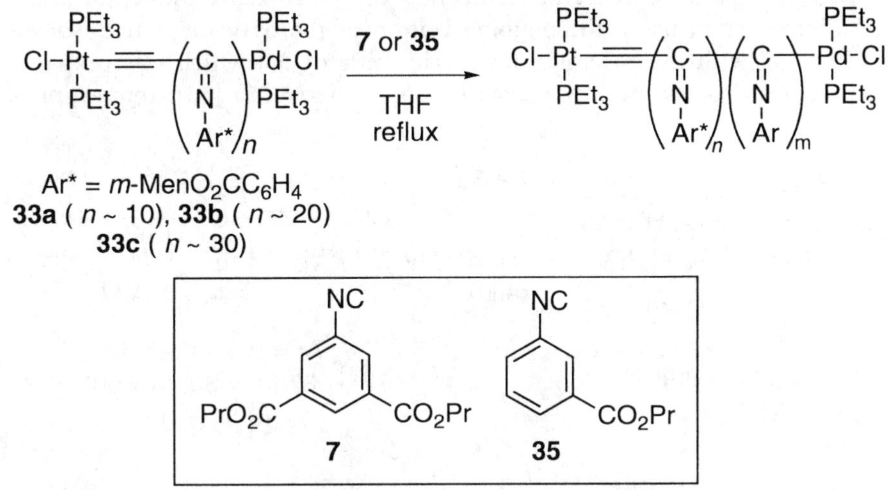

Scheme 31

Table 4 Block coplymerization of achiral aryl isocyanides 7 and 35 with oligomer initiators 33a–c

Run	Initiator	ArNC (equiv)	$M_w/10^3$	$[\alpha]^{20}_D$
1	33a (P_n=10, M_n=3720)	7(20)	13.5	126
2	33a	7(50)	21.0	92
3	33b (P_n=20, M_n=6580)	7(20)	14.5	288
4	33b	7(50)	23.5	246
5	33c (P_n=30, M_n=10,080)	7(20)	17.5	280
6	33c	7(50)	25.0	272
7	33c	7(70)	29.0	272
8	33c	35(30)	24.0	205
9	33c	35(50)	30.0	157

upon (1) the degree of oligomerization of the initiator used, (2) the degree of polymerization of the achiral isocyanide, and (3) the substitution pattern of the achiral isocyanide used. Almost constant optical rotations were obtained in the polymerization of 7 with the oligomer initiator **33c** ($P_n=30$), regardless of the degree of polymerization of the achiral isocyanide (Entries 5–7). Polymerization of 7 with initiator **33a** ($P_n=10$) resulted in the formation of polymers with lower optical rotations (Entries 1 and 2). The initiator **33b** ($P_n=20$) had a marginal effect, since an optical rotation similar to that of initiator **33c** was obtained for lower degrees of polymerization (Entry 3), whereas a decrease in optical rotation was observed with increasing degree of polymerization (Entry 4). The use of the sterically less demanding achiral isocyanide **35** with initiator **33c** resulted in a decrease in optical rotation for a high degree of polymerization.

An investigation into the mechanism of the screw-sense induction in the palladium system using the oligomerization of the optically active isocyanide (−)-**32** with the racemic oligomeric initiator **36**, which was prepared from achiral isocyanide **7**, was carried out (Scheme 32) [53]. Interestingly,

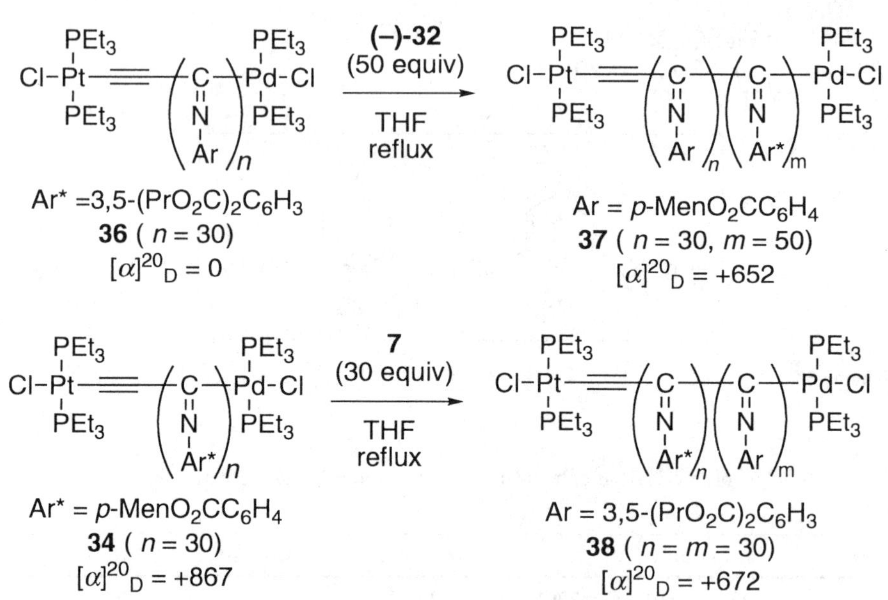

Scheme 32

the polymer **37** obtained showed a specific rotation of +652, which is similar to that of polymer **38** (+672) obtained from the polymerization of **7** (30 equiv) with the non-racemic oligomeric initiator **34** (see Scheme 30 for the synthesis of **34**). These experiments clearly demonstrate that the screw-sense is controlled thermodynamically.

The thermodynamic origin of the screw-sense induction is further supported by the results of the random copolymerization of achiral and optically active isocyanides in the presence of **23** (Scheme 33) [54]. Thus, the

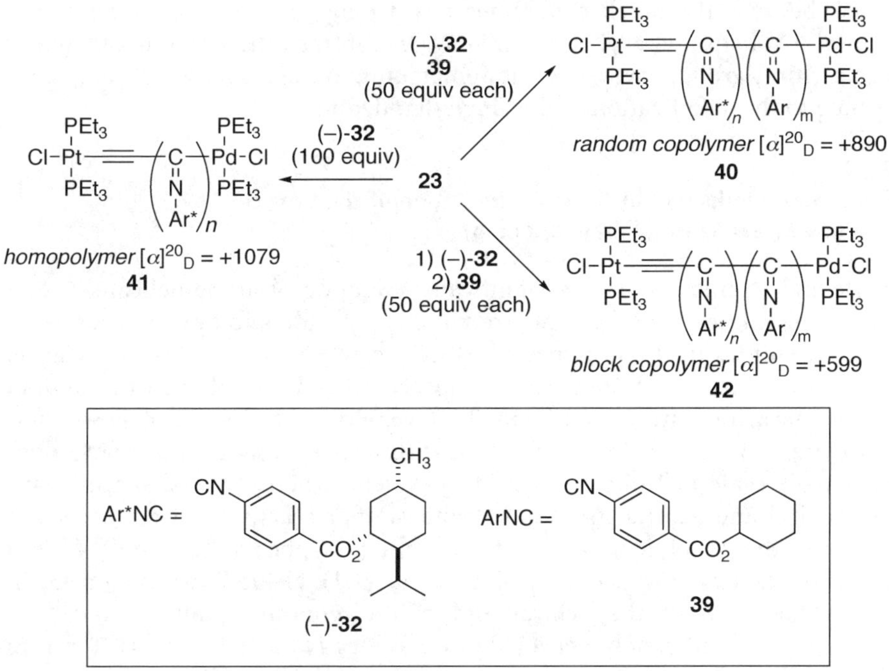

Scheme 33

optically active isocyanide (−)-**32** and the achiral p-cyclohexyloxycarbonylphenyl isocyanide (**39**) were copolymerized with **23** under reflux in THF for 15 h. The isocyanides were completely consumed, yielding the optically active copolymer **40** ($[\alpha]^{20}_D=+890$). The CD spectrum was very similar to that of homopolymer **41** derived from (−)-**32**, indicating that the copolymer adopted a helical structure with a preferred one screw-sense. The sign and the intensity of the specific rotation of **40** were compared with those of **41**. The same sign (positive) of the rotation suggested that the first copolymerization preferred the same screw-sense as that of the second copolymerization, although a lower screw-sense induction on the polymer main-chain was assumed during the comparison of the intensities. On the other hand, the 1:1 block copolymer **42** of (−)-**32** and **39** showed a specific rotation of $[\alpha]^{20}_D=+599$, which is significantly smaller than that of the random copolymer **40**. This relation, in which the random polymer shows a larger specific rotation than the block copolymer, holds for various (−)-**32/39** ratios. Data plots revealed a positive non-linear increase in the specific rotation with increasing content of the chiral isocyanide. The generally larger intensities of the random copolymers versus the block copolymers may arise from the relatively small steric hindrance of the *p*-substituted aryl isocyanides in the polymer backbone. Randomization of the main-chain structure can take place, because the block copolymer has a long segment consisting of the achiral *p*-substituted aryl isocyanide **39**. In contrast, the random copolymer has a chiral group, every two monomer units on average, which may contribute to the stabilization of the single-handed helix.

2.2.2
Screw-Sense Induction by Selective Inhibition of the Growth of a One Screw-Sense (Chiral Poisoning)

Nolte and Drenth reported a significant induction of a one helical sense in the copolymerization of achiral aromatic isocyanide and optically active alkyl isocyanides in the presence of a $NiCl_2$ catalyst [55, 56]. At first glance, this polymerization system seems closely related to Takahashi's random copolymerization system mentioned above. However, the use of slow-reacting chiral isocyanides in combination with fast-reacting achiral isocyanides is characteristic of Nolte and Drenth's system, whereas in Takahashi's system, chiral and achiral isocyanides with similar reactivities were used. For example, the copolymerization of the 4-methoxyphenyl isocyanide **43** and (*S*)-2-isocyanovaleric acid methyl ester **44** (1:1) yielded copolymer **45**, in which the ratio of the achiral and chiral monomer units was ca. 3:1 (Scheme 34). The copolymer **45** showed a negative couplet at $\lambda=270$ nm in the CD spectrum, along with a large negative optical rotation, which was assigned to the right-handed (*P*) helical structure. The formation of the *P*-helix is in sharp contrast to the formation of the *M*-helical polymer in the

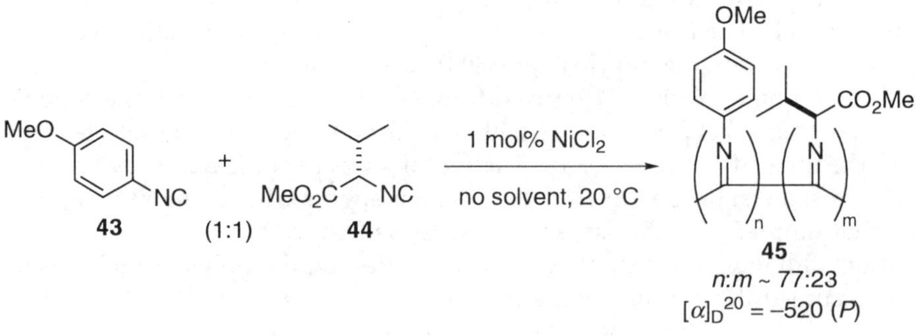

Scheme 34

homopolymerization of **44** (Table 3, Entry 6). Indeed, in Takahashi's system, the same screw-sense was preferred in the homopolymerization and random-copolymerization.

The proposed mechanism for this unusual screw-sense selection involves the selective inhibition of the growth of a one screw-sense by the chiral isocyanide [55, 56]. Thus, in the initial stage, both the right- and left-handed helices are derived from the fast-reacting achiral isocyanide (**43**), and these begin to grow. The chiral isocyanide (**44**) then selectively reacts with the left-handed helical oligomer, resulting in a slowing down of the growth of the left-handed helix. As a result, a faster growth of the right-handed (*P*) helical polymer was observed. This phenomena may be regarded as "chiral poisoning", and has been the topic of much attention in recent years in research looking for new catalyst systems for asymmetric synthesis [57]. According to the proposed chiral poisoning mechanism, a polymer with a higher degree of polymerization should show a higher screw-sense excess for the *P*-helix, and a lower incorporation of chiral isocyanide. This was confirmed by experiment, in which the resulting copolymer was fractionated by preparative GPC, and each fraction was then analyzed using optical rotation and ^{1}H NMR. The results clearly demonstrated that copolymers with low molecular weights exhibited low optical rotations and a high incorporation of the chiral isocyanide, whereas high molecular weight polymers showed high optical rotations and low chiral isocyanide incorporation.

The random copolymerization system utilizing chiral poisoning was extended to the asymmetric polymerization of a series of achiral aryl isocyanides with (*S*)-2-isocyanovaleric esters [56]. The mechanism for the asymmetric induction was interesting, but still needs further investigation, because there is some ambiguity in the assignment of the preferred screw-sense, as well as in the determination of the screw-sense excess using the CD spectra and optical rotation, which can be highly sensitive to polymer composition.

2.2.3
Screw-sense Induction by a Chiral Chain-end Group at the Non-Propagating Terminus

In Nolte and Drenth's nickel catalyzed system, the polymerization was believed to be initiated by a nucleophilic attack by the alcohol used as a solvent or the halide on the starting complex on the coordinated isocyanides. Successful asymmetric polymerization was achieved using a dicationic tetrakis(isocyanide)nickel(II) complex **46** with enantiopure primary amines, which served as a chiral nucleophile in the initial step (Scheme 35) [58, 59]. In a typical experiment, a catalyst prepared from $(t\text{-BuNC})_4\text{Ni(II)}(\text{ClO}_4)_2$ (**46a**) (1 mol%) and an optically active amine (1 mol%), was used for polymerization of isocyanides with, or without a solvent, such as *n*-hexane, in

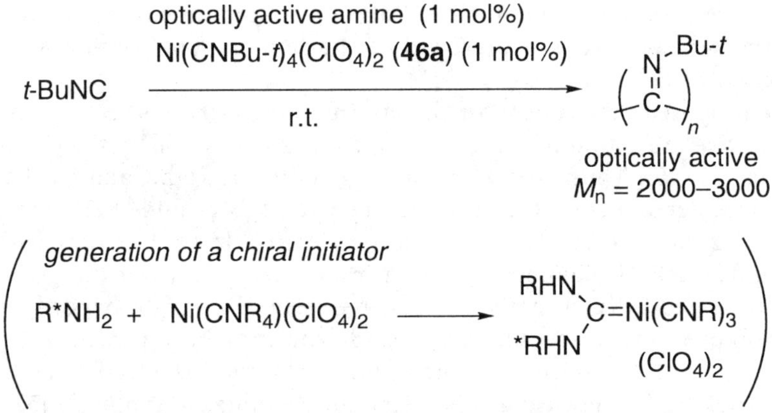

Scheme 35

the case of solid isocyanides. The polymerization was allowed to run for 1–5 days at ambient temperature. Typically, the molecular weights of the resultant polymers range from 2000 to 3000. The screw-sense excess is determined by comparing the intensities of the CD spectra with those of a completely resolved sample.

Among the polymerizations of *tert*-butyl isocyanide using a series of optically active amines, 1-phenylethylamine (**47**) served as the most stereoselective initiator, leading to the formation of optically active poly(*tert*-butyl isocyanide) with a 61–62% screw-sense excess (Scheme 36) [59]. A *P*-helical

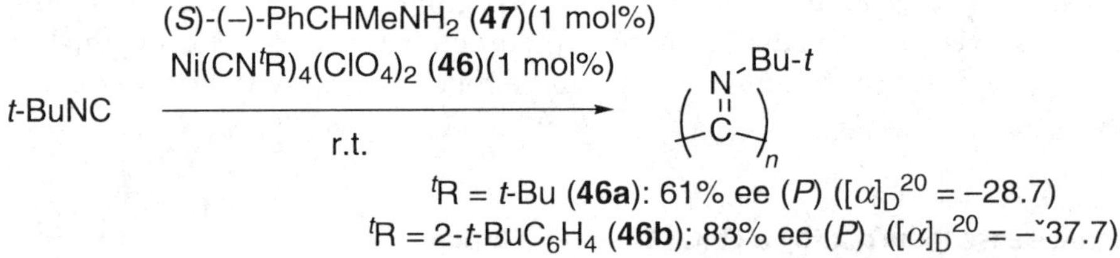

Scheme 36

sense was preferred by the (*S*)-enantiomer of **47**, and an *M*-helix sense by the (*R*)-enantiomer of **47**. The screw-sense selectivity also depended on the substituent of the isocyanide ligand on the dicationic nickel complex used as a catalyst precursor. Thus, among the four catalyst precursors tested, the tetrakis(2-*tert*-butylphenyl isocyanide)nickel(II) complex **46b** resulted in the highest screw-sense selectivity (83% for the *P*-helix from, (*S*)-1-phenylethylamine). In contrast, a nickel precursor bearing the more sterically demanding 2,6-diisopropylphenyl isocyanide group resulted only in a 22% excess of the P screw-sense.

A series of achiral isocyanides was polymerized using a catalyst prepared from $(t\text{-BuNC})_4\text{Ni(II)}(\text{ClO}_4)_2$ (**46a**) and (S)-**47** [59]. It turned out, however, that only a few isocyanides, including 1,1-dimethylpropyl, 1-methyl-1-phenylethyl, and 2,6-dichlorophenyl isocyanides, yielded optically active polymers. All primary and secondary aliphatic isocyanides, and aromatic aldehydes, except for a 2,6-dichlorophenyl derivatives, yielded optically inactive polymers. There appears to be a tendency that only sterically congested isocyanides permit the induction of a screw-sense.

A mechanism for asymmetric induction has been discussed based on the "merry-go-round" mechanism proposed for Ni(II)-catalyzed polymerization, and discussed earlier. In the initial stages, a chiral amine attacks one of the four isocyanide ligands (C^1) on the nickel to form a diaminocarbene complex **I**, whose structure has been elucidated in detail [60]. Among the possible conformations, the one that involves weak interaction between the Ph group and Ni was presumed to have the most favorable conformation (Scheme 37). The structure of the intermediate **I** has also been discussed in

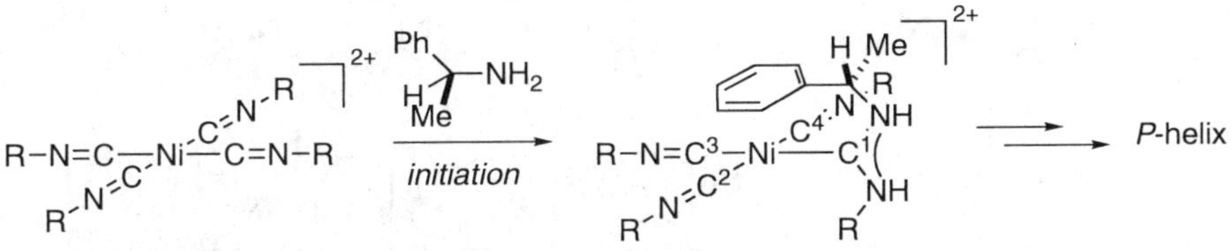

Scheme 37

detail [61]. The polymerization begins by the migration of the carbene-like group (C^1) to the isocyanide carbon marked "C^2" in Scheme 37, thus avoiding the possible steric repulsion that would be expected if migration to the "C^4" isocyano carbon atom occurred. The direction of the migration holds for each migratory insertion step throughout the polymerization process, leading to the formation of one helical sense with the retention of the screw-sense generated at the initial stage.

The details of this mechanism are still unclear, and need to be clarified. However, this asymmetric polymerization system using a nickel catalyst with optically active amines seems to be unique, in that the chiral elements that become apart from the propagating termini control the helical sense of the entire polymer main chain. A similar, but more stereoselective system is discussed below for the Pd-mediated polymerization of diisocyanobenzenes, which is discussed later.

2.2.4
Screw-sense Induction by a Chiral Ligand at the Propagating Terminus

Deming and Novak focused their attention on π-allylnickel complexes **48** and **49**, bearing optically active chiral carboxylato ligands, which would be retained on the nickel center during polymerization (Scheme 38) [62]. The

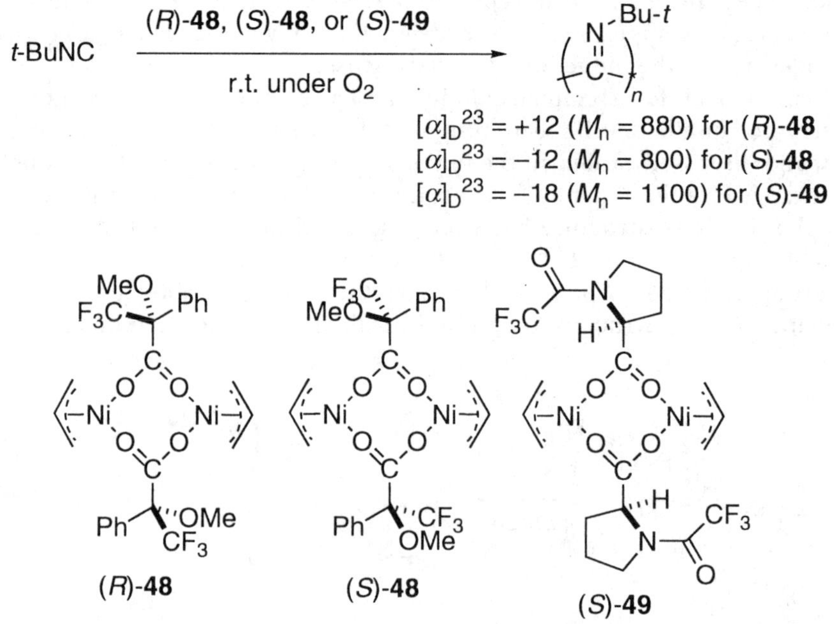

$[\alpha]_D^{23} = +12$ ($M_n = 880$) for (*R*)-**48**
$[\alpha]_D^{23} = -12$ ($M_n = 800$) for (*S*)-**48**
$[\alpha]_D^{23} = -18$ ($M_n = 1100$) for (*S*)-**49**

Scheme 38

use of chiral phosphine ligands was not considered, because of the high probability of the dissociation of the chiral ligand by coordination of the isocyanides existing as excess reactant. The polymerization of *tert*-butyl isocyanide in the presence of the complex (*R*)-**48** proceeded at 25 °C in air. Poly(*tert*-butyl isocyanide) (M_n=880) was obtained in the reaction, and had a specific rotation of $[\alpha]_D^{20}$=+12, which was assignable to excess *M*-helix based on the use of previous assumptions. The use of the enantiomeric initiator (*S*)-**48** produced a *P*-helical polymer exhibiting a specific rotation of $[\alpha]_D^{20}$=−12. In the polymerization of *tert*-butyl isocyanide using (*S*)-**49**, a polymer having a specific rotation of $[\alpha]_D^{20}$=−18 (M_n=1100) was obtained. Based on the specific rotation of the completely resolved poly(*tert*-butyl isocyanide), a minimum screw-sense excess was set at 46% ee for **48**, and 69% ee for **49**. It was also found that the coordination of cyanide ligands to the nickel may have a critical effect on the screw-sense selectivity. Thus, a bis-cyanide derivative of (R)-**48** afforded an optically active polymer of diphenylmethyl isocyanide (molecular weight=193) having a specific rotation of $[\alpha]_D^{23}=+23$, whereas an $[\alpha]_D^{23}$ of +10 was observed for polymerization in the absence of the cyanide source (Scheme 39). However, the low M_n values, some as low as 550, obtained in these polymerizations should be remarked on here.

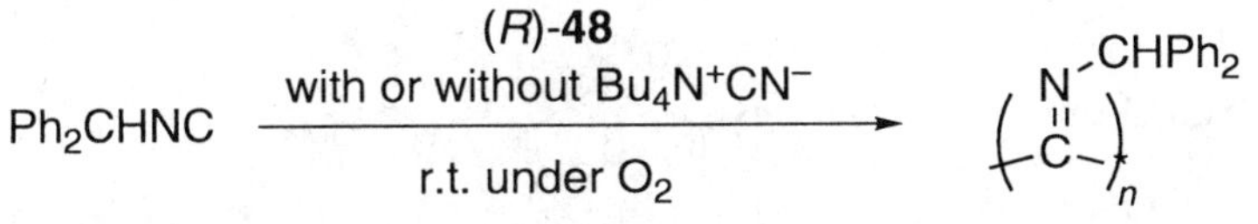

$$Ph_2CHNC \xrightarrow[\text{r.t. under } O_2]{\substack{(R)\text{-}\mathbf{48} \\ \text{with or without } Bu_4N^+CN^-}} \left(\underset{\underset{n}{\overset{\|}{C}}}{N}{-}CHPh_2 \right)$$

$[\alpha]_D^{23} = +10$ ($M_n = 880$) for (R)-**48**
$[\alpha]_D^{23} = +13$ ($M_n = 680$) for (R)-**48** +1 CN$^-$
$[\alpha]_D^{23} = +23$ ($M_n = 550$) for (R)-**48** + 2 CN$^-$

Scheme 39

2.3
Polymerization of Isocyanides Having Functional Groups

Because of the unique main chain conformation that is believed to be of a highly rigid rod shape, considerable effort has been devoted to the synthesis of poly(isocyanide)s bearing functional groups. It was expected that the helical arrangement of those functional groups around the rod-like polymer main-chain might confer special properties to the resulting polymers. Representative isocyanide monomers used for the polymerization reactions are shown in Figs. 4 and 5.

The peptide-based isocyanides **50a–e** were successfully polymerized by Ni(II) catalysts under an inert atmosphere [63–66]. An alanine-derived isocyanide, whose isocyano group was labeled using ^{13}C and ^{15}N, was prepared and successfully polymerized for structure elucidation [67]. Several solvent systems were used in the polymerization of the peptide-based isocyanides, depending upon the solubility of the isocyanides. As mentioned earlier, the use of alcohols as a solvent or co-solvent can accelerate the polymerization. The protective groups on the ester, hydroxy, and imidazole groups were removed after polymerization by treatment with aqueous NaOH to yield poly(isocyanide)s bearing unprotected peptide side chains.

Interesting arrangements of the side-chains in peptide-based poly(isocyanide)s have been reported. The polymers obtained from **50a** and **50e** exhibit β-helix structures, in which the Ala-Ala as well as Ala-Ala-Ala side-chains are organized in a β-sheet-like fashion (Scheme 40) [68, 69]. In this β-helix-

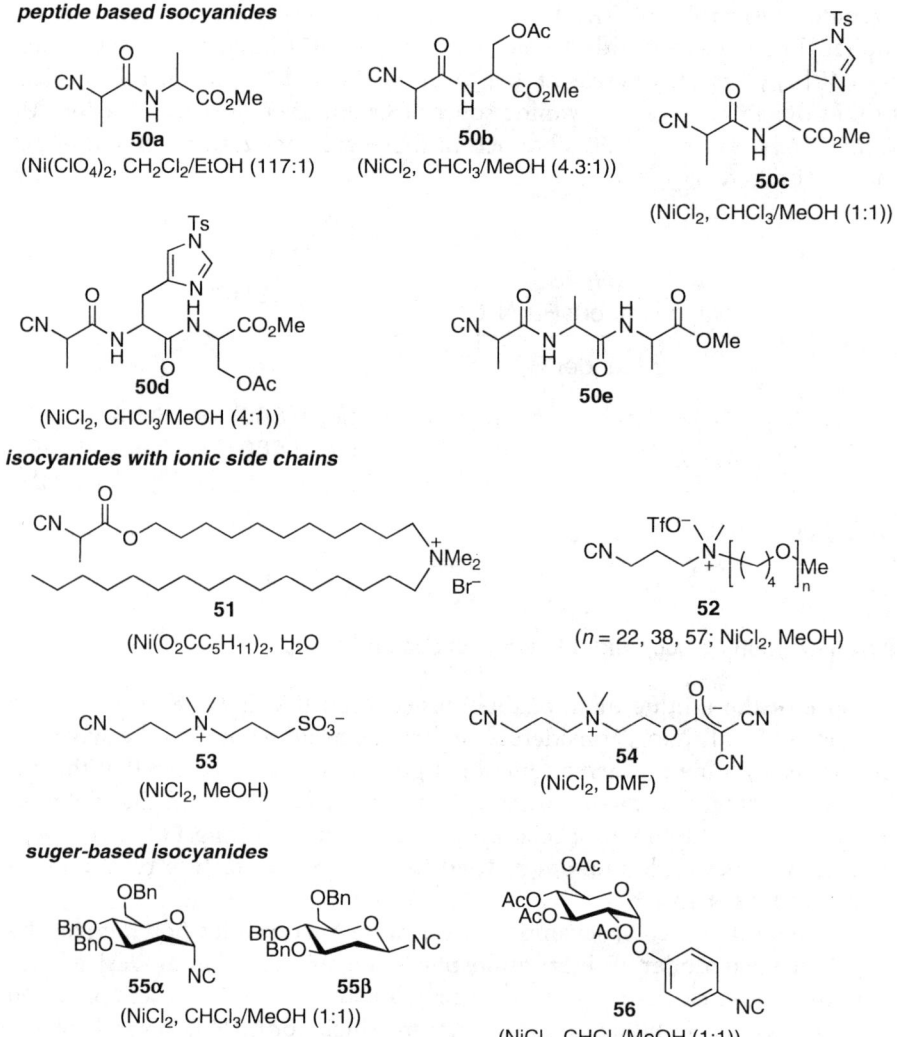

Fig. 4

like structure, the rigid helical backbone of the poly(isocyanide)s may be acting as a director for the suitable spatial arrangement of the side chains.

Peptide-based polymers **62**, containing imidazole, carboxyl, and hydroxymethyl functionalities, have been prepared from optically active **50d** and tested as mimics of enzymes, such as chymotrypsin, which have the same functionalities (Scheme 41) [70]. These polymers exhibit markedly higher activities than the corresponding low molecular weight compounds in the hydrolysis of nitrophenyl and dinitrophenyl esters. Increased activities were

miscellaneous

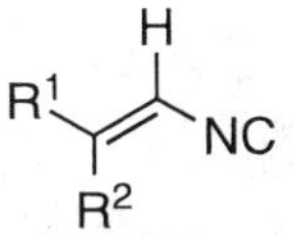

57
(Ni(acac)$_2$, EtOH or hexane)

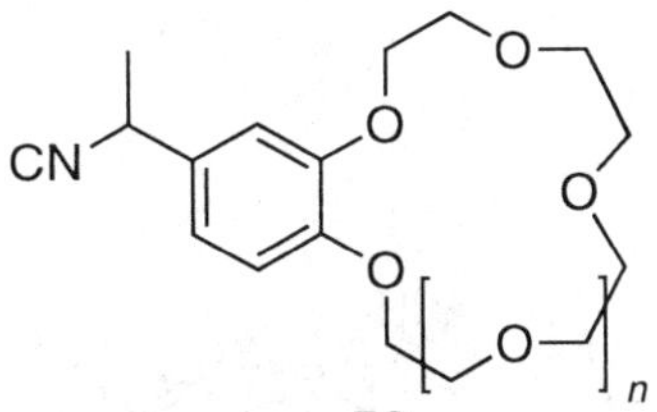

58
(NiCl$_2$, neat, 60 °C)

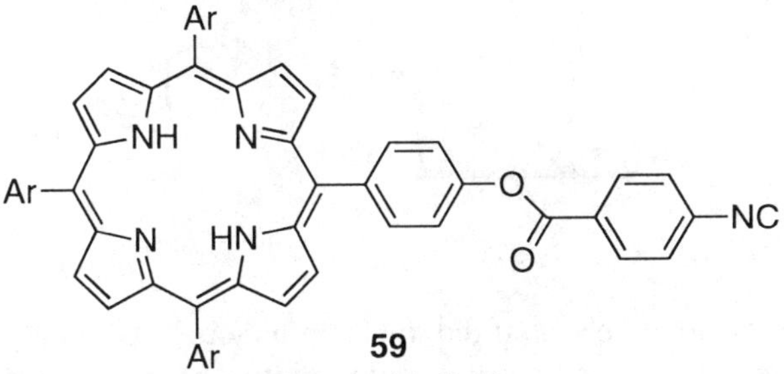

59
Ar = *p-t*-BuC$_6$H$_4$
((Et$_3$P)$_2$ClPt-CC-Pd(PEt$_3$)$_2$Cl, THF)

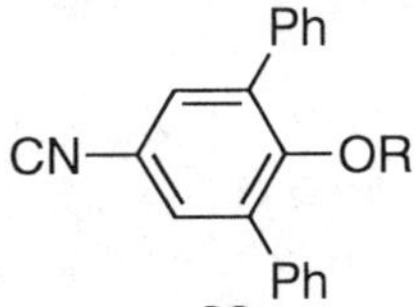

60
R = H, CH$_3$CO
(NiCl$_2$, CH$_2$Cl$_2$)

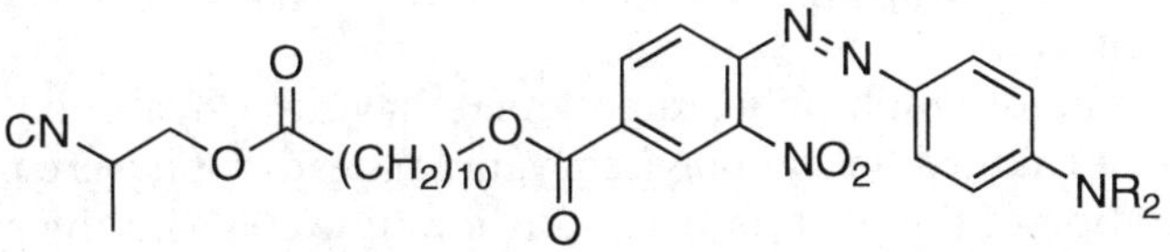

61 (R = Bn or morpholino)

Fig. 5

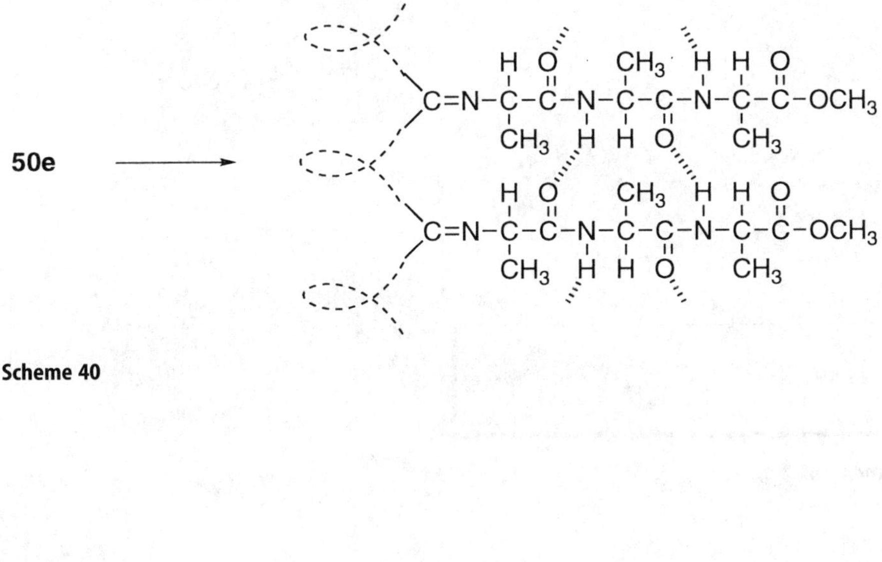

50e ⟶

Scheme 40

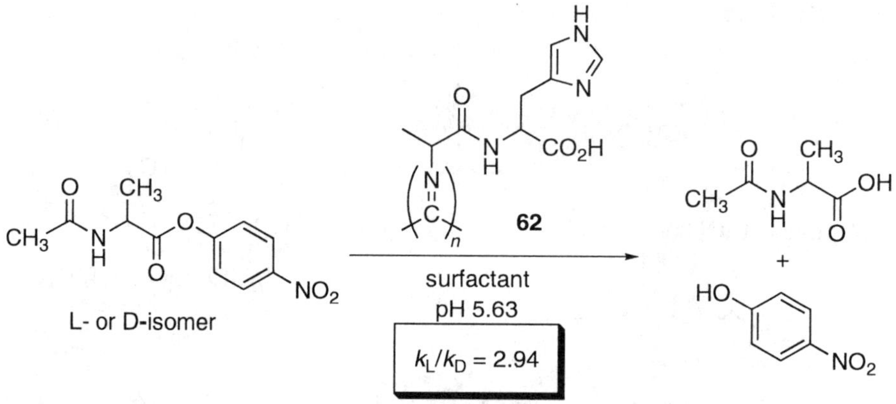

Scheme 41

observed in the presence of positively charged surfactants, such as N-cetylpyridinium chloride. The high activity was attributed to the cooperative action of the functionalities on the polymer side chains. In kinetic measurements on the hydrolysis of d- and l-4-nitrophenyl N-acetyl-2-aminopropionate with **62**, a k_L/k_D value of 2.94 was observed in the presence of the surfactant.

Synthesis of amphiphilic copolymers having a hydrophobic polystyrene tail with a charged helical poly(isocyanide) headgroup were prepared by polymerization of the peptide-derived isocyanide **50a** with the macromolecular nickel initiator **63** (Scheme 42) [71]. The macromolecular initiator was prepared by reaction of $(t\text{-BuNC})_4\text{Ni(II)(ClO}_4)_2$ (**46a**) with polystyrene, which was end-capped with an amino group. The block copolymer **64** thus ob-

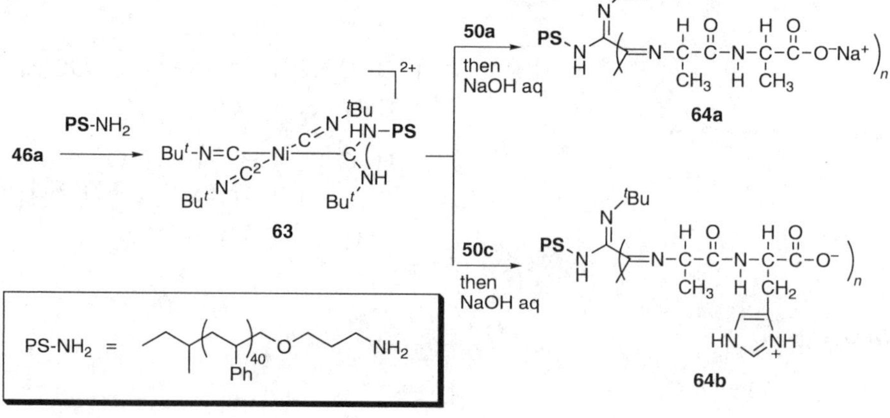

Scheme 42

tained exhibited a variety of morphologies, which depended upon the length of the poly(styrene) block, the pH of the solution, and the anion–headgroup interactions. These morphologies were supposed to arise from aggregation of the poly(styrene) block via hydrophobic interactions. Interestingly, the aggregation of the non-racemic helical copolymer resulted in the formation of helical superstructures, the screw-sense of which was opposite to that of the copolymer.

Isocyanides bearing ammonium side-chains **51** and **52** have been polymerized in the presence of nickel catalysts [72, 73]. The amphiphilic isocyanide **51** forms vesicles on dispersion in water. The isocyano groups located in the vesicle bilayers were polymerized by nickel capronate to form polymerized vesicles. The isocyanide **51** was also used in the preparation of polymerized vesicles containing metalloporphyrin components within the bilayer membrane [74]. The redox behavior of this membrane-bound cytochrome P-450 mimic has been investigated in detail. In addition to those bearing cationic side chains, isocyanides **53** and **54** bearing zwitterionic side chains were successfully used [75].

α- and β-D-Glucopyranosyl isocyanide fully protected by benzoyl groups (**55α** and **55β**) were polymerized in the presence of an $NiCl_2$ catalyst [76]. Although the polymerizations were found to be sluggish, presumably because of steric hindrance, the corresponding polymers were obtained, and their molecular weights were determined to be around 6000. Under the polymerization conditions used, the β-isocyanide underwent isomerization at the anomeric position. From optical rotation measurements, the formation of left- and right-handed helices from the α- and β-isocyanides, respectively, was presumed, which is in agreement with the empirical prediction mentioned above.

Phenyl isocyanides carrying fully acetylated α-glucose (**56**), β-glucose, β-galactose, and β-lactose were polymerized by a $NiCl_2$ catalyst system [77]. The resultant polymers were soluble in common organic solvents, and were deacetylated with sodium methoxide, yielding water- and DMSO-soluble polymers. Their CD spectra indicated that left-handed helical structures were dominant in the fully acetylated glycopolyisocyanides, except for those derived from **56**. After hydrolysis, all the deprotected glycopolyisocyanides possessed left-handed helical conformation in excess, indicating that an inversion of the helical sense took place during the hydrolysis of the polyisocyanide derived from **56**. The poly(isocyanide)s having glycosyl functions showed little specific interactions with lectins, whereas flexible glycosylated phenylacrylamide showed highly specific interactions. On the other hand, the glycosylated polymer underwent highly organized multilayer adsorption onto the hydrophilic surface of a mica substrate [78]. This phenomenon has been explained by the formation of densely crowded saccharide arrays along the rigid rod polymer backbone, which is characteristic for poly(isocyanide)s.

N-Vinyl-substituted isocyanides **57** have been polymerized in the presence of nickel catalysts, yielding poly(vinyl isocyanide)s [79]. The solubility and stability of these polymers very much depends upon the substituent on the vinyl group. Substituted vinyl isocyanide, $(CH_3)_2C=CHNC$, afforded polymers that were soluble in chloroform when freshly prepared, although they became insoluble on standing for several days, even at temperatures of –10 °C. The polymerization of vinyl isocyanide proceeded in hydrocarbon solvents, unlike aryl or alkyl isocyanides, which required an alcoholic solvent for efficient polymerization.

Polymerization of the isocyanide **58** having crown ether moieties in the presence of a nickel catalyst afforded poly(isocyanide)s that can be regarded as a synthetic model for a channel-forming ionophore. Examination of the incorporation ability of polymers bearing variable-sized rings of the crown ether moiety to the vesicle bilayers revealed that those bearing 18-crown-6 ($n=2$) and 21-crown-7 ($n=3$) type rings were successfully incorporated into the vesicle bilayers of dihexadecyl phosphate. A polymer with a 24-crown-8 ($n=4$) moiety was incorporated a little less effectively, and no incorporation was observed for a polymer with a 15-crown-5 ring ($n=1$) moiety. It was established that the incorporated crown ether-substituted polymer served as a channel for the Co^{2+} ions [80].

The nickel-mediated polymerization of *p*-hydroxy-substituted phenyl isocyanide **60** (R=H) afforded the corresponding poly(isocyanide) **65** in a 43% yield after 65 h at room temperature [81]. The polyhydroxylated polymer **65** was oxidized with NaOCl in THF (Scheme 43). The resultant black solid **66**,

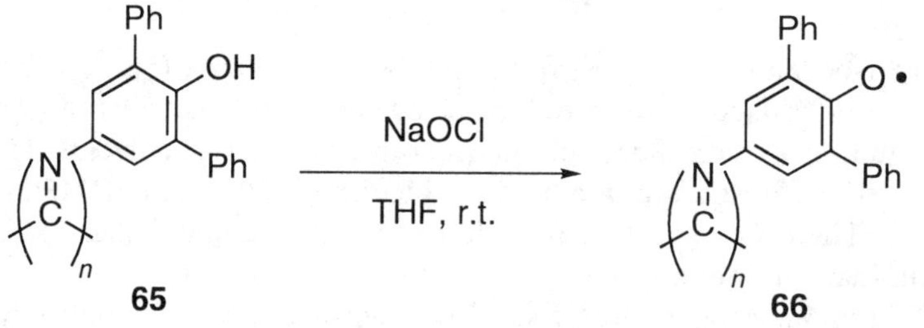

Scheme 43

which was insoluble in common solvents, is paramagnetic at both 4 and 320 K. A spin density of ca. 0.8 per repeating unit was observed, although no evidence for ferromagnetism or any other short-range ordering among the unpaired spins in the temperature region 77–320 K was observed. Related poly(isocyanide)s bearing pendant piperidyl-*N*-oxyl groups [82] and verdazyl radicals [83] have been prepared, although no bulk paramagnetic behavior has been found.

The porphyrin-substituted aryl isocyanide **59** has been polymerized using the Pd–Pt heterobinuclear initiator **23** in refluxing THF, yielding the corresponding poly(isocyanide) in good yield [84]. The degree of polymerization of the polymer could be varied from 20 to 100 within a generally narrow molecular weight distribution ($M_w/M_n \leq 1.14$) by changing the monomer-to-initiator ratio. The porphyrin-substituted polymer exhibited split Soret bands, which indicates that the porphyrin groups in the side chains are regularly arranged by virtue of the rigid rod polymer main chain.

Polymerization of isocyanide **61** having an NLO-active, azobenzene-based chromophore with, and without 2-isocyanopropyl acetate as a co-monomer afforded homopolymer or random copolymers, which form a stable monolayer on the water's surface [85–87]. Langmuir-Blodgett films of the poly(isocyanide)s generated second-harmonic optical radiation. The efficiency of the second harmonic light generation depended mainly on the degree of co-monomer incorporation and the spacer groups linking the NLO-active groups to the polymer backbone.

Deming and Novak's living polymerization system using allylnickel(II) trifluoroacetate (**14**) has been applied to the synthesis of block polymers, which were designed to serve as new light harvesting systems [88, 89]. In a comparison with the related homopolymers, diblock copolymers having acceptor and donor blocks showed enhanced fluorescence quenching, with the formation of radical ion pairs having lifetimes of about 1.1 µs. Moreover, the triblock copolymer **67**, which has a long intervening block in the middle section, exhibited an inhibited exciplex formation, and an increase in fluores-

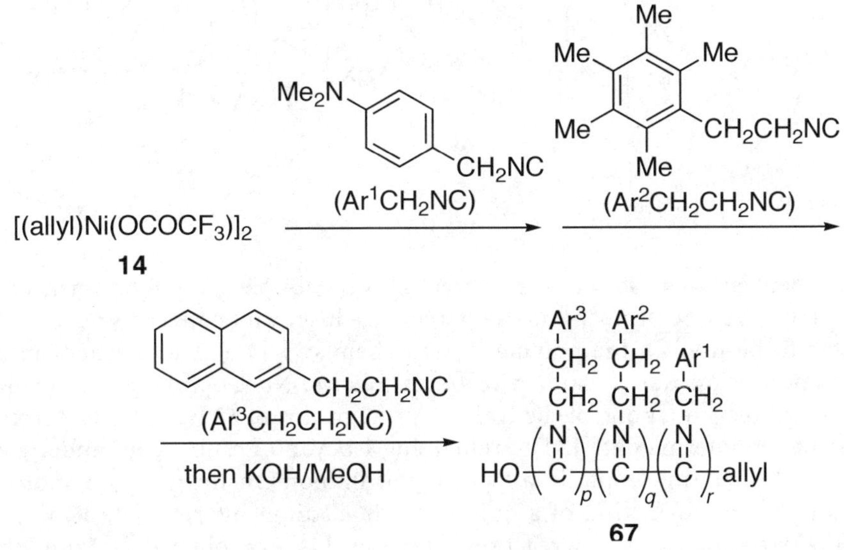

Scheme 44

cence quantum yield (Scheme 44). These results suggest that the attachment of chromophores and quencher groups on the rigid rod polymer backbone offers a new strategy for the construction of an efficient light harvesting system with a long-lived charge separation.

3
Polymerization of Diisocyanobenzenes

In 1990, Ito et al. reported on the oligomerization of a 1,2-diisocyanobenzene derivative using Grignard reagents [90]. When the reaction of **68** was carried out in dioxane at 0 °C with 0.5 equivalent of i-PrMgBr, oligoquinoxaline **69** with degrees of oligomerization in the range 1–4 were obtained in a 74% yield after hydrolysis (Scheme 45). The 3,6-substituents of 1,2-diiso-

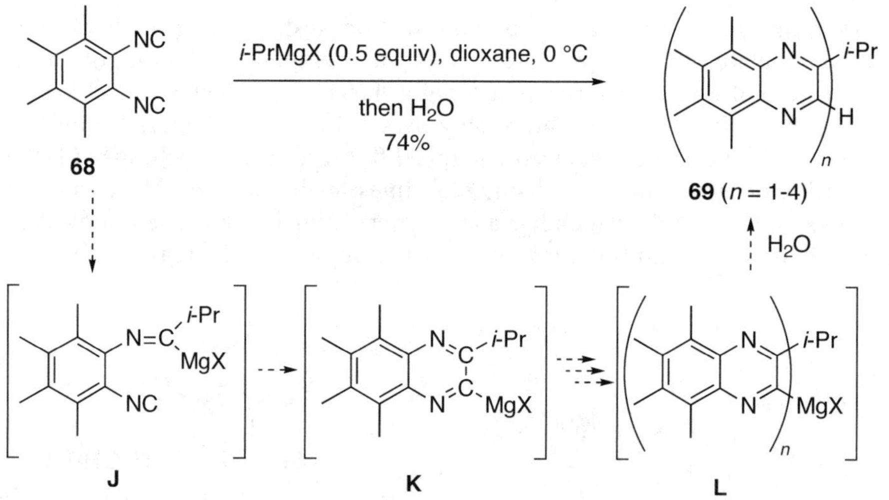

Scheme 45

cyanobenzene were found to be essential for stabilizing the monomer, presumably by steric protection. The parent 1,2-diisocyanobenzene was too unstable to be used in the polymerization chemistry [41]. The mechanism of oligomerization was presumed to be as follows. Nucleophilic attack of the Grignard reagent at one of the two isocyano groups of **68** leads to the formation of the intermediate **J**. The remaining isocyano group easily undergoes intramolecular nucleophilic attack of the iminomethylmagnesium moiety, leading to the formation of a quinoxalinylpalladium intermediate **K**, which again participates in the reaction with the diisocyanobenzene. From the mechanistic point of view, the polymerization of diisocyanobenzene is quite similar to that of monoisocyanides, which barely undergo any polymerization with carboanionic initiators, including Grignard reagents. As the successive incorporation of two isocyano groups produces a quinoxaline ring in each of the propagation steps, the reaction can be regarded as being an "aromatizing oligomerization".

The use of Grignard reagents as initiators in the reaction of 1,2-diisocyanobenzenes, however, did not lead to the formation of high molecular weight polymers. Labeling experiments using D_2O as a quenching agent suggested that the intermediate organomagnesium species are not stable enough to yield high molecular weight polymers [90]. The use of organo transition metal complexes was then examined to realize efficient polymerization.

3.1
Synthesis of Racemic Poly(quinoxaline-2,3-diyl)s

The oligomerization of diisocyanobenzene **68** was more efficiently catalyzed by organopalladium complexes than by Grignard reagents. In oligomerization using *trans*-MePdBr(PMe$_2$Ph)$_2$ (**70**), the oligomer distribution changed linearly with the ratio of **68** to palladium initiator **70** (Scheme 46) [91]. The

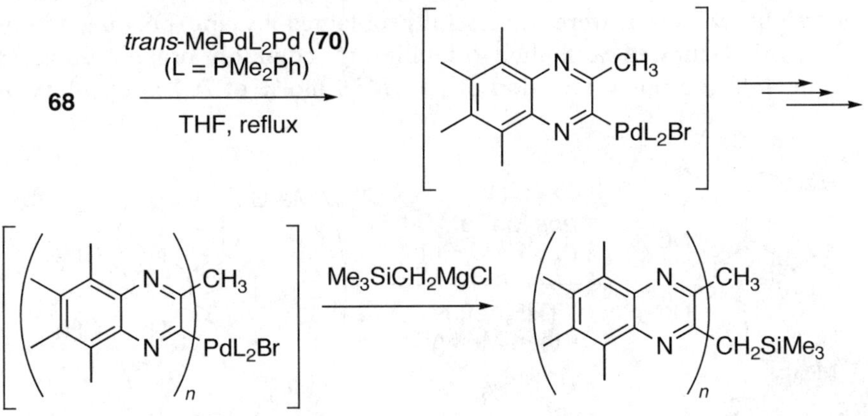

entry	68/70	%yield of oligomers							total
		$n=2$	$n=3$	$n=4$	$n=5$	$n=6$	$n=7$	$n>7$	
1	2	37	27	6	0	0	0	0	70
2	3	20	49	16	2	0	0	0	87
3	5	0	9	20	22	17	9	2	79
4	7	0	0	6	9	20	5	6	56

Scheme 46

oligomers bearing living palladium termini were quite stable, and could be isolated by preparative GPC, or even by chromatography on silica gel. The living oligomers were found to be active for further polymerization. For example, a bi(quinoxalinyl)palladium complex **71** reacted with two equivalents of diisocyanobenzene **68**, to yield the higher oligomers **72** as a mixture of trimers, tetramers, pentamers, and hexamers that still possessed living palladium termini (Scheme 47). Upon the polymerization of **68**, a low solubility

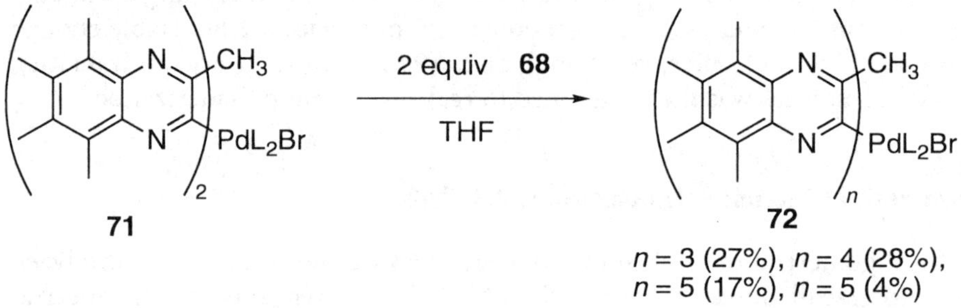

$n = 3\ (27\%), n = 4\ (28\%),$
$n = 5\ (17\%), n = 5\ (4\%)$

Scheme 47

hampered the formation of higher molecular weight polymers. High molecular weight polymers were successfully obtained by the polymerization of **73**, which showed a higher solubility in common organic solvents (Scheme 48). The polymerization of **73** with 5 mol% of **70** under reflux in

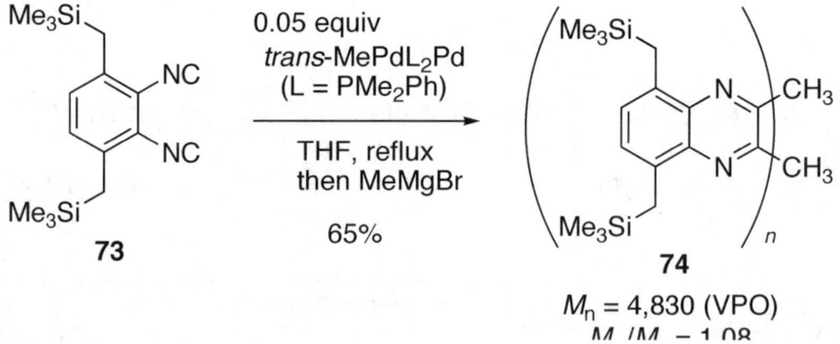

$M_n = 4,830$ (VPO)
$M_w/M_n = 1.08$

Scheme 48

THF followed by treatment with MeMgBr yielded polyquinoxaline **74** with a molecular weight of 4,830 (VPO) and a polydispersity index of 1.08.

Polymerization is also promoted by organonickel complexes [92]. A quinoxalinylnickel complex, prepared by the stoichiometric reaction of trans-o-TolNi(PMe$_3$)$_2$Cl with 73, was used as an initiator in the polymerization of 73, and yielded a soluble polymer with a narrow molecular weight distribution (M_n/M_w=1.10) (Scheme 49).

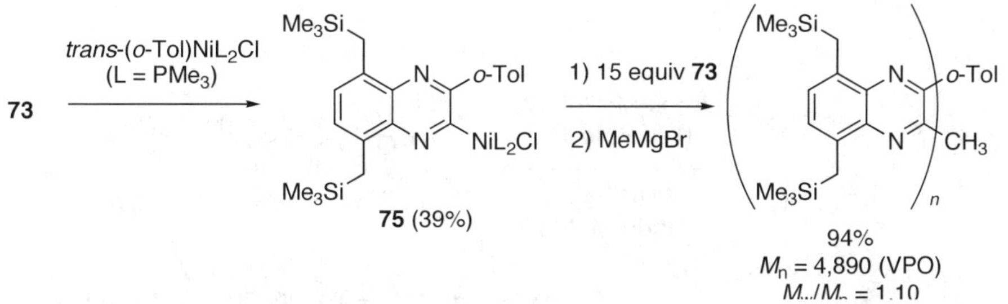

Scheme 49

One of the attractive features of the transition metal-mediated polymerization of 1,2-isocyanobenzenes is its ability for modifying the polymer termini by virtue of the remaining palladium (and nickel) groups at the living ends of the polymers. Thus, the introduction of ethoxycarbonyl groups, hydrogen, and methyl groups have been demonstrated using terquinoxalinylpalladium (76) and gunoxalinylnickel (77) complexes as model compounds (Schemes 50 and 51) [93].

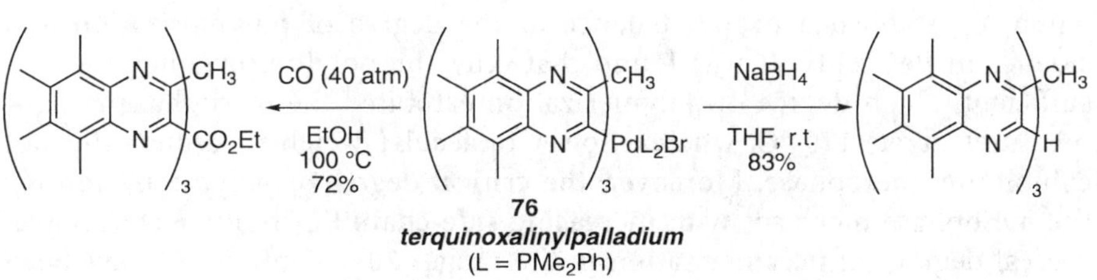

Scheme 50

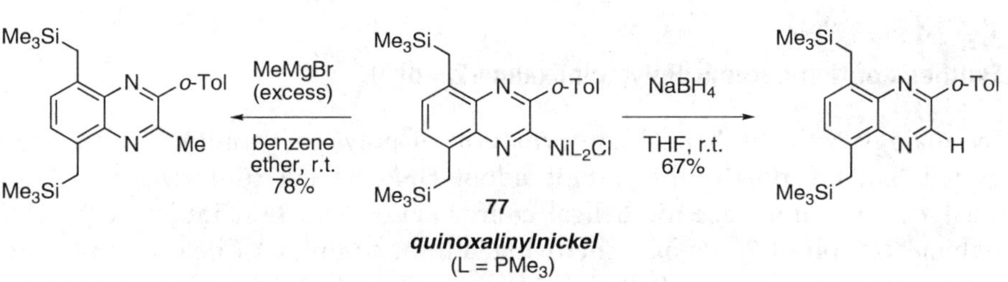

Scheme 51

Poly(quinoxaline)s showing liquid crystallinity have been prepared by palladium-mediated polymerization of 1,2-diisocyanobenzenes **78–80** bearing a series of alkoxymethyl substituents at the 4 and 5 positions (Scheme 52) [94]. Poly(quinoxaline)s can be classified as rigid-rod liquid

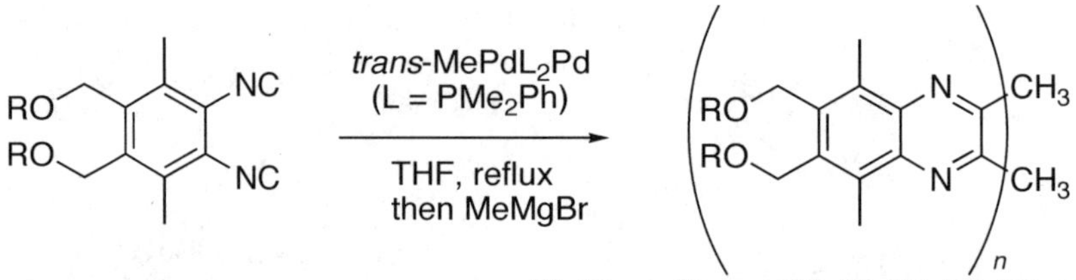

78 (R = n-Pr)
79 (R = n-C$_5$H$_{11}$)
80 (R = n-C$_7$H$_{15}$)

81 (R = n-Pr, n = 10, 20, 30, 50, 70, 100)): mesophase observed for n ≥ 30
82 (R = n-C$_5$H$_{11}$, n = 30, 50, 70, 100): mesophase observed for n ≥ 50
83 (R = n-C$_7$H$_{15}$, n = 30, 40, 50, 70, 100): mesophase observed for n ≥ 70

Scheme 52

crystalline polymers in which the mesogenic rigidity is derived from the rod-like structure of the main chain, and balanced with surrounding flexible side-chains of the polymer. A series of poly(quinoxaline)s **81–83** was prepared to establish a proper balance of the degree of polymerization and side-chain flexibility. It was found that only the poly(quinoxaline)s with a sufficiently high degree of polymerization exhibited the mesophase at temperatures above 120 °C, whereas lower molecular weight polymers did not exhibit any mesophase. Moreover, the critical degree of polymerization for the mesophase increases with increasing side-chain flexibility. For example, critical degrees of polymerization in the ranges 20–30 and 50–70 have been observed for poly(quinoxaline)s bearing propoxymethyl (**81**) and heptyloxymethyl (**83**) side-chains, respectively.

3.2
Synthesis of Non-Racemic Poly(quinoxaline-2,3-diyl)s

By analogy with the helical conformation of poly(isocyanide)s, it was suggested that the poly(quinoxaline)s adopt rigid helical conformations. The existence of a non-racemic helical conformation was first indicated by the polymerization of **73** using a methylpalladium complex **84** bearing optically active phosphine ligands (Scheme 53) [95]. The polymerization, on quenching with methylmagnesium bromide, yielded the optically active polymer **85** with an $[\alpha]_D = +7.2$. Because the optically active phosphine moiety had been

Scheme 53

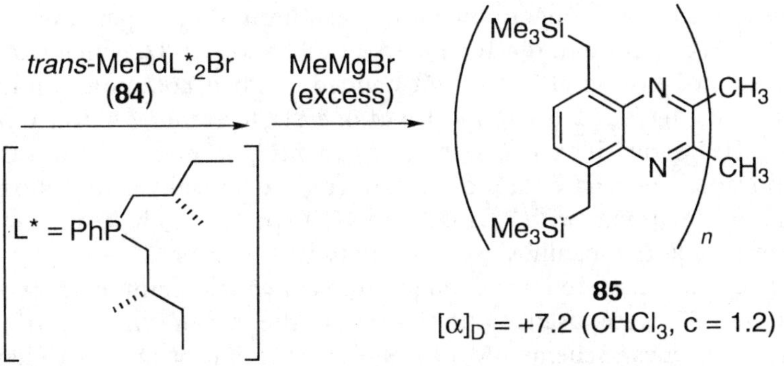

removed by the treatment with a Grignard reagent, the optical activity must solely rely on the helical chirality of the polymer backbone. However, the optical activity was gradually lost at room temperature in solution. Effort then focused on the asymmetric synthesis of poly(quinoxaline)s with a stable single-handed helical structure.

3.2.1
Asymmetric Polymerization via Optical Resolution of Living Oligomers

It was presumed from molecular models that the 3,6-substituents of quinoxaline units may play an important role in the stabilization of optically active helical structures. 1,2-Diisocyanobenzene **86** (4 equiv) bearing *p*-tolyl groups at the 3- and 6-positions was subjected to oligomerization in the presence of the chiral palladium complex **84** (Scheme 54) [96]. The resulting mixture of oligomers was separated by preparative GPC, leading to the isola-

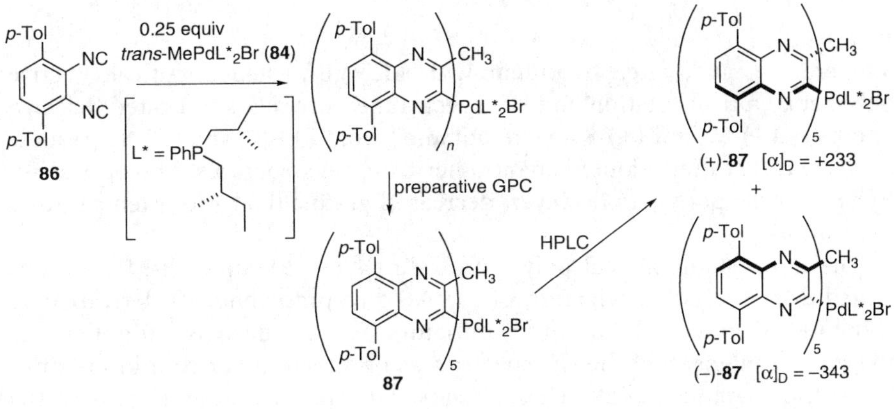

Scheme 54

tion of pentameric oligoquinoxalinylpalladium **87**. The pentamer was found to be a 3:4 mixture of diastereomers (+)-**87** and (−)-**87**, which had an opposite screw-sense (right- and left-handed) with a common chirality at the phosphine ligand. The CD spectra of one diastereomer, separated by preparative HPLC on silica gel, was almost a mirror image of the other. Upon removal of the palladium moiety carrying the optically active phosphine ligands by treatment with MeMgBr, the CD spectra of the resultant two pentameric oligoquinoxalines were found to be the reverse.

The two diastereomeric oligoquinoxalinylpalladium complexes (+)-**87** and (−)-**87** were used as initiators in the polymerization of 1,2-diisocyanobenzenes (Scheme 55) [96]. Polymerization of **88** proceeded at room

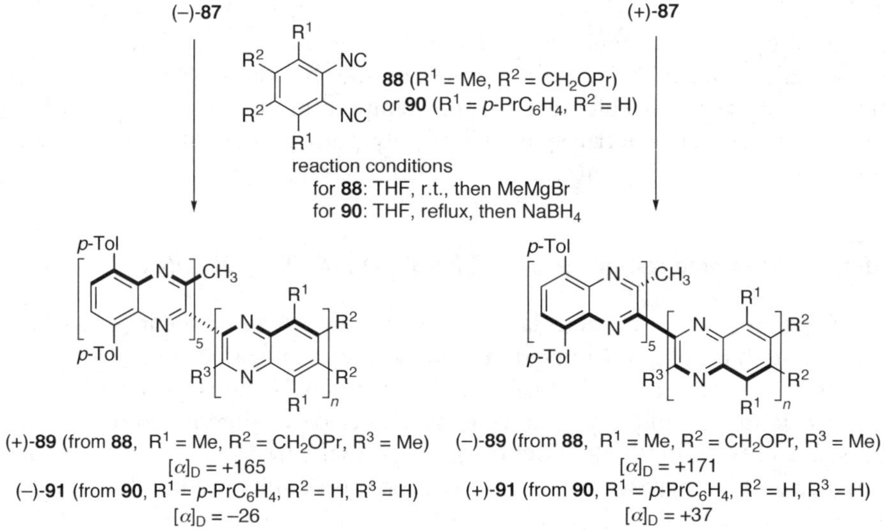

Scheme 55

temperature, and, after treatment with MeMgBr, yielded optically active polymers. Optical rotation and CD spectra measurements indicated that two polymers (−)-**89** and (+)-**89**, were obtained from (+)-**87** and (−)-**87**, respectively, and that they adopted enantiomeric helical structures. The optical activity of these polymers however, decreased gradually at room temperature in solution.

The more stable helical poly(quinoxaline)s (+)-**91** and (−)-**91** were obtained after the polymerization of the 3,6-di(p-propylphenyl) derivative **90** using (+)-**87** and (−)-**87** as chiral initiators [96]. No decrease in optical activity or in intensity of the CD spectra was observed under conditions similar to those applied to **89**. These results indicate that the bulkiness of the

3,6-substituent participates in the stabilization of optically active helical structures of poly(quinoxaline)s. A related asymmetric polymerization of 1,2-diisocyanobenzenes has been reported using diastereomerically enriched hexameric oligoquinoxalinylpalladium complexes [97].

The racemization process of oligo- and polyquinoxalines has been investigated in detail [98]. Thus, the resolved pentameric and hexameric (oligoquinoxalinyl)palladium complexes **92**, **93**, and **94** were treated with MeMgBr, leading to the isolation of the optically active oligoquinoxalines **95**, **96**, and **97**, respectively, in which both the chain ends are terminated with methyl groups (Scheme 56). Among the three optically active oligomers, only **97**

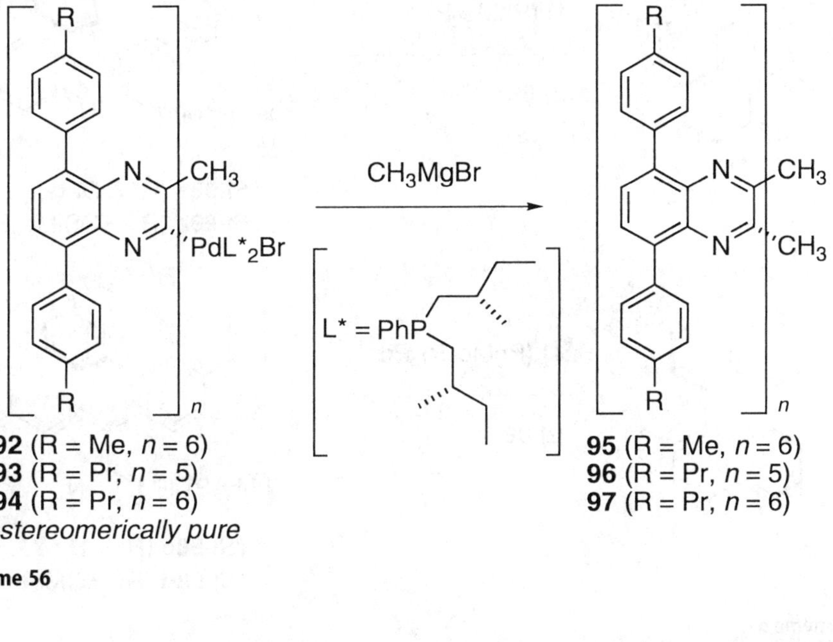

92 (R = Me, n = 6)
93 (R = Pr, n = 5)
94 (R = Pr, n = 6)
diastereomerically pure

95 (R = Me, n = 6)
96 (R = Pr, n = 5)
97 (R = Pr, n = 6)

Scheme 56

was tolerant towards racemization at room temperature, whereas the others underwent racemization at room temperature. The racemization process obeyed first-order kinetics. The activation energies for the racemization of **95**, **96**, and **97** were calculated to be 106, 100, and 133 KJ mol^{-1}, respectively. These results and the fact that no intermediary conformers were detected using ^{1}H NMR and in the CD spectra throughout the racemization process suggest that the racemization, i.e., helical inversion, takes place in a single step. It was suggested that the stability of the helical conformation depended significantly on the chain length and on the 5,8-substituent of the oligomer.

3.2.2
Asymmetric Polymerization by Chiral Initiators

Chiral organopalladium complexes have been designed as initiators in the asymmetric polymerization of 1,2-diisocyanobenzenes instead of the resolved oligoquinoxalinylpalladium complexes. The first series of chiral initiators examined were the organopalladium complexes **98a–d** bearing chiral 1,1'-binaphthyl groups (Scheme 57) [99, 100]. They were prepared from

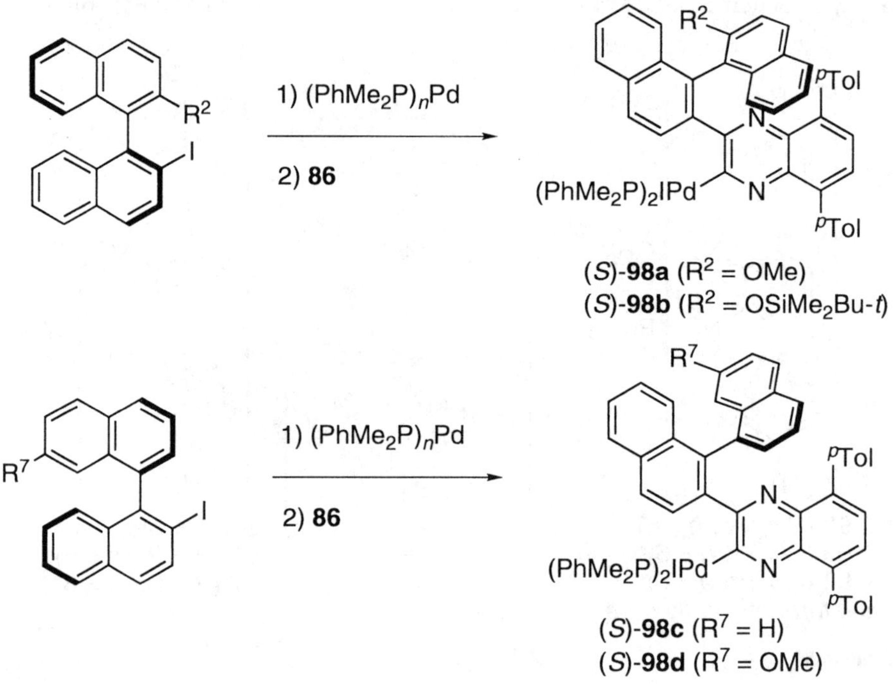

Scheme 57

enantiopure binaphthyl iodide derivatives. Thus, the iodides were treated with $(PhMe_2P)_nPd(0)$, yielding bis(dimethylphenylphosphine)(1,1'-binaphth-2-yl)iodopalladium(II) complexes. The binaphthylpalladium complexes were reacted further with 3,6-di-(p-tolyl)-1,2-diisocyanobenzene to afford stable quinoxaline complexes **98a–d**, which were used as chiral initiators for polymerization.

The polymerizations of the diisocyanides **86** and **88** were attempted using four binaphthyl-based initiators **98a–d**, which differed only in the substituents on the binaphthyl group (Scheme 58). All the polymerizations produced poly(quinoxaline)s **99** and **100** that showed CD spectra characteristic

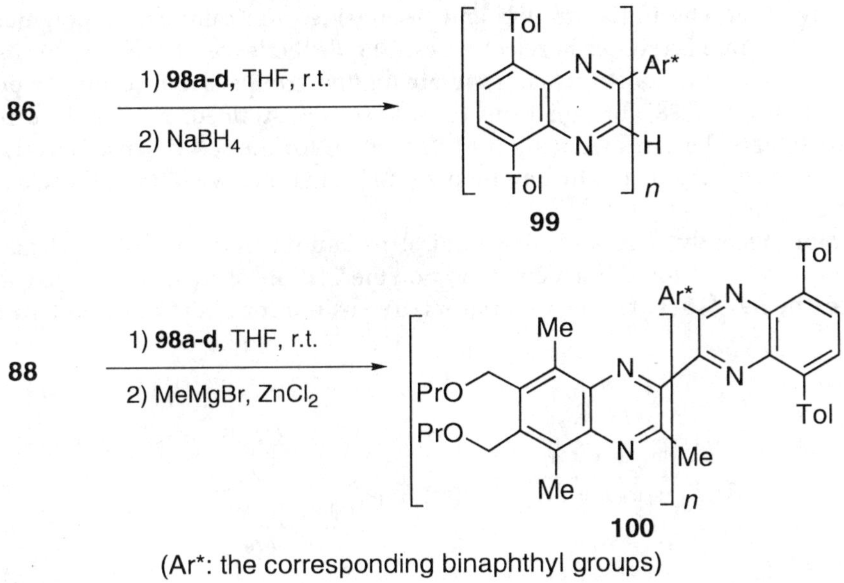

86 $\xrightarrow{\text{1) 98a-d, THF, r.t.}}{\text{2) NaBH}_4}$

[structure **99** with Tol, Ar*, N, H groups]

88 $\xrightarrow{\text{1) 98a-d, THF, r.t.}}{\text{2) MeMgBr, ZnCl}_2}$

[structure **100** with PrO, Me, Tol, Ar* groups]

(Ar*: the corresponding binaphthyl groups)

Scheme 58

of optically active helical structures (see Table 5). However, the screw-sense selectivity depended markedly on the substituent of the binaphthyl group. In the polymerizations using **86**, the selectivities ranged between 16% and 82% screw-sense excess (se) using the four chiral initiators. The highest selectivity (82% se) was recorded for the 7′-methoxy-substituted binaphthyl derivative **98d**, whereas only a 16% se was attained with the 2′-methoxy de-

Table 5 Asymmetric polymerization of diisocyanobenzenes **86** and **88** using chiral initiators **98a-d**

Entry	Isocyanide	Initiator	S/I ratio	Yield (%)	$M_n/10^3$	M_w/M_n	se % (config.)
1	86	(S)-98a	40	69	5.81	1.29	16 (P)
2	86	(S)-98b	40	35	10.2	1.90	52 (P)
3	86	(S)-98c	40	78	4.23	1.30	69 (P)
4	86	(S)-98d	40	64	8.92	1.47	82 (P)
5	86	(R)-98d	40	70	6.39	1.28	82 (M)
6	88	(S)-98a	40	77	7.3	1.48	<1
7	88	(S)-98b	40	76	20.0	2.43	65 (P)
8	88	(S)-98c	40	74	8.3	1.30	69 (P)
9	88	(S)-98d	40	79	11.1	1.39	87 (P)
10	88	(R)-98d	40	78	8.8	1.15	87 (M)
11	88	(S)-98d	60	87	15.1	1.33	87 (P)
12	88	(S)-98d	100	70	23.8	1.21	87 (P)

rivative **98a**. The initiators **98b** and **98c** yielded optically active polymers with moderate screw-sense selectivities. The dependence of the screw-sense selectivity on the substituents of the binaphthyl groups also held for the polymerization of **88**, for which up to 87% se was attained. It should be remarked here that the originally reported values for the screw-sense selectivity were corrected in a following report [101]. This review refers to those corrected values.

More recently, a series of new chiral palladium initiators **101–103** have been developed for the asymmetric polymerization of diisocyanobenzenes (Scheme 59) [101]. The new initiators are easily prepared from o-iodoben-

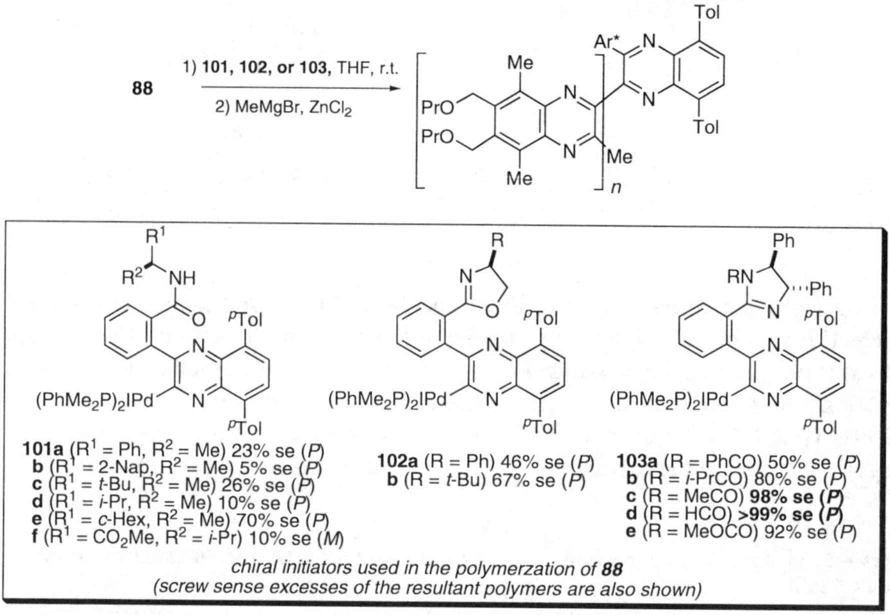

Scheme 59

zoic acid with commercially available enantiopure compounds. The amide (**101**) and oxazoline-based (**102**) chiral initiators achieved only moderate screw-sense selectivities, although the screw-sense excess significantly varied with the structure of the chiral groups. The highest selectivity has been attained with imidazoline-based initiators **103**, whose substituent on the imidazoline nitrogen had a marked influence on the selectivity. Acetyl- (**103c**) and formyl- (**103d**) substituted imidazoline derivatives yielded >98% se, in contrast to the corresponding benzoyl and isobutyryl derivatives, which gave significantly lower selectivities.

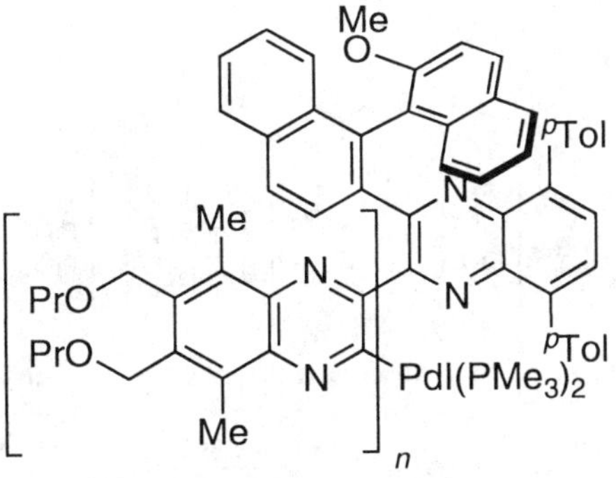

104a (*n* = 1)
105a (*n* = 2) 52:48 (^{1}H NMR)

Fig. 6

The mechanism of asymmetric induction by the binaphthyl-based initiators was investigated based on the structural analyses of isolated living oligomers [100]. (Oligoquinoxalinyl)palladium complexes **104** and **105** bearing two and three quinoxaline units, respectively, were synthesized by reaction of enantiopure 2′-methoxy-1,1′-binaphthyl derivative **98a** with **86**, and isolated by preparative GPC (Fig. 6). The ^{1}H NMR spectrum of the terquinoxalinyl complex **105a** showed the presence of two diastereomers (ratio=52:48), whereas the biquinoxalinyl complex **104a** did not. The diastereoisomerism observed for **105a** may arise from the helical chirality of the terquinoxaline moiety, in conjunction with the axial chirality of the binaphthyl moiety. ^{1}H NMR measurements revealed that the two diastereomers of **105a** were in equilibrium at room temperature. The activation energy for the helical inversion was estimated to be 16.1–16.3 kcal/mol from temperature-dependent ^{1}H NMR measurements.

Terquinoxalinylpalladium complexes **105b**, **105c**, and **105d** bearing various substituents on the binaphthyl ring were prepared from **98b–d** to gain an insight into the mechanism of screw-sense selection in asymmetric polymerization (Fig. 7) [100]. It has been established that the diastereomeric ratios of the terquinoxalinyl complexes correlate well with the screw-sense selectivities of the polymerization. Thus, terquinoxalinyl complex **105d** bearing a 7′-methoxy group adopts an almost single helical sense, as evidenced by its ^{1}H NMR spectrum. Higher oligomers **106d** and **107d** bearing the 7′-methoxybinaphthyl group also show the presence of a single helical sense. On the other hand, the two isomers exist as **105b** and **105c** in ratios of 89:11 and 84:16, respectively. From these results, it is presumed that the screw-sense of the polymerization is determined from the formation of a terquinoxaline intermediate, whose diastereomeric ratio holds during the course of further polymerization.

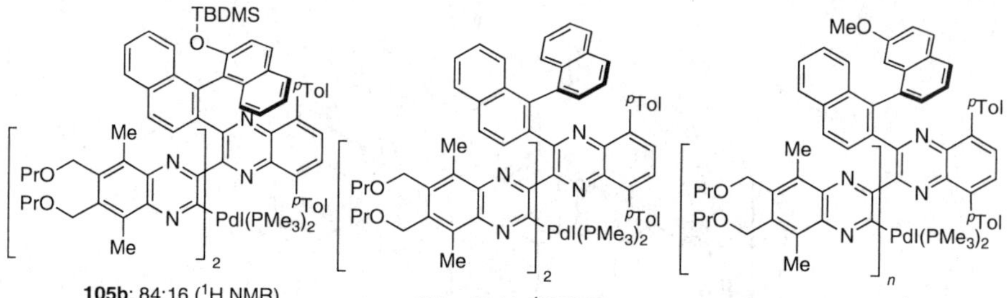

[Structures of 105b, 105c, 105d/106d/107d shown]

105b: 84:16 (^{1}H NMR)

105c: 89:11 (^{1}H NMR)

105d: (n = 2) >99:1 (^{1}H NMR)
106d: (n = 3) >99:1 (^{1}H NMR)
107d: (n = 4) >99:1 (^{1}H NMR)

Fig. 7

In this polymerization system, the screw-sense excess of the polymers was determined from a comparison of their CD spectra with that of an authentic polymer prepared from conformationally stable (quinquequinoxalinyl)palladium complexes [99, 100]. Initially, the authentic polymer was obtained from diastereomerically pure **108**, whose right-handed helical structure was determined from single-crystal X-ray analysis (Fig. 8). However, it was later shown that the polymerization was accompanied by partial racemization [101]. The use of a (quinquequinoxalinyl)palladium complex **109** led to the formation of polyquinoxaline showing CD spectrum that was more intense (ca. 10%) than the original authentic polymer. These results indicate that helical inversion may still be possible for oligomers higher than the pentamer, unless the propagating polymer has an effective chiral group, such as the imidazoline group, at the polymer terminus. The slight racemization observed in the synthesis of the former authentic polymer may be

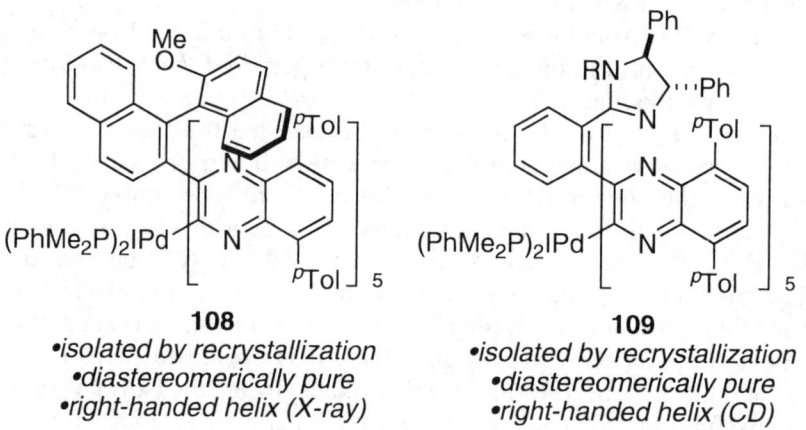

108
•isolated by recrystallization
•diastereomerically pure
•right-handed helix (X-ray)

109
•isolated by recrystallization
•diastereomerically pure
•right-handed helix (CD)

Fig. 8

due to the terminal 2'-methoxybinaphthyl group, which has a small effect on the control of the screw-sense, as shown in polymerization using **98a** as an initiator.

3.2.3
Asymmetric Block Copolymerization

By virtue of the living polymerization mechanism, the palladium-mediated polymerization of 1,2-diisocyanobenzene has been applied to the synthesis of block copolymers. A block copolymer **110** was synthesized by successive polymerization of diisocyanobenzene **88** and **111** using the chiral binaphthyl initiator (S)-**98d** (Scheme 60) [102]. The copolymer **110** showed a high solu-

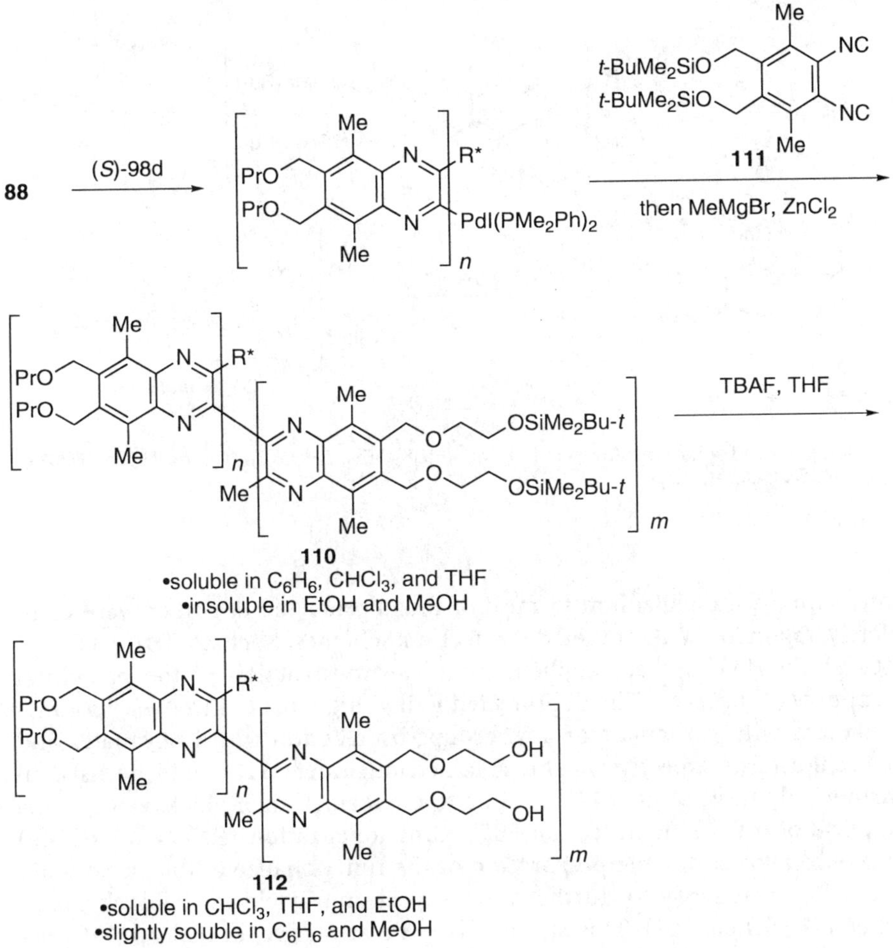

110
- soluble in C_6H_6, $CHCl_3$, and THF
- insoluble in EtOH and MeOH

112
- soluble in $CHCl_3$, THF, and EtOH
- slightly soluble in C_6H_6 and MeOH

Scheme 60

bility in benzene, chloroform, and THF, but was not soluble in ethanol or methanol. When the TBDMS groups were deprotected with TBAF, a remarkable change in solubility was observed. Thus, the deprotected polymer **112** was only slightly soluble in benzene, but was highly soluble in ethanol. The CD spectra of **112** in $CHCl_3$ and ethanol were quite similar in shape, although they showed significantly different intensities. The similarity in CD shape supports the highly rigid nature of the poly(quinoxaline) backbone, which is hardly affected by solvation.

The high stability of the living oligomers enables their isolation, and their chemical modification with no loss of activity toward further polymerization [103]. For example, living oligomer **113** was prepared from the TBDMS-protected monomer **111** and treated with boron trifluoride etherate at 0 °C in Et_2O, affording the TBDMS-free oligomer **114** quantitatively (Scheme 61).

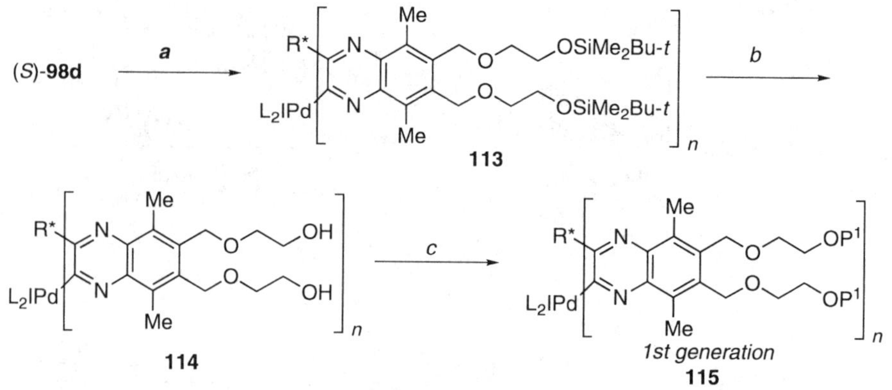

Reagents and conditions: (a) **111**, THF, r.t.; (b)BF_3OEt_2, CH_2Cl_2, 0 °C.; (c)RNCO, DMAP or RCOCl, DMAP, Et_3N, r.t.

Scheme 61

Interestingly, the palladium termini of the deprotected oligomer were completely retained. When other deprotection schemes, such as TBAF/THF or HCl aq/EtOH–THF were applied, significant destruction of the palladium termini was observed. The deprotected living oligomer **114** were easily functionalized with carbamate or acyl groups by reaction with isocyanates and acyl chlorides, respectively. The obtained oligomers **115** could be used in further polymerization of **111**, leading to the formation of block copolymers. Application of the deprotection (BF_3)-functionalization (RNCO or RCOCl) sequence allowed for the preparation of the functionalized diblock polymer **116**, which was ready for further polymerization to yield the triblock copolymer **117** (Scheme 62). This strategy is quite attractive, in that optically ac-

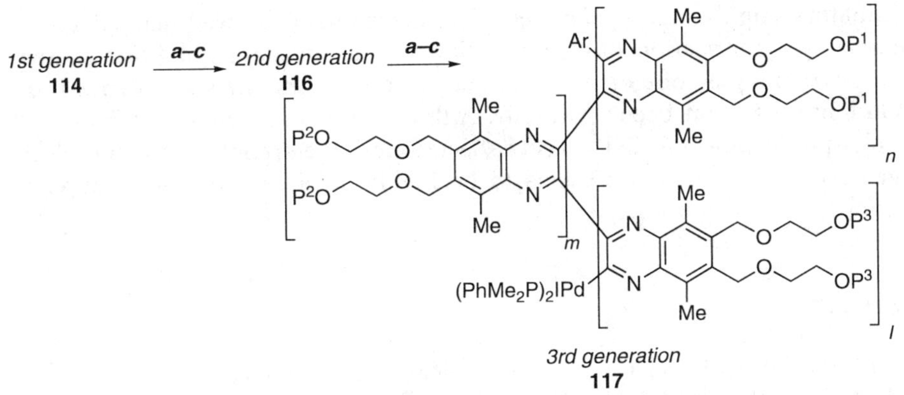

Scheme 62

tive block copolymers bearing various side-chains are accessible from a single TBDMS-protected monomer 111.

4
Concluding Remarks

The development of a variety of transition metal catalyst systems for the polymerization of isocyanides has opened up new possibilities for poly(isocyanide) chemistry. This strategy has achieved the synthesis of structurally well-defined isocyanide polymers, including those bearing various functionalities, and the asymmetric construction of helical main-chains of poly(isocyanide)s using chiral transition metal complexes as initiators. The rigid rod helical backbones of poly(isocyanide)s have received much attention in attempts at building up new macromolecular functional systems, in which the well organized spatial arrangement of multiple functionalities is necessary. Transition-metal systems also provide new access to block copolymers, which seems to be important for the development of functional materials. These new findings will undoubtedly contribute to the development of materials science in the future. Such developments will be promoted by further basic research on polymerization mechanisms, conformational analysis, and exploration of more efficient chiral initiators for asymmetric polymerization.

The chemistry of monoisocyanide polymers has been successfully applied to the aromatizing polymerization of 1,2-diisocyanobenzenes. The mechanism is essentially the same as that for monoisocyanides, involving successive insertion of neighboring isocyano groups, which cyclize to afford a quinoxaline ring. The polymerization is characteristic in that the living oligomers or polymers are stable enough to be isolated. Using chiral palladium initiators, highly screw-sense-selective, mechanistically well-defined asymmetric polymerization has been achieved. The resultant poly(quinoxaline-2,3-diyl)s may be one of the most stable non-racemic helical polymers, including those that occur naturally. Although no remarkable application of poly(quinoxaline-2,3-diyl)s has yet been found, the highly rigid and well-organized helical structure seems to be highly attractive for future applications.

References

1. Ugi I (ed) (1971) Isonitrile chemistry. Academic Press, London
2. Ugi I, Lohberger S, Karl R (1991) In: Comprehensive organic synthesis, vol 2. Pergamon, Oxford, p 1083
3. Millich F (1972) Chem Rev 72:101
4. Millich F (1980) J Polym Sci Macromol Rev 15:207
5. Yamamoto Y, Takizawa T, Hagihara N (1966) Nippon Kagaku Zasshi 87:1355
6. Yamamoto Y, Hagihara N. (1968) Nippon Kagaku Zasshi 89:898
7. Stackman RW (1968) J Macromol Sci Chem A2:225
8. Millich F, Sinclair RG (1968) Polym Sci Part C 22:33
9. Millich F, Sinclair RG (1968) J Polymer Sci Part A-1 6:1417
10. Ugi I, Fritzer V (1961) Chem Ber 94:2239
11. For a review on the migratory insertion of isocyanides, see: Durfee LD, Rothwell IP (1988) Chem Rev 88:1059
12. Yamamoto Y, Yamazaki H (1972) Coord Chem Rev 8:225
13. Saegusa T, Ito Y, Kobayashi S, Hirota K (1967) Tetrahedron Lett 8:521
14. Otsuka S, Nakamura A, Yoshida T (1969) J Am Chem Soc 91:7196
15. Nolte RJM, Stephany RW, Drenth W (1973) Recl Trav Chim Pays-Bas 92:83
16. Deming TJ, Novak BM (1991) Macromolecules 24:326
17. Yamamoto Y, Yamazaki H, Hagihara N (1968) Bull Chem Soc Jpn 41:532
18. Yamamoto Y, Yamazaki H, Hagihara N (1969) J Organomet Chem 18:189
19. Takei F, Tung S, Yanai K, Onitsuka K, Takahashi S (1998) J Organomet Chem 559:91
20. Otsuka, S, Nakamura A, Yoshida T, Naruto M, Ataka K (1973) J Am Chem Soc 95:3180
21. Nolte RJM, Drenth W (1973) Redl Trav Chim Pays-Bas 92:788
22. King RB, Greene MJ (1987) J Polym Sci Part A Polym Chem 25:907
23. Drenth E, Nolte RJM (1979) Acc Chem Res 12:30
24. Deming TJ, Novak BM (1991) Macromolecules 24:6043
25. Deming TJ, Novak BM (1993) J Am Chem Soc 115:9101
26. Deming TJ, Novak BM (1993) Macromolecules 26:7092
27. Deming TJ, Novak BM (1991) Macromolecules 24:5478; Deming TJ, Novak BM (1991) Polym Mater Sci Eng 65:148
28. Deming TJ, Novak BM, Ziller JW (1994) J Am Chem Soc 116:2366
29. Tomita I, Taguchi M, Takagi K, Endo T (1997) J Polym Sci Part A Polym Chem 35:431
30. Yamamoto Y, Yamazaki H (1970) Bull Chem Soc Jpn 43:2653
31. Yamamoto Y, Yamazaki H (1974) Inorg Chem 13:438
32. Ogawa H, Joh T, Takahashi S (1988) J Chem Soc Chem Commun 561

33. Onitsuka K, Ogawa H, Joh T, Takahashi S, Yamamoto Y, Yamazaki H (1991) J Chem Soc Dalton Trans 1531
34. Onitsuka K, Yanai K, Takei F, Joh T, Takahashi S (1994) Organometallics 13:3862
35. Takei F, Onitsuka K, Kobayashi N, Takahashi S (2000) Chem Lett 914
36. Yamamoto M, Onitsuka K, Takahashi S (2000) Organometallics 19:4669
37. Millich F, Baker GK (1969) Macromolecules 2:122
38. Okamoto Y, Nakano T (1994) Chem Rev 94:349
39. Nakano T, Okamoto Y (2001) Chem Rev 101:4013
40. Cornelissen JJLM, Rowan AE, Nolte RJM, Sommerdijk NAJM (2001) Chem Rev 101:4039
41. Nolte RJM (1994) Chem Soc Rev 11
42. Rowan AE, Nolte RJM (1998) Angew Chem Int Ed 37:63
43. Pu L (1997) Acta Polym 48:116
44. Nolte RJM, van Beijinen AJM, Drenth BW (1974) J Am Chem Soc 96:5932
45. Beijinen AJM, Nolte RJM, Drenth W, Hezemans AMF (1976) Tetrahedron 32:2017
46. Beijnen AJM, Nolte RJM, Zwikker JW, Drenth W (1978) J Mol Catal 4:427
47. Beijnen AJM, Nolte RJM, Drenth W, Hezemans AMF, van de Coolwijk PJFM (1980) Macromolecules 13:1386
48. Beijnen AJM, Nolte RJM, Naaktgeboren AJ, Zwikker JW, Drenth W, Hezemans AMF (1983) Macromolecules 16:1679
49. Deming TJ, Novak BM (1992) J Am Chem Soc 114:4400
50. Green MM, Gross RA, Schilling FC, Zero K, Crosby C III (1988) Macromolecules 21:1839
51. Ramos E, Bosch J, Serrano JL, Sierra T, Veciana J (1996) J Am Chem Soc 118:4703; Amabilino DB, Ramos E, Serrano JL, Sierra T, Veciana J (1998) J Am Chem Soc 120:9126
52. Takei F, Yanai K, Onituska K, Takahashi S (1996) Angew Chem Int Ed 35:1554
53. Takei F, Yanai K, Onitsuka K, Takahashi S (2000) Chem, Eur J 6:983
54. Takei F, Onitsuka K, Takahashi S (2000) Polym J 32:524
55. Harada T, Cleij MC, Nolte RJM, Hazemans AMF, Drenth W (1984) J Chem Soc Chem Commun 726
56. Kamer PCJ, Cleij MC, Nolte RJM, Harada T, Hezemans AMF, Drenth W (1988) J Am Chem Soc 110:1581
57. Faller JW, Lavoie AR, Parr J (2003) Chem Rev 103: ASAP article
58. Kamer PCJ, Nolte RJM, Drenth W (1986) J Chem Soc Chem Commun 1789
59. Kamer PCJ, Nolte RJM, Drenth W (1988) J Am Chem Soc 110:6818
60. Kamer PCJ, Nolte RJM, Drenth W (1988) Recl Trav Chim Pays-Bas 107:175
61. 61 Kamer PCJ, Nolte RJM, Drenth W, Nijs HLLM, Kanters JA (1988) J Mol Catal 49:21
62. Deming TJ, Novak BM (1992) J Am Chem Soc 114:7926
63. van der Eijk JM, Nolte RJM, Drenth W, Hezemans AMF (1980) Macromolecules 13:1391
64. Visser HGJ, Nolte RJM, Zwikker JW, Drenth W (1985) J Org Chem 50:3133
65. Visser HGJ, Nolte RJM, Zwikker JW, Drenth W (1985) J Org Chem 50:3138
66. van der Eijk JM, Richters VEM, Nolte RJM, Drenth W (1984) Recl Trav Chim Pays-Bas 103:46
67. Vlietstra E, Nolte RJM, Zwikker JW, Drenth W, Jansen RHAM (1982) Recl Trav Chim Pays-Bas 101:183
68. Cornelissen JJLM, Graswinckel WS, Adams PJHM, Nachtegaal GH, Kentgens APM, Sommerdijk NAJM, Nolte RJM (2001) J Polym Sci Part A Polym Chem 39:4255

69. Cornelissen JJLM, Donners JJJM, De Gelder R, Graswinckel WS, Metselaar GA, Rowan AE, Sommerdijk NAJM, Nolte RJM (2001) Science 293:676
70. Visser HGJ, Nolte RJM, Drenth W (1985) Macromolecules 18:1818
71. Cornelissen JJLM, Fischer M, Sommerdijk NAJM, Nolte RJM (1998) Science 280:1427
72. Roks MFM, Visser HGJ, Zwikker JW, Verkley AJ, Nolte RJM (1983) J Am Chem Soc 105:4507
73. Grassl B, Rempp S, Galin JC (1998) Macromol Chem Phys 199:239
74. van Esch J, Roks MFM, Nolte RJM (1986) J Am Chem Soc 108:6093
75. Bieglé A, Mathis A, Galin JC (2000) Macromol Chem Phys 201:113
76. Nolte RJM, Zomeren JAJ, Zwikker JW (1978) J Org Chem 43:1972
77. Hasegawa T, Kondoh S, Matsuura K, Kobayashi K (1999) Macromolecules 32:6595
78. Hasegawa T, Matsuura K, Ariga K, Kobayashi K (2000) Macromolecules 33:2772
79. King RB, Borodinsky L (1985) Macromolecules 18:2117
80. Roks MFM, Nolte RJM (1992) Macromolecules 25:5398
81. Abdelkader M, Drenth W, Meijer EW (1991) Chem Mater. 3:598
82. Vlietstra EJ, Nolte RJM, Zwikker JW, Drenth W, Meijer EW (1990) Macromolecules 23:946
83. Bosch J, Rovira C, Veciana J, Castro C, Palacio F (1993) Synth Met 55:1141
84. Takei F, Onitsuka K, Kobayashi N, Takahashi S (2000) Chem Lett 914
85. Kauranen M, Verbiest T, Boutton C, Teerenstra MN, Clays K, Schouten AJ, Nolte RJM, Persoons A (1995) Science 270:966
86. Teerenstra MN, Klap RD, Bijl MJ, Schouten AJ, Nolte RJM, Verbiest T, Persoons A (1996) Macromolecules 29:4871
87. Teerenstra MN, Hagting JG, Schouten AJ, Nolte RJM, Kauranen M, Verbiest T, Persoons A (1996) Macromolecules 29:4876
88. Hong B, Fox MA (1994) Macromolecules 27:5311
89. Hong B, Fox MA (1995) Can J Chem 73:2101
90. Ito Y, Ihara E, Hirai H, Ohsaki A, Ohnishi M, Murakami M (1990) J Chem Soc Chem Commun 403
91. Ito Y, Ihara E, Murakami M, Shiro M (1990) J Am Chem Soc 112:6446
92. Ito Y, Ihara E, Murakami M (1992) Polym J 24:297
93. Ito Y, Kojima Y, Suginome M, Murakami M (1996) Heterocycles 42:597
94. Ito Y, Ihara E, Uesaka T, Murakami M (1992) Macromolecules 25:6711
95. Ihara E (1992) PhD thesis, Kyoto University
96. Ito Y, Ihara E, Murakami M (1992) Angew Chem Int Ed Engl 31:1509
97. Ito Y, Kojima Y, Murakami M (1993) Tetrahedron Lett 34:8279
98. Ito Y, Kojima Y, Murakami M, Suginome M (1997) Bull Chem Soc Jpn 70:2801
99. Ito Y, Ohara T, Shima R, Suginome M (1996) J Am Chem Soc 118:9188
100. Ito Y, Miyake T, Hatano R, Shima R, Ohara T, Suginome M (1998) J Am Chem Soc 120:11880
101. Suginome M, Collet S, Ito Y (2002) Org Lett 4:351
102. Ito Y, Miyake T, Ohara T, Suginome M (1998) Macromolecules 31:1697
103. Ito Y, Miyake T, Suginome M (2000) Macromolecules 33:4034

Coordination Polymerization of Dienes, Allenes, and Methylenecycloalkanes

Kohtaro Osakada (✉) · Daisuke Takeuchi

Chemical Resources Laboratory, Tokyo Institute of Technology,
4259 Nagatsuta, Midori-ku, Yokohama 226-8503, Japan
kosakada@res.titech.ac.jp

1	Introduction	139
2	Polymerization of Conjugated Dienes	139
2.1	Polymerization Catalyzed by Early Transition Metal Complexes	143
2.1.1	Polymerization Catalyzed by Half Titanocene and Zirconocene	143
2.1.2	Polymerization Catalyzed by V and Nb Complexes	145
2.1.3	Polymerization Catalyzed by Cr Complexes	146
2.1.4	Theoretical Insights of the Reaction Mechanism	147
2.2	Polymerization Catalyzed by Lanthanides	148
2.2.1	Polymerization Catalyzed by Halogeno and Carboxylato Lanthanides	148
2.2.2	Polymerization Catalyzed by Organolanthanide Complexes	150
2.3	Polymerization Catalyzed by Late Transition Metals	152
2.3.1	Polymerization Catalyzed by Ni Complexes	153
2.3.2	Polymerization Catalyzed by Co and Fe Complexes	155
2.4	Copolymerization of Conjugated Dienes	156
3	Polymerization of Nonconjugated Dienes	159
3.1	Polymerization by Early Transition Metal Complexes	159
3.2	Copolymerization of Nonconjugated Dienes with Alkenes	162
4	Polymerization of 1,2-Dienes (Allenes)	165
4.1	Living Polymerization of Allenes Catalyzed by Ni and Pd Complexes	165
4.1.1	Polymerization of Aryl- and Alkylallenes	165
4.1.2	Polymerization of Alkoxyallenes	166
4.1.3	Polymerization of Functionalized Allenes	169
4.2	Copolymerization of Allenes with 1,3-Butadiene, Propyne, and Isocyanides	171
4.3	Polymerization Catalyzed by Co and Rh Complexes	172
5	Polymerization and Copolymerization of Methylenecyclopropanes	172
5.1	Ring-Opening Polymerization of Methylenecycloalkanes Catalyzed by Early Transition Metal Complexes	173
5.2	Ring-Opening Polymerization of Methylenecyclopropanes Catalyzed by Pd Complexes	175
5.3	Addition Polymerization of Methylenecyclopropanes by Ni Complexes	178
6	Copolymerization of Dienes and Methylenecyclopropanes with CO	180
6.1	Copolymerization of Nonconjugated Dienes with CO Catalyzed by Pd Complexes	180

6.2	Copolymerization of Alkyl- and Arylallenes with CO Catalyzed by Pd and Rh Complexes	181
6.3	Ring-Opening Copolymerization of Methylenecyclopropanes with CO by Pd Complexes	184
7	Conclusion	188
	References	189

Abstract This review describes recent progress in the polymerization of conjugated and nonconjugated dienes catalyzed by transition metal complexes. Coordination polymerization of 1,3-butadiene produces *cis*-1,4-, *trans*-1,4-, syndio-1,2-, and iso-1,2-polymers which are obtained selectively by choosing the catalysts and the reaction conditions. Lanthanides catalyze polymerization of 1,3-butadiene to produce the *cis*-1,4-polymer with excellent selectivity. The lanthanocenes, $CoCl_2$/MAO, $CpTiCl_3$/MAO, and (π-allyl)Ni($OCOCF_3$)(L) promote living polymerization of 1,3-dienes as well as copolymerization of 1,3-dienes with vinyl monomers. Nonconjugated dienes polymerize to give products with vinylic pendant groups or those with five- to six-membered rings. Late transition metals such as Ni, Pd, and Rh catalyze polymerization of various allene derivatives. Living polymerization of allenes with polar functional groups is used to prepare new copolymers. Methylenecycloalkanes are polymerized by transition metal complexes to produce polymers via addition polymerization or ring-opening polymerization. Copolymerization of dienes with CO is promoted by late transition metal complexes to give polyketones with vinyl groups or cyclic units.

Keywords Coordination polymerization · Diene · Allene · Methylenecyclopropane · Copolymerization

Abbreviations
ADMET	acyclic diene metathesis polymerization
BARF	$[B\{C_6H_3(CF_3)_2\text{-}3,5\}_4]^-$
(R,S)-BINAPHOS	(*R*)-2-(diphenylphosphino)-1,1'-binaphthalen-2'-yl (*S*)-1,1'-binaphthalen-2,2'-diyl phosphite
i-Bu	isobutyl
CD	circular dichroism
CGC	constrained geometry catalyst
Cp*	permethylcyclopentadienyl
DFT	density functional theory
DEPT	distortionless enhancement by polarization transfer
dppp	1,3-bis(diphenylphosphino)propane
dipp	1,3-bis(diisopropylphosphino)propane
dmpe	1,2-bis(dimethylphosphino)ethane
DSC	differential scanning calorimetry
EPDM	ethylene propylene diene methylene linkage
Flu	fluorenyl
Hex	hexyl
HIBAO	hexaisobutylaluminoxane
Ind	indenyl
naph	naphthenate
NMP	1-methyl-2-pyrrolidinone
MAO	methylaluminoxane
MBIC	1-methyl-3-*n*-butylimidazolium chloride

phen	1,10-phenanthroline
TCBQ	tetrachloro-1,4-benzoquinone
TG	thermogravimetry

1
Introduction

Coordination polymerization of dienes has long attracted significant attention both from academic and industrial aspects. 1,4-Polymerization of conjugated dienes results in a unique polymer having C=C double bonds in the main chain, while 1,2-polymerization of conjugated and nonconjugated dienes forms polymers with olefin functionality in the side chain group. The structures of the polymers with olefin functionality in the repeating units are closely related to their physical and chemical properties. Since there are many possible structures for the monomer units of the polymer of a diene, selective preparation of the polymers with a single structure is a demanding and important subject.

Recent progress in diene polymerization catalyzed by metal complexes, such as half-titanocene, lanthanide complexes, and late transition metal complexes, has enabled highly selective synthesis of the polymers of conjugated and nonconjugated dienes. Many metal compounds catalyze living polymerization of dienes and their copolymerization with vinyl monomers. Experimental and theoretical studies of the polymerization revealed the relevance of the mechanism of polymer growth promoted by the metal complexes to the structures of the polymers produced. Several books and reviews, published in the 1980s–1990s, cover earlier reports on the coordination polymerization of dienes [1].

This article will focus on studies of more recent developments in the polymerization of conjugated and nonconjugated dienes promoted by transition metal complexes, as well as the new catalytic polymerization of allenes and methylenecyclopropanes, the latter of which is an isomer of diene and causes ring-opening polymerization to afford new polymers composed of four-carbon repeating units. Metathesis polymerization of dienes, including ADMET polymerization [1d, 1e], is not described here because it is usually discussed together with ring-opening metathesis polymerization of cyclic alkenes rather than with addition polymerization of the unsaturated compounds promoted by metal complexes.

2
Polymerization of Conjugated Dienes

The polymerization of 1,3-dienes has been investigated using various initiators such as radicals, anions, and transition metal compounds because choice of the initiators and the reaction conditions influences the polymer structure. Scheme 1 depicts the structure of four fundamental repeating

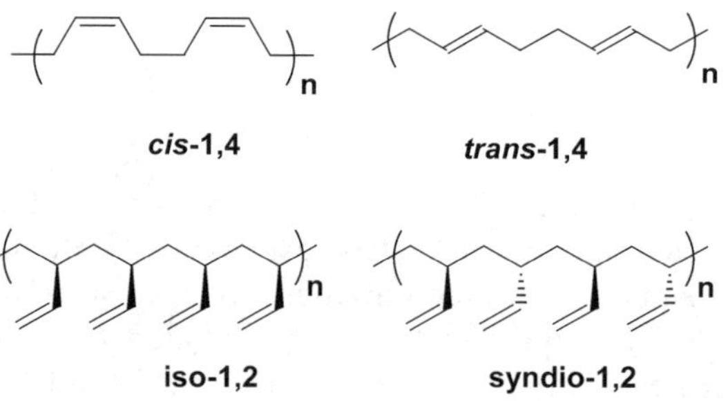

Scheme 1 Structures of 1,3-butadiene polymer

units, *cis*-1,4, *trans*-1,4, iso-1,2, and syndio-1,2, of poly(1,3-butadiene). The former two structural units are unique to the diene polymer, while the latter two can be formally regarded as the polymer of "vinylated ethylene". The synthesis of *cis*-1,4-polybutadiene that contains a *cis*-vinylene group in the structural unit has been the most important issue of diene polymerization because of its industrial use for synthetic rubber. Discovery of selective polymerization of 1,3-butadiene by a Ziegler–Natta catalyst, giving the *cis*-1,4-polymer, prompted studies of coordination polymerization of the conjugated dienes.

Table 1 summarizes typical results of the polymerization of 1,3-butadiene catalyzed by Ziegler-type catalysts and transition metal complexes. Selective formation of the four isomeric polybutadienes is enabled by choosing the catalyst. Polymerization catalyzed by $TiI_4/Al(i-Bu)_3$, $CoCl_2/AlEt_2Cl$, $[(\pi-allyl)Ni(OCOCF_3)]_2$, and $Nd(OCOR)_3/AlEt_2Cl/Al(i-Bu)_3$ produces *cis*-1,4-polybutadiene selectively, while *trans*-1,4-polybutadiene is formed from the reactions catalyzed by $VOCl_3/AlEt_2Cl$, $Ti(OBu)_4/AlEtCl_2$, and $RhCl_3 \cdot 3H_2O$. The reactions of butadiene catalyzed by $Co(acac)_3/AlEt_3/CS_2$ and by $Cr(acac)_3/AlR_3$ afford the syndio-1,2-polybutadiene and the iso-rich 1,2-polymer, respectively.

Porri et al. summarized the reports of chemo-, regio-, and stereoselective coordination polymerization of 1,3-dienes, including their own, and suggested a relationship of the selectivity to the reaction mechanism and the structure of the reaction intermediates [1b]. Scheme 2 depicts the general pathways of the growth of poly(1,3-butadiene) promoted by transition metal complexes. Three types of coordination of the monomer, η^4-*cis*, η^4-*trans*, and η^2-*trans*, are possible. The two former coordination modes differ from each other in the geometry about the C–C single bond of the ligand. Yasuda and Erker independently found higher stability of the η^4-*cis* coordination of

Table 1 Polymerization of 1,3-butadiene

Catalyst	Selectivity, %			References
	cis-1,4	trans-1,4	1,2	
$TiI_4/Al(i-Bu)_3$	92–93	2–3	4–6	a
$CoCl_2/AlEtCl_2/H_2O$	98	1	1	b
$[(\pi\text{-allyl})Ni(OCOCF_3)]_2$	97–99			c
$NdCl_3/EtOH/AlEt_3$	98	0	2	d
$VOCl_3/AlEt_2Cl$		97–98	2–3	e
$Ti(OBu)_4/AlEtCl_2$		93–94		f
$RhCl_3\cdot 3H_2O$	<1	99	0.2	g
$V(acac)_3/AlEt_3$			80–86 (syndio)	h
$Co(acac)_3AlR_3/CS_2$			>99 (syndio)	i
$Cr(acac)_3AlR_3$			80–85 (iso)	j

[a] Saltman WM, Link TH (1964) Ind Eng Chem Prod Res Dev 3:199
[b] Gippin M (1962) Ind Eng Chem Prod Res Dev 1:32
[c] Dawans F, Teyssié P (1967) Ind Eng Chem Prod Res Dev 10:261
[d] Shen Z, Ouyang J, Wang F, Hu Z, Yu F, Qian B (1980) J Polym Sci Polym Chem Ed 18:3345
[e] Natta G, Porri L, Mazzei A (1959) Chim Ind 41:1959
[f] Cucinella S, Mazzei A, Marconi W, Busetto C (1970) J Macromol Sci Chem Part A Chem 4:1549
[g] Rinehart RE, Smith HP, Witt HS, Romeyn H Jr (1961) J Am Chem Soc 83:4864
[h] Natta G, Porri L, Zanini G, Fiore L (1959) Chim Ind 41:526
[i] Ashitaka H, Ishikawa H, Ueno H, Nagasaka A (1983) J Polym Sci Polym Chem Ed 21:1853

1,3-butadiene to Zr than the η^4-*trans* bonding [2]. η^4-*Trans* coordination is a simple π-bond between a C=C double bond and the metal center.

Insertion of the monomer, bonded to the metal in η^4-*cis* fashion, into the metal–polymer bond forms a new π-allyl polymer end with the substituent at the *anti*-position. Successive insertion of the new monomer with η^4-*cis* coordination would produce *cis*-1,4-polybutadiene. Insertion of η^4-*trans*-coordinated monomer into metal–polymer bond leads to *trans*-1,4-polybutadiene via *syn* π-allyl intermediates. The above *anti* π-allyl polymer end is often equilibrated with the thermodynamically more favorable *syn* π-allyl structure via π–σ–π rearrangement. Thus, the ratio of *cis*-1,4 and *trans*-1,4 repeating units of the polymer produced depends on the relative rates of the two reactions: C–C bond formation between the monomer and the polymer end, and *anti* to *syn* isomerization of the π-allyl end of the growing polymer. If the *anti*–*syn* isomerization of the *anti* π-allyl polymer end occurs more rapidly than the insertion of a new monomer, the polymer with *trans*-1,4 units is formed even from η^4-*cis*-coordinated monomer. The polymerization catalyzed by Ti, Co, or Ni complexes shows high *cis*-1,4 selectivity, while that with low monomer concentration results in increase of the *trans* content of

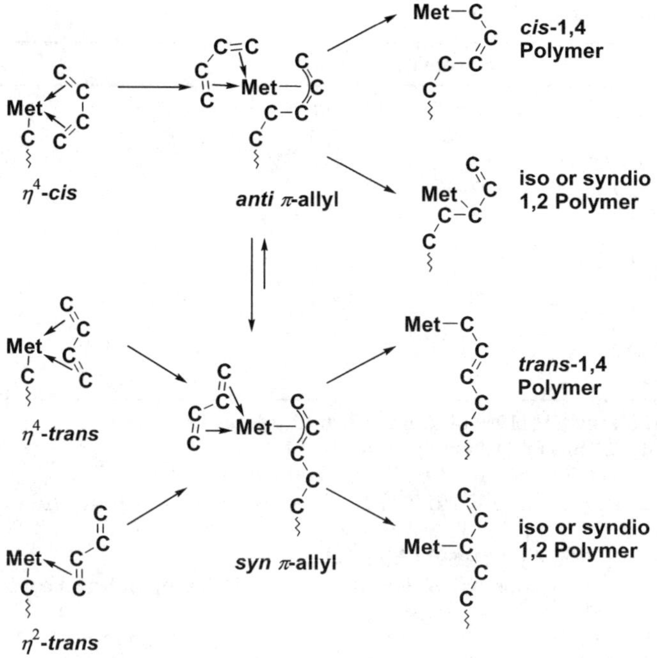

Scheme 2 Pathways for formation of isomeric polybutadiene

the polymer due to slower polymer growth than the *anti–syn* isomerization of the π-allylic polymer end.

The 1,2-polymerization of diene may be accounted for by assuming rearrangement of the π-allyl polymer into a σ-allyl structure having a single bond between the metal and a CH allyl carbon of the polymer, and insertion of the diene monomer into the M–C σ bond. A more convincing and generally accepted rationalization for the smooth 1,2-polymerization of 1,3-dienes is shown in Scheme 3. The diene monomer coordinates to the metal at the

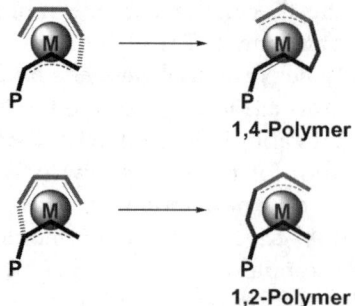

Scheme 3 1,2- vs 1,4-polymer formation

opposite site to the π-allyl end of the polymer. A new C–C bond is formed between the terminal CH_2 carbon of the π-allyl ligand and one CH_2= carbon of the monomer or between the internal CH π-allyl carbon and the other CH_2= carbon of the monomer. Although the latter bond-forming reaction is sterically less favorable than the former due to the bulky polymer chain at the π-allyl end, it could also take place smoothly with little change of the coordination structure of the monomer and polymer end. Both 1,2- and 1,4-polymerization can occur depending on the catalyst, because the two polymerization reactions via the common intermediate in Scheme 3 have small difference in their activation energies.

2.1
Polymerization Catalyzed by Early Transition Metal Complexes

Ziegler–Natta catalysts containing Ti compounds, such as TiX_4/AlR_3 (X=halogen) or $Ti(OBu)_4/AlR_3$, have been studied extensively for the polymerization of 1,3-dienes. The catalysts TiI_4/AlR_3 and $TiCl_4/I_2/AlR_3$ exhibit higher selectivity for cis-1,4 polymerization of 1,3-butadiene than the $TiCl_4/AlR_3$ catalyst. The discovery of alkene polymerization by homogeneous Kaminsky catalyst [3] and of syndiospecific styrene polymerization by $CpTiCl_3$/MAO [4, 5] has shifted the research interest into homogenous catalysts composed of titanocene or zirconocene and MAO for the polymerization of 1,3-dienes.

2.1.1
Polymerization Catalyzed by Half Titanocene and Zirconocene

$CpTiCl_3$/MAO exhibits high catalytic activity not only for syndiospecific styrene polymerization but also for the polymerization of conjugated dienes, 1,3-butadiene [6, 7], isoprene [6, 7], 1,3-pentadienes [6, 7, 8], and dimethyl-1,3-butadienes [6, 7, 9]. The polymerization of 1,3-butadiene catalyzed by $CpTiCl_3$/MAO produces mainly the cis-1,4-polymer (Eq. 1), although the cis–trans selectivity (79% cis) is lower than that obtained by conventional Ziegler–Natta catalysts (up to 93% cis).

The catalytic activity of the titanocene derivatives increases in the order $Cp_2TiC_2 < Cp_2TiCl < CpTiCl_3$ for the polymerization of 1,3-butadiene and 4-methyl-1,3-pentadiene [10]. Substituted 1,3-dienes also polymerize in the presence of $CpTiCl_3$/MAO catalyst with the selectivity depending on the monomer structure. The polymerization of (Z)-1,3-pentadiene forms cis-

1,4-polymer at 20° C, whereas 1,2-syndiotactic polymer is obtained at −20° C (Eq. 2) [8, 11]. The polymerization of 4-methyl-1,3-pentadiene by CpTiCl$_3$/MAO gives 1,2-syndiotactic polymer [9],

the mechanism of which is similar to that of the syndiotactic polymerization of styrene. The polymerization of 2-phenyl-1,3-butadiene and (E)-1-phenyl-1,3-butadiene by CpTiCl$_3$/MAO gives cis-1,4-polymer (cis-1,4 >99%) and 3,4-polymer (3,4=76%), respectively [12].

In 1999, Miyazawa reported living polymerization of 1,3-butadiene by using CpTiCl$_3$/MAO catalyst at −25 °C [13]. The molecular weight distribution of the polymer is reasonable as the living polymerization (M_w/M_n=1.04) although the complex initiates the reaction with moderate efficiency (~50%). The reaction rate and cis-selectivity of the 1,4-polymerization are largely affected by the substituent on the cyclopentadienyl ligand of the Ti complex. The catalyst composed of C$_5$H$_4$(t-Bu)TiCl$_3$ and MAO produces polybutadiene with high cis content (93.5%) and narrow molecular weight distribution (M_w/M_n=1.04) (Eq. 3) [14].

Nonmetallocene Ti-based catalyst, Ti(CH$_2$Ph)$_4$/MAO, polymerizes butadiene and isoprene giving the cis-1,4-polymers, while the polymerization of 4-methyl-1,3-butadiene produces the syndio-1,2-polymer.

Zirconocene complexes show much lower activity as the catalyst of 1,3-butadiene polymerization than the titanocene catalysts [15]. The catalyst composed of rac-[CH$_2$(3-tert-butyl-1-indenyl)$_2$]ZrCl$_2$ and MAO promotes the 1,4-polymerization of 1,3-butadiene or (Z)-1,3-pentadiene and 1,2-polymerization of (E)-1,3-pentadiene and 4-methyl-1,3-pentadiene (Eq. 4) [16]. The bulky tert-butyl group of the ligand is essential for smooth polymerization because the *ansa*-zirconocenes without bulky substituents show much lower catalytic activity.

$$\text{CH}_2=\text{CHCH}=\text{CH}_2 \xrightarrow[\text{toluene, 15-50 °C}]{\text{Zr(Cl)}_2/\text{MAO}} \left(\text{polybutadiene}\right)_n \quad \text{eq 4}$$

cis:trans = 50:50

2.1.2
Polymerization Catalyzed by V and Nb Complexes

V(acac)$_3$/MAO was reported initially by Ricci to be the catalyst of 1,3-butadiene polymerization which produces *trans*-1,4-polymer as shown in Eq. 5 [17]. *Cis–trans* selectivity of the

$$\text{CH}_2=\text{CHCH}=\text{CH}_2 \xrightarrow[\text{toluene, 15 °C}]{\text{V(acac)}_3 \text{ MAO}} \left(\text{polybutadiene}\right)_n \quad \text{eq 5}$$

trans-1,4 (90.9-97.3 %)

reaction varies depending on the reaction conditions. Detailed studies by Endo showed that the temperature and molar ratio of Al to V change the stereoselectivity of the polymerization. An increase in the relative amount of MAO to V(acac)$_3$ causes an increase of molecular weight and of *cis* content of the polymer [18]. The selectivity of the polymerization of isoprene catalyzed by V(acac)$_3$/MAO depends on the reaction temperature; the reaction produces predominantly the *trans*-1,4 form at −78 °C and *cis*-1,4-polymer as a major structure at room temperature. This indicates that the isomerization of the *anti* π-allyl polymer end into the *syn* π-allyl form occurs effectively even at −78 °C, and that the insertion of the monomer is more rapid than the isomerization to keep the *anti* π-allyl structure of the polymer end during the reaction at room temperature [19].

Nb-based catalyst [NbO(C$_{16}$H$_{11}$O$_6^-$)(C$_2$O$_4^{2-}$)] (C$_{16}$H$_{12}$O$_6$=hemateine), prepared from the reaction of hemateine with [NH$_4$·H$_2$NbO(C$_2$O$_4^{2-}$)$_3$·3H$_2$O], also shows different selectivity depending on the reaction conditions. 1,3-Butadiene polymerization occurs specifically to produce *cis*-1,4-polymer in toluene, whereas *trans*-1,4-polymer is obtained in the molten salt, AlCl$_3$/MBIC (Eq. 6) [20, 21].

$(C_{16}H_{11}O_6)^- =$ [structure]

$$\text{CH}_2=\text{CH-CH=CH}_2 \xrightarrow[\text{toluene, 50 °C}]{[\text{NbO}(C_{16}H_{11}O_6^-)(C_2O_4^{2-})], \text{AlEt}_2\text{Cl}} \text{cis-1,4 (> 95 \%)}$$

$$\xrightarrow[\text{MBIC, AlCl}_3, -5 \text{ °C}]{[\text{NbO}(C_{16}H_{11}O_6^-)(C_2O_4^{2-})], \text{AlEt}_2\text{Cl}} \text{trans-1,4 (> 95 \%)}$$

Vanadocene catalysts, $(C_5H_4Me)VCl_2(PEt_3)_2$/MAO and Cp_2VCl/MAO, produce polybutadiene with a cis-1,4 content of 80–89% [22]. Amino group-functionalized vanadocenes bring about the polymerization of 1,3-butadiene in the presence of MAO. Complexes $\{C_5H_4CH(CH_2)_4NMe\}_2VCl$, $\{C_5H_4(CH_2)_2N(CH_2)_5\}_2VCl$, Cp_2VCl_2, and $(C_5H_4Me)(PEt_3)_2VCl_2$ give cis-1,4-polymer (cis-1,4=80–91%), which is in contrast to the reaction by $\{C_5H_4CH(CH_2)_4NMe\}(PMe_3)_2VCl_2$/MAO and $\{C_5H_4(CH_2)_nNR_2\}(PMe_3)_2VCl_2$/MAO to form the polymer with trans-1,4 structure (trans-1,4=45–62%) [23].

2.1.3
Polymerization Catalyzed by Cr Complexes

Mixtures of organoaluminum compounds with Cr compounds such as Cr(acac)$_3$, Cr(CO)$_5$(py), and Cr(NCPh)$_6$ were employed as the catalyst for the polymerization of 1,3-butadiene. The reactions afford the 1,2-polymer rather than the 1,4-polymer. CrCl$_2$(dmpe)/MAO and CrMe$_2$(dmpe)/MAO give crystalline 1,2-syndiotactic polybutadiene at temperatures ranging from −30 to +20 °C (Eq. 7) [24]. The selectivity does not vary with

$$\xrightarrow[\text{toluene 20 °C}]{\text{CrCl}_2(\text{dmpe}) \text{MAO}} \quad \text{1,2-Syndio, 1,2 = 97\%, [rrrr] up to 90\%} \qquad \text{eq 7}$$

the polymerization temperature, in contrast to the polymerization catalyzed by CpTiCl$_3$/MAO to form cis-1,4-polymer at +20 °C and 1,2-syndio product at −20 °C. The polymerization of isoprene catalyzed by CrCl$_2$(dmpe)/MAO gives atactic 3,4-polymer via preferential insertion of the less sterically hindered C=C bond into the Cr–polymer bond. The copolymerization of butadiene and isoprene by CrCl$_2$(dmpe)/MAO produces polymers whose melting points vary from 49 to 162 °C by changing the relative ratio of the two monomers [25]. The 1,2-polymerization of 1,3-butadiene catalyzed by Cr(acac)$_2$/alkylaluminum was studied in detail [26].

2.1.4
Theoretical Insights of the Reaction Mechanism

As described above, the Ti-based catalysts, including CpTiCl$_3$/MAO, tend to produce *cis*-1,4-polybutadiene, whereas V(acac)$_3$/MAO catalyst forms *trans*-1,4-polybutadiene. The reaction often gives 1,2-polymer as the minor product. Formation of the *cis*- and *trans*-1,4-polymers depending on the kind of catalyst was discussed, based on the structure of the coordinated monomer and the π-allylic polymer end as depicted in Scheme 2. The yield and *cis* selectivity of the polymerization catalyzed by V(acac)$_3$/MAO increase by raising the reaction temperature or by increasing the Al/V ratio of the catalyst [18]. A high reaction temperature and high Al cocatalyst concentration enhance the insertion of the monomer into the metal–polymer bond, which make it faster than the isomerization of the π-allylic polymer end from *anti* to *syn* form [27].

The pathways and the intermediates of 1,3-butadiene polymerization catalyzed by CpTiCl$_3$/MAO (Scheme 2) were recently reinvestigated from theoretical aspects using molecular dynamics, ab initio, and DFT calculations [28]. Two complexes, CpTi(η^3-CH$_2$CHCHCH$_3$)(η^4-CH$_2$=CHCH=CH$_2$) and CpTi(η^3, η^1-CH$_2$CHCH–(CH$_2$)$_2$–CH=CHCH$_3$) (and CpTi(η^3, η^2-CH$_2$CHCHCH$_2$CH(CH$_3$)CH=CH$_2$)), were calculated as the model compounds of the monomer-coordinated intermediate and the monomer-free intermediate of the polymerization, respectively (Scheme 4). The rate-deter-

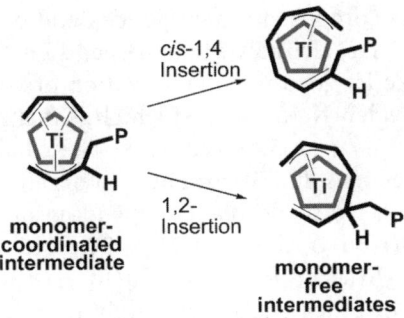

Scheme 4 The pathway and intermediates of the polymerization catalyzed by CpTiCl$_3$/MAO (P=polymer chain although the structures were calculated for P=H)

mining step of the reaction is included in conversion of the monomer-free intermediate into the monomer-coordinated intermediate, accompanied by elimination of the coordinated vinylene group of the polymer chain and by coordination of a new monomer according to the results of DFT calculation [28]. Monomer insertion into the Ti-π-allyl bond of the monomer-coordinated intermediates, giving the monomer-free intermediate, takes place more easily. Thus, selectivity between 1,4-insertion and 1,2-insertion of 1,3-butadiene is governed not only by the insertion of the monomer but also by the coordination of a new monomer. DFT investigation showed that one of the stereochemically different pathways for 1,4-insertion is more favorable than any of those for 1,2-insertion. It is consistent with the experimental results that CpTiCl$_3$/MAO catalyzes 1,3-butadiene polymerization to produce mostly the *cis*-1,4 product.

2.2
Polymerization Catalyzed by Lanthanides

Lanthanides, which are d-electron deficient metals similar to early transition metals such as Ti and Zr, polymerize ethylene and acrylic esters. Although studies of coordination polymerization of 1,3-butadiene catalyzed by lanthanide compounds started as early as 1964 by discovery of the Ce catalysts [29], the number of reports on lanthanide catalysis for the polymerization of diene had been small until the 1980s. Recently, fast and selective polymerization reactions have been found and have attracted growing attention. Three types of lanthanide catalysts are employed for diene polymerization: mixtures with organoaluminum compounds like Ziegler–Natta-type catalysts, allyl complexes, and metallocenes.

2.2.1
Polymerization Catalyzed by Halogeno and Carboxylato Lanthanides

Two kinds of catalysts composed of lanthanides and organoaluminum compounds, LnCl$_3$·nL (L=THF, alcohols, etc.)/AlR$_3$ and Ln(OCOR)$_3$ (or Ln(OR)$_3$/AlR$_3$/AlR$_2$Cl), are effective for the polymerization of 1,3-butadiene. The former resembles the Ziegler–Natta catalyst which is composed of a mixture of TiX$_m$ (X=halogen, OR etc.; m=3,4) and AlR$_3$. The catalysts are usually heterogeneous due to the low solubility of LnCl$_3$ in organic solvents. The activity of the catalysts from NdCl$_3$ decreases depending on the cocatalyst in the order Al(i-Bu)$_3$=AlH(i-Bu)$_2$>AlEt$_3$>>AlMe$_3$. The Nd catalysts, containing NdCl$_3$, NdCl$_3$/P(OBu)$_3$/Al(i-Bu)$_3$ and NdCl$_3$/(ROH)$_3$/AlEt$_3$, also polymerize 1,3-butadiene [30]. The silica-supported catalyst, prepared by the reaction of Nd(η^6-C$_6$H$_5$Me)(AlCl$_4$)$_3$ with Lewis-acid-masked silica promotes *cis*-1,4-polymerization of 1,3-butadiene in the presence of AlEt$_3$ [31].

The soluble Ln(OCOR)$_3$ catalyzes the polymerization of 1,3-butadiene in the presence of the cocatalysts AlR$_3$ and AlR$_2$Cl. A representative catalyst is prepared by mixing Nd(OCOR)$_3$ (OCOR=ethylhexanoate (OCOCH

(C_2H_5)(CH_2)$_3CH_3$), naphthenate ($C_{11}H_{10}O_2$), versatate, etc.) and $AlEt_2Cl$ in solution, followed by addition of $Al(i\text{-}Bu)_3$ [32]. The lanthanides commonly used are Nd, Gd, Ce, and Pr, in which the Nd catalysts exhibit higher activity than the other metal catalysts (Eq. 8). The catalysts,

$$\text{CH}_2=\text{CH-CH=CH}_2 \xrightarrow[\text{(Ln = Nd, Gd, Ce, Pr)}]{\text{Ln(OCOR)}_3,\ \text{Al(i-Bu)}_3,\ \text{AlEt}_2\text{Cl} \atop \text{toluene, 20-50 °C}} \left(\text{CH}_2\text{-CH=CH-CH}_2\right)_n \quad \text{eq 8}$$

cis-1,4 (97-98%)

$Nd(OCOR)_3/Al(i\text{-}Bu)_3/AlEt_2Cl$, and $Nd(OCOR)_3/Al_2Et_3Cl_3/AlH(i\text{-}Bu)_2$, polymerize 1,3-butadiene to produce the polymer with cis-1,4 structure [33]. The chlorinated organoaluminum compounds are indispensable for smooth polymerization. Addition of water and carboxylic acid improves the reactivity and selectivity.

The kinetics of the polymerization of 1,3-butadiene by $Nd(OCOR)_3$ (OCOR=versatate)/$AlH(i\text{-}Bu)_2/Et_3Al_2Cl_3$ were studied [34]. The molecular weight increases linearly with increasing monomer conversion. The broad molecular weight distribution of the polymer at the initial stage of polymerization (M_w/M_n=2.5–3.5) decreases with conversion of the monomer to M_w/M_n=2.0–2.5. The catalyst with a higher $AlH(i\text{-}Bu)_2/Nd(OCOR)_3$ ratio produces a larger number of polymer chains, while the molecular weight distribution of the polymer is not changed by the Al/Nd ratio. This is accounted for by the fast and reversible transfer of the growing polymer chain between Nd and Al. Increase of the Al/Nd ratio of the catalyst does not affect the structure of the active-site-containing Nd center but increases the number of the dormant-site-containing Al centers.

$Nd(OCOR)_3/AlEt_2Cl/Al(i\text{-}Bu)_3$ promotes pseudo-living polymerization of 1,3-butadiene. Analyses of the catalyst by synchrotron X-ray absorption and UV visible spectroscopy enabled the observation of the catalyst structure in situ [35]. The $Nd(OCOR)_3$ used in the polymerization catalyst, such as $Nd(OCOR)_3/AlEt_2Cl/Al(i\text{-}Bu)_3$, was prepared in situ from the reaction of $Nd(OAc)_3$ or Nd_2O_3. Evans prepared and purified the carboxylates of Nd and La carefully and characterized them by X-ray crystallography. The catalysts prepared by mixing $AlEt_2Cl$ with the well-defined carboxylate of La contain $LaCl_3$ whose chlorides are derived from the $AlEt_2Cl$ [36].

The polymerization of 1,3-butadiene is also catalyzed by $Nd\{N(SiMe_3)_2\}_3/Al(i\text{-}Bu)_3/AlEt_2Cl$ [37] and by catalyst containing $Nd(OCOR)_3$ (OCOR=versatate) and $SiCl_4$ cocatalyst [38]. The polymerization of 1,3-butadiene catalyzed by $Nd(O\text{-}i\text{-}Pr)_3/MAO/t\text{-}BuCl$ takes place faster than that by the conventional $Nd(O\text{-}i\text{-}Pr)_3/AlH(i\text{-}Bu)_2/t\text{-}BuCl$ catalyst and produces a polymer with higher molecular weight as well as better molecular weight distribution and cis-1,4 stereospecificity [39]. A homogeneous catalyst, $Nd(OR)_3/MAO$, is soluble in organic solvents and results in high cis-1,4 content and narrower molecular weight distribution of the polymer (M_w/M_n=2.0) than the ternary catalysts, Nd(OCOR)$_3$/AlR$_3$/Cl-containing compound, and the binary catalyst, NdCl$_3$/AlR$_3$. The larger molecular weight distribution of the polymer product obtained by the latter two catalysts is ascribed to low solubility of the complexes and the presence of multiple active sites [40].

The 1,4-polymerization of (E)- and (Z)-1,3-pentadiene can give four stereoregular polymers originating from the chiral carbon atom in the main chain as shown in Scheme 5. Three of these polymers, *trans*-1,4-isotactic

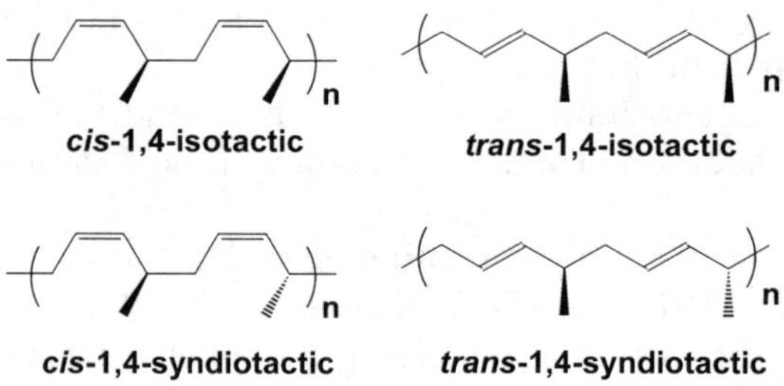

cis-1,4-isotactic   *trans*-1,4-isotactic

cis-1,4-syndiotactic   *trans*-1,4-syndiotactic

Scheme 5 Structures of 1,4-polymer of 1,3-pentadiene

[41], *cis*-1,4-isotactic [42], and *cis*-1,4-syndiotactic [43], were actually synthesized. *Cis*-1,4-isotactic polypentadiene with high stereoregularity (*cis*=95%) was prepared by the catalyst Nd(OCOC$_7$H$_{15}$)$_3$/AlEt$_2$Cl/Al(i-Bu)$_3$, and was analyzed by NMR spectroscopy [44]. The crystalline polymer showed the presence of two-fold helical conformation, revealed by the fiber X-ray diffraction.

2.2.2
Polymerization Catalyzed by Organolanthanide Complexes

Polymerization of 1,3-butadiene is catalyzed by (π-allyl)$_3$M (M=La, Nd) and Li[Nd(π-allyl)$_4$] to produce *trans*-1,4-polybutadiene, whereas the complexes (π-allyl)$_3$Nd and Nd, in the presence of the MAO cocatalyst, give *cis*-1,4-polybutadiene [45] (Eq. 9). The reaction without MAO renders π-coordination of

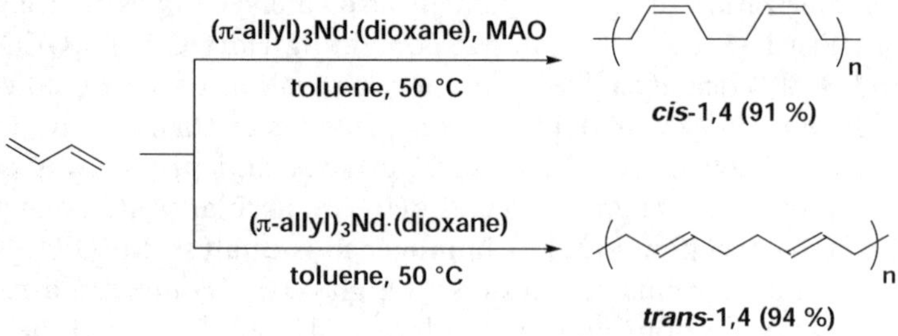

(π-allyl)$_3$Nd·(dioxane), MAO
toluene, 50 °C
cis-1,4 (91 %)

(π-allyl)$_3$Nd·(dioxane)
toluene, 50 °C
trans-1,4 (94 %)

the s-*trans* monomer to Nd more dominant. Insertion of the s-*trans* diene into the *syn* π-allyl–metal bond produces the *trans*-1,4-polymer. (π-Allyl)$_3$Nd/AlMe$_2$Cl/AlMe$_3$ catalyzes the polymerization of 1,3-butadiene to produce the *cis*-1,4-polymer [46]. The kinetics of the reaction suggest the formation of an intermediate complex with coordinated butadiene in the s-*cis* form. Insertion of the monomer into the polymer–Nd bond forms the *anti* π-allyl polymer end similarly to the polymerization promoted by CpTiCl$_3$/MAO (Scheme 4). π-Allylneodymium chloro complexes, (π-allyl)NdCl$_2$ and (π-allyl)$_2$NdCl, also catalyze polymerization of 1,3-butadiene in the presence of MAO to give the *cis*-1,4-polymer [47]. Polymerization promoted by mixtures of the chloro(π-allyl)neodymium complexes with HIBAO or MAO proceeds via the intermediate with s-*cis*-coordinated butadiene to produce *cis*-1,4-polybutadiene [48].

Lanthanocene catalysis for the polymerization of conjugated dienes [49] has a very short history and great impact on this field. The first report by Jin in 1998 demonstrated high utility of mixtures of the lanthanocenes

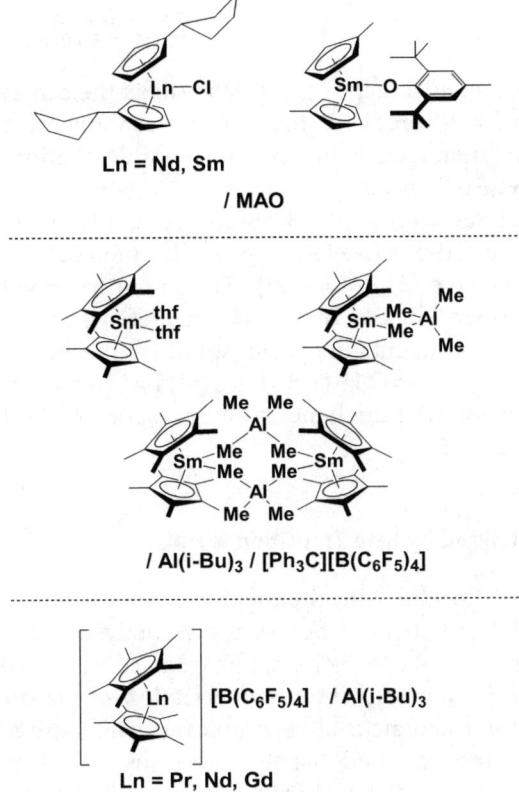

Scheme 6 Lanthanocene catalysts for 1,3-butadiene polymerization [49]

(C_5H_9Cp)$_2$NdCl, (C_5H_9Cp)$_2$SmCl, and (MeCp)$_2$SmOAr (Ar=C_6H_2-2,6-(t-Bu)$_2$-4-Me) with MAO, for the polymerization of 1,3-butadiene [50]. Hou and Wakatsuki found that (C_5Me_5)$_2$Sm(THF)$_2$/Al(i-Bu)$_3$/[Ph$_3$C][B(C_6F_5)$_4$] promotes living polymerization of 1,3-butadiene to produce a polymer with high cis-1,4 structure content (94.2–95.0%) and relatively narrow molecular weight distribution (M_w/M_n=1.31–1.39) [51]. Scheme 6 summarizes a series of the important lanthanocene catalysts. (C_5Me_5)$_2$Sm[(μ-Me)AlMe$_2$(μ-Me)]$_2$Sm(C_5Me_5)$_2$/Al(i-Bu)$_3$/[Ph$_3$C][B(C_6F_5)$_4$] shows higher activity than the above catalyst. (C_5Me_5)$_2$Sm(μ-Me)$_2$AlMe$_2$/Al(i-Bu)$_3$/[Ph$_3$C][B(C_6F_5)$_4$] leads to excellent cis-1,4-selectivity of the reaction (99.0%) and narrow molecular weight distribution of the product (M_w/M_n=1.20–1.23) (Eq. 10) [52].

$$\text{butadiene} \xrightarrow[\text{toluene}\quad -20\,°C]{\text{Sm/Al catalyst, Al(iBu)}_3,\ [Ph_3C][B(C_6F_5)_4]} \text{cis-1,4-polybutadiene} \quad \text{eq 10}$$

cis-1,4 (99.0%)
M_w/M_n = 1.20-1.23

Another important advantage of this catalysis is the achievement of living polymerization of 1,3-butadiene that makes random and block copolymerization of 1,3-butadiene with styrene possible. Cationic Gd complex, [(C_5Me_5)$_2$Gd][B(C_6F_5)$_4$] in combination with Al(i-Bu)$_3$, was recently found to catalyze the polymerization of 1,3-butadiene to afford cis-1,4-polybutadiene with perfect selectivity (>99.9%) [53]. The molecular weight distribution is relatively narrow (M_w/M_n=1.45). The samarocene with a substituent at the cyclopentadienyl ligand brings about stereospecific polymerization of isoprene [54]. A cyclopentadienyl-phosphido samarium complex, [Me$_2$Si(C_5Me_4)(PAr)Sm(thf)] (Ar=C_6H_2(t-Bu)$_3$-2,4,6), has been used as the catalyst for polymerization of 1,3-butadiene in the presence of MMAO to produce cis-1,4-rich polymer [55].

2.3
Polymerization Catalyzed by Late Transition Metals

Mixtures of Ni and Co salts with organoaluminum reagents catalyze the polymerization of 1,3-butadiene, whereas these catalysts are not effective for alkene polymerization due to facile chain transfer by β-hydrogen elimination of the growing polymer. The polymerization of 1,3-butadiene usually proceeds via the intermediate π-allyl complexes, which are more stable than the alkyl metal complexes and hardly cause any β-hydrogen elimination [56]. The late transition metal compounds polymerize 1,3-butadiene even in the absence of organoaluminum cocatalyst. As early as the 1960s–1970s, $RhCl_3 \cdot 3H_2O$ [57] and [(π-allyl)NiX]$_2$ (X=halide, trifluoroacetate) [58] were found to catalyze the polymerization of 1,3-butadiene to produce the *trans*-1,4-polymer and the *cis*- or *trans*-1,4-polymer, respectively. Recent progress in this field has enabled living polymerization, highly selective polymerization, and elucidation of the reaction mechanism.

2.3.1
Polymerization Catalyzed by Ni Complexes

Ni(naph)$_2$/AlEt$_3$/organic activator and Ni(naph)$_2$/AlEt$_2$F/H$_2$O catalysts polymerize 1,3-butadiene to produce the *cis*-1,4-polymer with high selectivity (>95%) [59]. The catalytic activity and molecular weight of the polymer are controlled by the composition of the catalyst. Ni(acac)$_2$/MAO and CpTiCl$_3$/MAO catalyze polymerization of 1,3-butadiene to give the *cis*-1,4-polymer and of styrene to give syndiotactic polystyrene [6, 12]. Ni(acac)$_2$/MAO is the most effective catalyst of 1,3-butadiene among the mixtures of MAO and the acetylacetonate compounds of transition metals such as Cr, V, Fe, and Co [60, 61], and shows higher catalytic activity than Ni(OCOR)$_2$/MAO or NiCp$_2$/MAO.

The polymer composed of 2-cyclohexenyl-1,4 units can exhibit high T_g (176 °C) when the tacticity is controlled well [62]. Recently, [(π-allyl)NiBr]$_2$, in conjunction with MAO, was found to initiate stereo- and regiospecific polymerization of 1,3-cyclohexadiene [63]. Although the polymer has too poor solubility in organic solvents to be analyzed by NMR spectroscopy, the copolymers of 1,3-cyclohexadiene with 1,3-butadiene and norbornene prepared by the Ni catalyst show NMR spectra that indicate the presence of the 2-cyclohexene-1,4-diyl unit formed via 1,4-polymerization of 1,3-cyclohexadiene (Eq. 11). X-ray diffraction analysis of the crystalline poly(1,3-cyclohexadiene) as well as studies by molecular dynamics confirmed the *cis*-syndiotactic structure of the polymer [64].

eq 11

cis-sydiotactic

π-Allylnickel complexes polymerize 1,3-butadiene without addition of the organoaluminum reagents. This single-component catalyst shows as high a catalytic activity as the catalysts containing organoaluminum. [(π-Allyl)Ni(OCOCF$_3$)]$_2$ was reported to initiate living polymerization of 1,3-butadiene in the presence of P(OPh)$_3$ or TCBQ to afford the *cis*-1,4-polymer with relatively narrow molecular weight distribution (M_w/M_n=1.2–2.0) [65]. The 1,3-butadiene polymerization catalyzed by [(π-allyl)Ni(OCOCF$_3$)]$_2$/TCBQ, however, was later suggested to involve chain transfer probably via β-hydrogen elimination [66]. Addition of P(OPh)$_3$ to [(π-allyl)Ni(OCOCF$_3$)]$_2$ changes the microstructure of the polymer to *trans*-1,4. The *cis* content of the polymer can be regulated in the range 2–93.5% by control of the amount of added P(OPh)$_3$ (Eq. 12) [65].

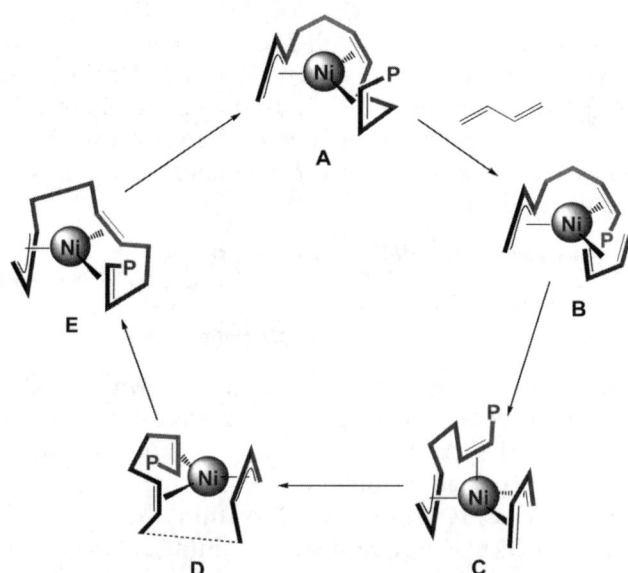

Cationic π-allylnickel complexes polymerize 1,3-butadiene to produce the *cis*-1,4-polymer. Taube investigated the polymer growth via smooth and selective insertion of the diene into the π-allyl–Ni bond of the growing polymer, both from experimental and theoretical aspects. The reaction catalyzed by the cationic C$_{12}$-allylnickel(II) complex shows kinetics that agree with a chain propagation transfer model [67]. The reaction mechanism of the *cis*-1,4-polymerization using technical Ni catalysts is also discussed [68]. He compared the mechanism of the reaction catalyzed by allylnickel complexes [69].

Scheme 7 The pathway and intermediates of the polymerization catalyzed by cationic Ni complexes

The theoretical approach to the mechanism of the reaction is free from the issue of the role of the cocatalyst such as MAO, which is present in the many transition metal-catalyzed polymerizations but is usually precluded from the theoretical studies. Tobisch compared the stereochemically different pathways and the related intermediates of polymer growth catalyzed by cationic Ni complexes. Scheme 7 depicts the plausible pathway of insertion of the s-*cis*-1,3-diene ligand into the π-allyl–Ni bond [70]. The polymer has a π-allylic end and two C=C bonds to coordinated Ni in A (the monomer-free intermediate). Coordination of a diene monomer to Ni forms the intermediates B and C. The two intermediates have the diene ligand and the growing polymer that is bonded to Ni both via the π-allyl end and via the C=C double bond. The dual coordination of the polymer even in the monomer-bonded intermediate differs from the CpTiCl$_3$/MAO catalysis, in which the monomer-bonded intermediate has no bonding between the C=C bond of the polymer and Ti (Scheme 4). Concerted bond formation between the terminal allyl carbon of the polymer and a CH$_2$ diene carbon (intermediate D) forms a new monomer-free intermediate (E) which undergoes structural change to regenerate A. The *cis*-C=C double bond in the polymer coordinates to the Ni center and stabilizes the intermediates throughout the reaction.

2.3.2
Polymerization Catalyzed by Co and Fe Complexes

Cobalt compounds are used as a mixture with organoaluminum in the catalytic 1,3-butadiene polymerization. CoCl$_2$/MAO initiates living polymerization of 1,3-butadiene to produce *cis*-1,4-polymer with 98–99% selectivity based on the ^{13}C NMR analyses (Eq. 13) [71]. The molecular weight of the polymer increases

$$\text{butadiene} \xrightarrow[\text{toluene} \quad 0-18\,°C]{\text{CoCl}_2 \quad \text{MAO}} \text{[-CH}_2\text{-CH=CH-CH}_2\text{-]}_n \qquad \text{eq 13}$$

cis-1,4 (98-99%)
$M_w/M_n = 1.3-1.5$

linearly with the reaction time, whereas increase of the polymer yield is not directly related to the change of the molecular weight. Further detailed study of the polymerization revealed that the initiation is slower than the propagation reaction that is free from chain transfer. The kinetic model based on this assumption agrees well with the experimental results including relative polymerization rate, molecular weight, and polymer yield. Use of the Co salt with the other halogeno ligands causes a decrease in the initiation efficiency and stereoselectivity in the order Cl>Br>I>F, although the propagation rate is not affected by the halogeno ligands [72]. Co(acac)$_3$/MAO is also effective for the *cis*-specific polymerization of 1,3-butadiene [73]. Although the molecular weight distribution of the polymer produced is somewhat broad ($M_w/M_n>1.8$), the molecular weight increases linearly according to the polymer yield. This enables copolymerization of 1,3-butadiene with styrene [74].

Co(OCOR)$_2$/AlEt$_3$/H$_2$O initiates polymerization of 1,3-butadiene. Addition of phosphine ligands influences the structure and molecular weight of the polymer [75]. Sterically bulky phosphines decrease catalytic activity. Co(OCOR)$_2$/MAO/t-BuCl is also effective for the polymerization of 1,3-butadiene to produce the cis-1,4-polymer [76]. The catalyst preparation procedure and aging time have a critical influence on the cis content and yield of the polymer. The reaction rate is reduced by addition of mesitylene or trimethoxybenzene to the reaction mixture [77]. The concentration of the active species of the catalyst is estimated based on the equilibrium between the Co and Al compounds in the reaction mixture [78].

Fe compounds have received much less significant attention than Ni or Co compounds as the diene polymerization catalyst. FeEt$_2$(bpy)$_2$ catalyzes cyclodimerization of 1,3-butadiene [79] and polymerization of vinyl monomers such as acyclic ester [80]. Recently, FeEt$_2$(bpy)$_2$/MAO was found to show high catalytic activity toward 1,2-polymerization of 1,3-butadiene and 3,4-polymerization of isoprene at −40 to +25 °C (Eq. 14) [81]. The crystalline polybutadiene prepared below 0 °C is composed of

$$\text{butadiene} \xrightarrow[\text{toluene, < 0 °C}]{\text{FeEt}_2(\text{bpy})_2 \quad \text{MAO}} \text{polymer} \quad \text{eq 14}$$

1,2-syndio
1,2 = 80–85%

syndiotactic 1,2-structural units and a smaller portion of 1,4-unit (10–15%). (E)-1,3-Pentadiene, isoprene, and 3-methyl-1,3-pentadiene give the 1,2-polymer, 3,4-polymer, and 1,2-polymer of the respective monomers. The latter two polymers were prepared for the first time.

2.4
Copolymerization of Conjugated Dienes

This section summarizes the copolymerization of conjugated dienes with other monomers catalyzed by transition metal complexes. Some of the reactions here were also mentioned in the previous section. The catalyst CpTiCl$_3$/MAO is active not only for the polymerization of 1,3-butadiene, isoprene, 1,3-pentadiene, and styrene but also for the copolymerization of these individual monomers [82].

The catalyst composed of rac-[CH$_2$(3-$tert$-butyl-1-indenyl)$_2$]ZrCl$_2$ and MAO initiates the copolymerization of 1,3-butadiene with ethylene [83]. The polymer produced contains repeating units having cyclopropane and cyclopentane groups, but not unsaturated group was observed (Eq. 15).

This is the first example of cyclopolymerization of butadiene and of copolymerization to form the cyclopropane-containing polymer. 1,2-Insertion of the monomer leads to the η^1-polymer end with a vinyl group at the β-position. Further intramolecular insertion of the C=C bond of the vinyl pendant group into the Zr–C bond forms the cyclopropane group. Cyclopolymerization in nonconjugated α,ω-dienes is much more common, and is mentioned in the following section.

The copolymerization of ethylene with nonconjugated dienes is of significant interest due to its applicability to ethylene–propylene–diene rubber (EPDM) [84]. The cyclic dienes such as 1,3-cyclopentadiene, dicyclopentadiene, and vinylcyclohexane copolymerize with ethylene catalyzed by rac-Et(Ind)$_2$ZrCl$_2$/MAO (Eq. 16) [85].

[Me$_2$Si(3-Me$_3$SiC$_5$H$_3$)$_2$NdCl/BuLi/AlH(i-Bu)$_2$] catalyzes the copolymerization of 1,3-butadiene with ethylene effectively. The content of the butadiene unit, whose microstructure is mainly trans-1,4, in the copolymer is in good agreement with the ratio of 1,3-butadiene to ethylene used in the polymerization [86].

(Me$_2$CC$_5$H$_4$)$_2$Sm(allyl)$_2$Li(dme) promotes the copolymerization of isoprene with α-olefins (Eq. 17) [87] and with nonconjugated dienes [88]. The isoprene units of the copolymers have a *trans*-1,4 structure. A similar allylsamarocene complex, prepared in situ from [(Me$_2$CC$_5$H$_4$)$_2$SmCl·MgCl·(THF)$_3$]$_2$ and allyl Li(dioxane), was reported to polymerize both isoprene and ϵ-caprolactone, and copolymerize these monomers to afford the block copolymer –{CH$_2$–CH= CMe–CH$_2$}–{C(=O)–(CH$_2$)$_4$–CH$_2$O}– [89]. The structure of the active species and the reaction pathway are discussed based on the results of the reactions catalyzed by the unhindered *ansa* samarocenes [90].

The catalyst composed of [(π-allyl)Ni(OCOCF$_3$)]$_2$ and hexafluoroacetone or hexachloroacetone polymerizes various monomers such as 1,3-butadiene, vinyl ethers, norbornene, isocyanide, styrene, and isoprene [91]. [(π-Allyl) Ni(OCOCF$_3$)]$_2$ is used to synthesize butadiene–isocyanide diblock [92] and triblock copolymers (Scheme 8) [93].

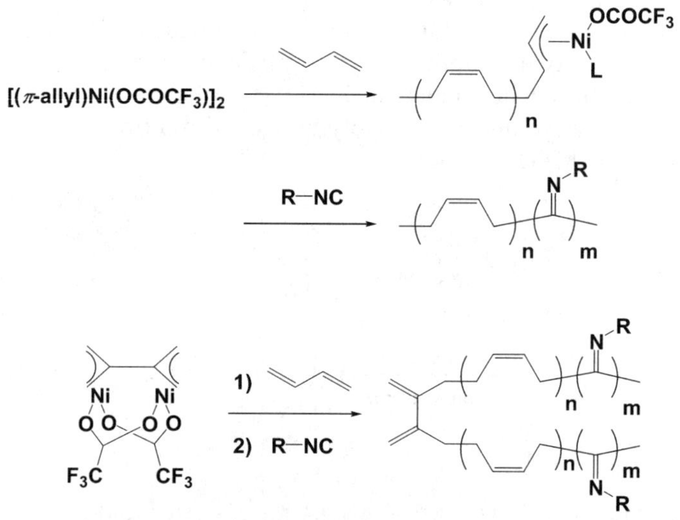

Scheme 8 Synthesis of butadiene–isocyanide diblock and triblock copolymers

Successive addition of 1,3-butadiene and isocyanide to the solution of the Ni catalyst forms a product with flexible polybutadiene blocks and rigid polyisocyanide blocks. Although the reaction of butadiene and isoprene in the presence of CoCl$_2$/MAO causes homopolymerization of butadiene, the reaction catalyzed by CoCl$_2$/MAO/PPh$_3$ affords a copolymer with 1,2-butadiene and 3,4-isoprene units [94]. The monomer reactivity ratios indicate higher reactivity of butadiene than isoprene.

3
Polymerization of Nonconjugated Dienes

3.1
Polymerization by Early Transition Metal Complexes

As shown above, conjugated dienes such as 1,3-butadiene polymerize in 1,4 or 1,2 fashion depending on the catalyst and the monomer structure. Polymerization of nonconjugated dienes also results in 1,2-polymers which have pendant vinyl groups, similar to those of the conjugated dienes.

Although 1,3-polymerization is not feasible for the nonconjugated dienes, they undergo cyclization polymerization via insertion of one C=C bond of the monomer into the metal–polymer bond, followed by intramolecular insertion of the remaining C=C bond of the growing end into the new metal–polymer bond. Scheme 9 depicts the cyclization of the polymer end to form *trans*- and *cis*-fused rings, depending on the relationship of the position of the coordinating olefin face with the polymer chain.

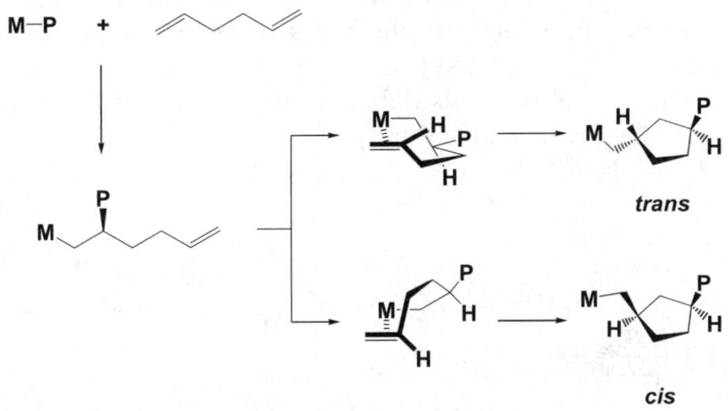

Scheme 9 Mechanism of cyclization polymerization of 1,5-hexadiene

Living polymerization of 1,5-hexadiene is catalyzed by V(acac)$_3$/AlEt$_2$Cl to give a polymer with alternating methylene-1,3-cyclopentylene and 1-vinyltetramethylene units (Eq. 18) [95]. Waymouth reported selective

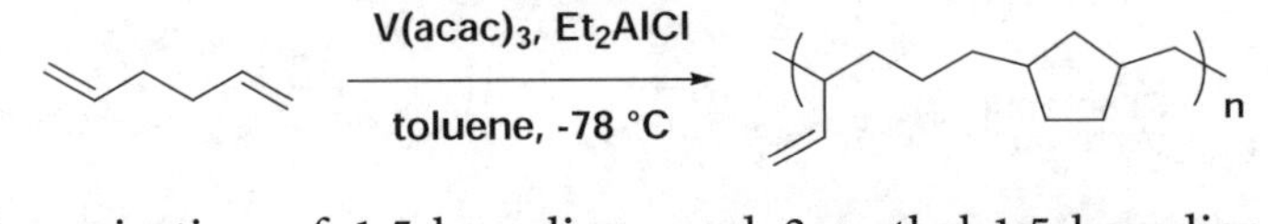

cyclopolymerization of 1,5-hexadiene and 2-methyl-1,5-hexadiene by zirconocene catalysts (Eq. 19) [96]. Cp$_2$ZrCl$_2$/MAO and Cp*$_2$ZrCl$_2$/MAO produce a polymer

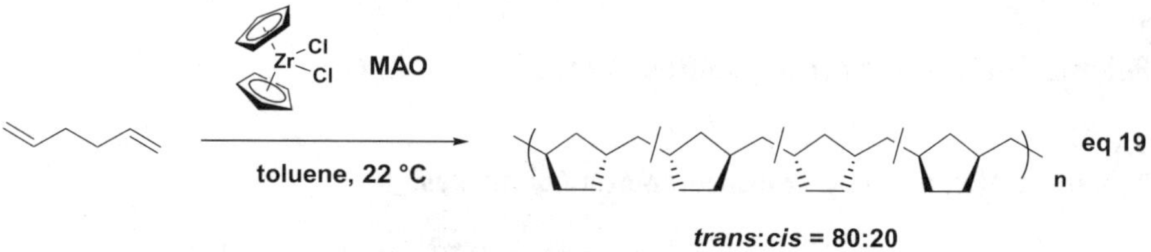

$$\text{(1,5-hexadiene)} \xrightarrow[\text{toluene, 22 °C}]{\text{Cp}_2\text{ZrCl}_2, \text{MAO}} \text{(cyclic polymer)}_n \qquad \text{eq 19}$$

trans:cis = 80:20

which contains the cyclic repeating unit with 99% selectivity. The molecular weight of the poly(1,5-hexadiene) (M_w>20,000) is much larger than that of poly(1-hexene) prepared by the same catalyst. This is ascribed to the slow termination of the growing polymer end whose β-hydrogen elimination should produce the strained and thermodynamically unstable methylenecycloalkane. Yttrium complexes, including [Cp*Y(OAr)H]$_2$ (Ar=C_6H_3-2,6-t-Bu) with bulky aryloxide and Cp* ligands, promote cyclopolymerization of 1,5-hexadiene [97].

The polymers with *trans*-fused five-membered rings linked with a diisotactic head-to-tail sequence have chirality, although the polymers composed of the *cis*-fused ring are achiral. Scheme 10 summarizes the structures of the stereoisomeric polymers. The optically active zirconocene complex with a C_2 symmetric structure catalyzes the enantioselective cyclopolymerization of 1,5-hexadiene (Eq. 20) [98, 99]. Although the polymer contains not only *trans*-fused ring but also *cis*-fused ring units (ca. 68:32), it shows optical rotation due to the main chain chirality.

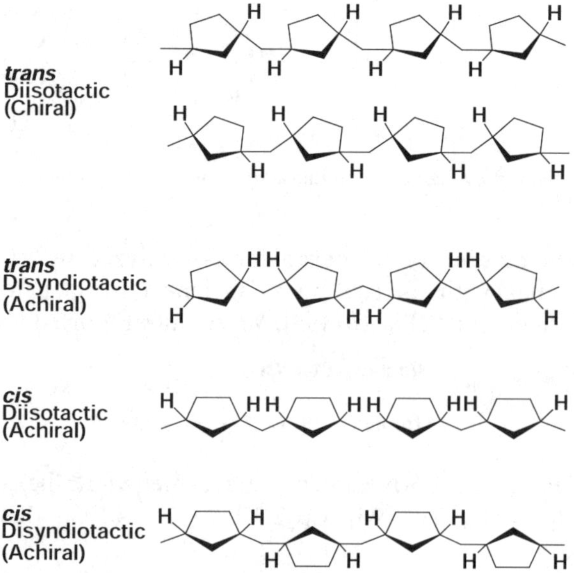

Scheme 10 Four stereoisomers of poly(1,5-hexadiene)

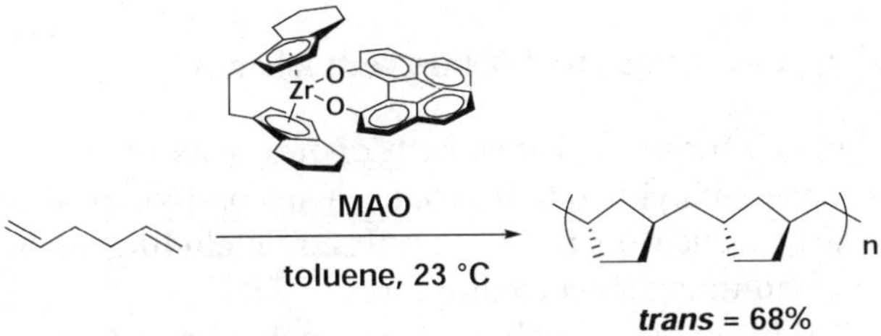

Coordination Polymerization of Dienes, Allenes, and Methylenecycloalkanes 161

Zirconocene complexes with ferrocenyl groups promote selective cyclopolymerization of 1,5-hexadiene to give a polymer with high content of the *trans*-unit (up to 98% trans selectivity) (Eq. 21) [100]. Sita reported living

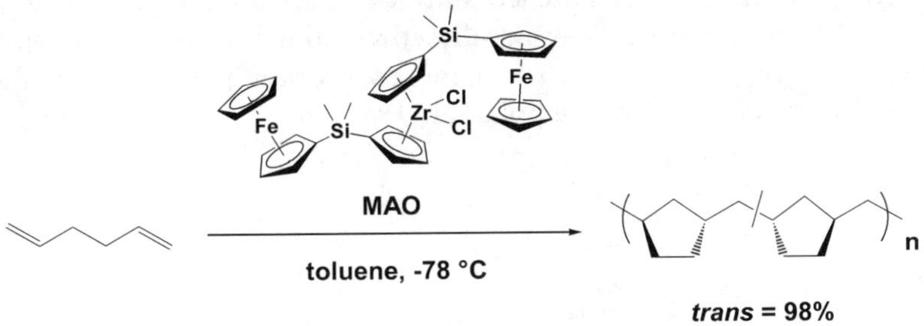

eq 21

cyclopolymerization of 1,5-hexadiene by zirconium amidinate complexes, which promote isospecific living polymerization of 1-hexene to produce polymer with a five-membered ring in the structural units (*trans*=64–82%) and with narrow molecular weight distribution (M_w/M_n<1.10) [101, 102]. Block copolymerization of 1,5-cyclohexadiene and 1-hexene forms the diblock and triblock copolymers. The diblock copolymer has a microphase separated into a cylindrical morphology consisting of hard cylinders of poly(methylenecycloalkane) surrounded by the more elastic poly(1-hexene) domains (Eq. 22).

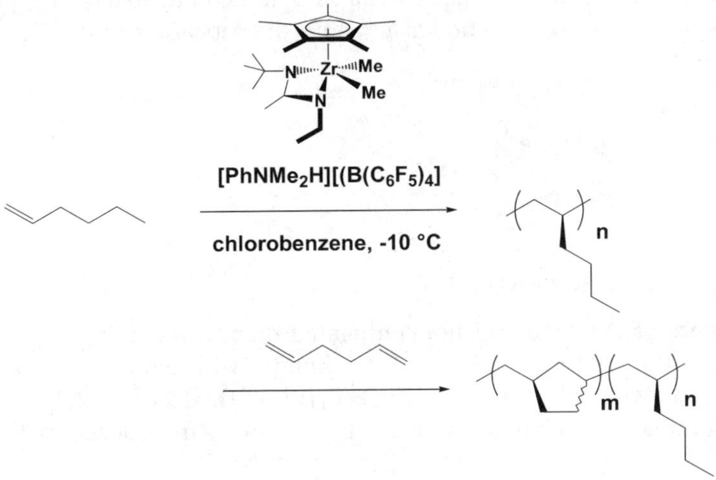

eq 22

3.2
Copolymerization of Nonconjugated Dienes with Alkenes

Smooth copolymerization of alkenes with dienes would produce polyolefins with cyclic groups or vinyl pendant groups. The copolymerization was studied by using early transition metal complexes, including metallocene, half metallocene, and nonmetallocene complexes of Zr.

Et(Ind)$_2$ZrCl$_2$/MAO gives copolymers of ethylene or propylene with nonconjugated dienes, such as 2-methyl-1,4-pentadiene, 7-methyl-1,6-octadiene and 1,7-octadiene, (Eq. 23) [103]. rac-Et(Ind)$_2$ZrCl$_2$/MAO also catalyzes copolymerizations of asymmetrically substituted linear dienes, 6-phenyl-1,5-hexadiene, 7-methyl-1,6-octadiene, and R-(+)-5,7-dimethyl-1,6-octadiene. The copolymerization of R-(+)-5,7-dimethyl-1,6-octadiene with propylene to give the polymer with ca. 15% diene incorporation. The ratio of the diene-derived part is ca. 15% of the polymer [104].

Copolymerization of ethylene and nonconjugated dienes catalyzed by Cp$_2$ZrCl$_2$/MAO has been reported by many research groups [105]. Cationic half titanocene catalyst, [Cp*TiMe$_2$][MeB(C$_6$F$_5$)$_3$], polymerizes 1,5-hexadiene to form a polymer containing both 1,2-polymerization units and cyclization polymerization units, while the reaction of norbornene leads to a polymer from vinyl polymerization and ring-opening polymerization [106]. Cyclopolymerization of 1,7-octadiene catalyzed by Me$_2$Si(Ind)$_2$ZrCl$_2$/MAO [107] and copolymerization of propylene and nonconjugated diene [108] produce polymers having a cyclic structure in the repeating units (Eq. 24). The structure of

the copolymers of ethylene and nonconjugated dienes, including degree of cross-linking, prepared by Cp$_2$ZrCl$_2$/MAO changes with temperature of the polymerization [109]. Zirconocene/Al(i-Bu)$_3$/[Ph$_3$C][B(C$_6$F$_5$)$_4$] catalyzes the copolymerization of propylene and 1,5-hexadiene. Zirconocene with C_2

symmetry shows a higher catalytic activity than the complex with C_s symmetry, whereas the latter catalyst introduces the hexadiene unit in higher content than the former [110]. Copolymerization of ethylene and the nonconjugated dienes produces long-chain-branched polyethylene [111]. The catalysts formed from various substituted zirconocene derivatives copolymerize ethylene with nonconjugated dienes such as 1,5-hexadiene, 1,7-octadiene, and 1,9-nonadiene. The ratios of the cyclized to uncyclized units and degree of cross-linking are altered by the kind of Zr complex [112]. Zirconium complex containing the 2-methylindenyl ligand promotes copolymerization of ethylene and nonconjugated diene [113].

Copolymerization of alkenes with nonconjugated dienes is conducted by using zirconocene, CGC catalyst, and the catalyst composed of iron diiminopyridine complexes (Eq. 25). The former two catalysts give the copolymer,

eq 25

whereas the Fe catalyst polymerizes ethylene even in the presence of the dienes. The Ni complex polymerizes neither ethylene nor diene, probably due to formation of the stable π-allyl complexes which are inactive for the polymerization of both monomers [114]. CGC type Ti catalyst copolymerizes propylene with 1,5-cyclohexadiene to produce polymer with a cyclic structure in the main chain. The copolymerization with 1,9-dodecadiene does not proceed smoothly and leaves unreacted propylene [115].

Polymerization of 1,6-heptadiene by Ti complexes with a phenoxyimine ligand produces a polymer with both methylene-1,3-cyclohexane and ethylene-2-cyclopentane groups in the main chain (Eq. 26) [116]. It contrasts with the reaction catalyzed by *rac*-ethylene-bis(indenyl)zirconium dichloride which produces polymer containing the methylene-1,3-cyclohexane unit only. The polymerization of 1,5-hexadiene and copolymerization of propylene and 1,5-hexadiene catalyzed by the phenoxyimine-Ti complex produces a polymer with vinyl pendant groups [117]. Scheme 11 depicts the pathways for the methylene-1,3-cyclopentane unit and for the vinyltetramethylene unit from the polymerization of 1,5-hexadiene. The latter pathway involves 2,1-insertion of a C=C double bond of the monomer into the polymer–Ti bond, followed by intramolecular 1,2-insertion of the remaining C=C double bond of the polymer to form the cyclobutylmethyl group attached to the Ti center.

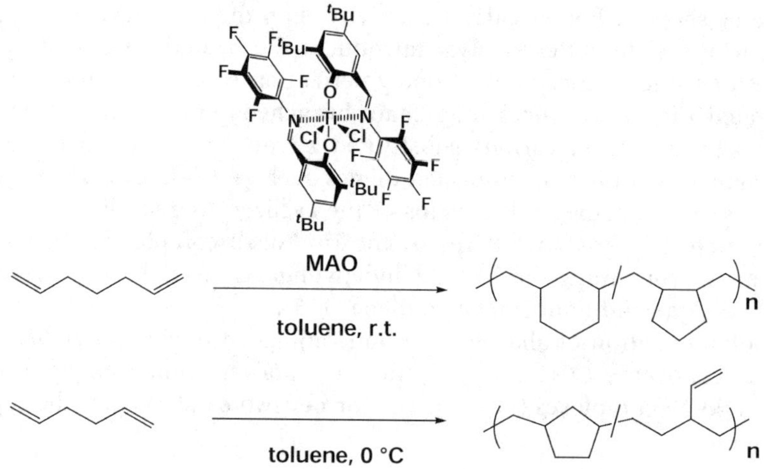

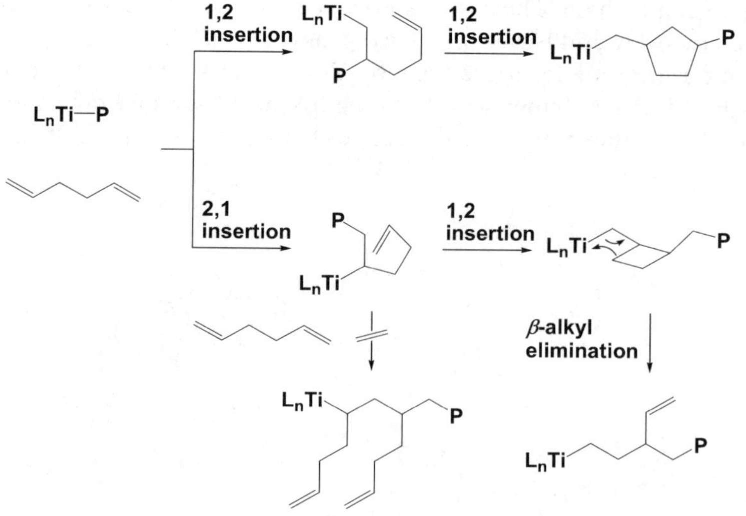

Scheme 11 Mechanism of polymerization of 1,5-hexadiene by phenoxyimine–Ti complex

This intermediate with a four-membered ring undergoes β-alkyl elimination to generate a new polymer–Ti bond with the vinyl group at the γ-position.

Propylene-b-poly(methylene-1,3-cyclopentane-co-propylene) was synthesized by the copolymerization of propylene with 1,5-hexadiene by MgCl$_2$-supported Ziegler–Natta catalyst using a stopped-flow technique [118].

4
Polymerization of 1,2-Dienes (Allenes)

Allene is an abundant resource easily available from petroleum, but its industrial usage is still limited. Transition metal-catalyzed polymerization of allene has been studied since the 1950s [119]. The coordination polymerization produces soluble polymers with clear structures in high yields. In the 1960s Otsuka and Nakamura reported Rh complex-catalyzed polymerization of allene as well as its cyclooligomerization reactions promoted by Ni(0) complexes. Monosubstituted allenes were polymerized by RhCl(CO)$_2$(PPh$_3$) catalyst [120]. Endo and Tomita found living polymerization of substituted allenes using Ni complexes and designed the synthesis of various functionalized polymers and copolymers containing the vinylidene group. Since the reports before 1997 are summarized in recent reviews [121], this section focuses on the more recent reports on this subject.

4.1
Living Polymerization of Allenes Catalyzed by Ni and Pd Complexes

4.1.1
Polymerization of Aryl- and Alkylallenes

[(π-Allyl)Ni(OCOCF$_3$)]$_2$ brings about the polymerization of phenylallene at room temperature to give a polymer consisting of the 2,3-polymerization unit, {–CH$_2$–C(=CHPh)–}, in high yield (Eq. 27) [122]. The molecular weight of the polymer produced

$$\text{=·=}\underset{R}{} \xrightarrow[\text{toluene, r.t.}]{[(\pi\text{-allyl})Ni(OCOCF_3)]_2} \left(\text{-CH}_2\text{-C(=CHR)-}\right)_n \quad \text{eq. 27}$$

R = Ph   0 : 100
R = n-Bu   9 : 91
R = t-Bu   0 : 100

increases with increase of the monomer-to-Ni ratio. The growing polymer shows a linear dependence of the molecular weight on the conversion of the monomer during polymerization, indicating that chain transfer or termination of this polymerization is insignificant. The polymerization of other arylallenes also takes place in the presence of [(π-allyl)Ni(OCOCF$_3$)]$_2$. Arylallenes with an electron-donating group, such as (4-methoxyphenyl)allene, polymerize more rapidly than monomers with less electron-donating substituents at the phenyl ring.

Monoalkylallenes undergo smooth living polymerization promoted by the Ni complex to give the corresponding polyallenes with controlled molecular weights and narrow molecular weight distribution [123]. The polymers produced are composed of –C(=CH$_2$)–CHR– units formed via 1,2-polymerization and –CH$_2$–C(=CHR)– units via 2,3-polymerization in approximately 1:9 ratio. The polymerization of *tert*-butylallene and 1,1-dialkylallenes affords the corresponding polymers which contain the structural unit via 2,3-polymerization, –CH$_2$–C(=CH-*t*-Bu)– or –CH$_2$–C(=CRR')–, respectively. 1,3-Dialkylallene polymerizes by the Ni catalyst more slowly, while 1,2-cyclononadiene polymerizes smoothly, due to release of the ring strain of the monomer caused by the polymerization (Eq. 28) [124].

eq 28

4.1.2
Polymerization of Alkoxyallenes

Polymerization of methoxyallene initiated by a mixture of [(π-allyl)Ni (OCOCF$_3$)]$_2$ and PPh$_3$ produces a polymer containing two repeating units, –C(=CH$_2$)–CH(OMe)– and –CH$_2$–C(=CHOMe)–, randomly in 32:68 ratio in the polymer chain. These structural units are derived from 1,2- and 2,3-polymerization of the monomer, respectively (Eq. 29) [125].

The molecular weight of the polymer increases linearly with increase of the monomer-to-Ni molar ratio in the polymerization with narrow molecular weight distribution (M_w/M_n<1.1), indicating a living character of the polymerization. Catalysts based on the allylnickel complex initiate the living polymerization of other alkoxyallenes [126]. The living poly(methoxyallene) in the reaction mixture is so stable that addition of monomer to the reaction mixture after consumption of the initially charged monomer resumes the polymerization. Allenes with a bulky alkoxy substituent (*t*-BuO-, PhO-) give mainly polymers containing the structural unit from 2,3-polymerization.

Scheme 12 depicts the proposed reaction pathways for the polymerization of allenes catalyzed by π-allylnickel complexes. Insertion of the 1,2-C=C

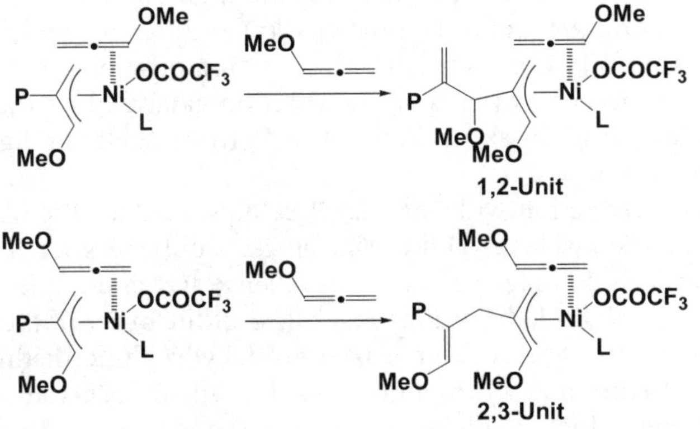

Scheme 12 Mechanism for formation of 1,2- and 2,3-units on Ni-catalyzed polymerization of substituted allenes

bond of the monomer into the π-allyl–Ni bond of the polymer end forms the –C(=CH$_2$)–CH(OMe)– unit, while the other unit –CH$_2$–C(=CHOMe)– is caused by the reaction of the 2,3-C=C bond. The monomer probably coordinates at the apical site of the square planar nickel center. The authors rationalize the presence of the 1,2- and 2,3-polymerization units in the polymer by assuming that the insertion of the monomer occurs into the Ni–σ-allyl bond rather than the Ni–π-allyl bond of the polymer end, as shown in Scheme 13. The bulky substituent of the monomer makes the σ-allylnickel complex with a CH$_2$–Ni bond more favorable than that with a CH(OMe)–Ni bond, leading to the 2,3-polymerization.

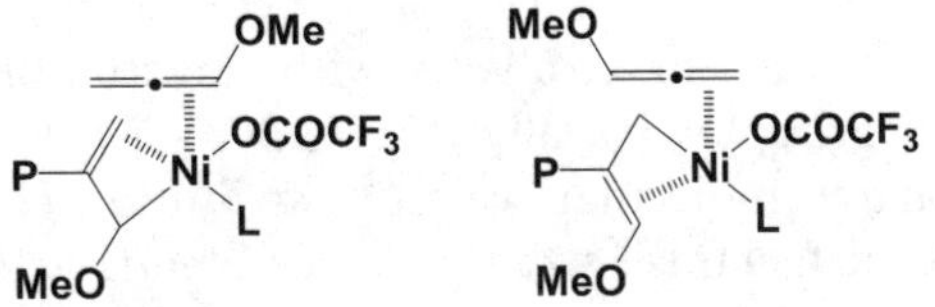

Scheme 13 σ-Allyl intermediates for 1,2- and 2,3-polymerization

The polymerization of (n-octyloxy)allene is influenced significantly by the anionic ligand of the π-allylnickel catalysts and by addition of PPh$_3$ [127]. [(π-Allyl)Ni(OCOCF$_3$)]$_2$ provides the living polymerization system that produces the polymers with controlled molecular weights and narrow molecular weight distribution. [(π-Allyl)Ni(OCOCH$_3$)]$_2$ forms oligomeric by-products from the alkoxyallene. The ratio of 1,2- to 2,3-polymerization units in the polymer increases with the electron-withdrawing character of the carboxylate ligands of the catalyst. [(π-Allyl)NiCl]$_2$/PPh$_3$ and [(π-allyl)NiI]$_2$/PPh$_3$ catalysts produce polymers with a high content of the unit from 2,3-polymerization (94 and 84%, respectively) and narrow molecular weight distribution (M_w/M_n=1.03 and 1.10). The reaction promoted by the chloronickel catalyst proceeds slower than the reaction catalyzed by the iodonickel complex. Addition of an excess amount of PPh$_3$ or chelating ligand retards the polymerization.

Living polymerization with the nickel catalyst enables the block copolymerization of alkoxyallenes [128]. The block copolymers from methoxyallene, ethoxyallene, butoxyallene, t-butoxyallene, and phenylallene have narrow molecular weight distributions regardless of the order of the addition of the two monomers. The copolymerization of hydrophilic diethylene glycol allenyl methyl ether and hydrophobic hexyloxyallene forms an amphiphilic block copolymer which is soluble in both water and hexane (Eq. 30).

eq 30

Polymerization of the alkoxyallene with macromonomers having a poly(ethyleneglycol) group by [(π-allyl)Ni(OCOCF$_3$)]$_2$/PPh$_3$ produces a graft copolymer with narrow molecular weight distribution [129]. The products serve as polymeric surfactants in the polymer blend system of polystyrene and poly(methyl methacrylate).

Block copolymer of alkoxyallenes and phenylallene is obtained by using the different π-allylnickel catalyst [130]. The alkoxyallenes and phenylallene polymerize catalyzed by [(π-allyl)Ni(OCOCF$_3$)]$_2$/PPh$_3$ and by [(π-allyl)Ni(OCOCF$_3$)]$_2$, respectively. Initial polymerization of phenylallene by [(π-allyl)Ni(OCOCF$_3$)]$_2$ and subsequent addition of PPh$_3$ and the alkoxyallene to the living polymer produce the block copolymer (Eq. 31). Preparation of the copolymer using alkoxyallene as the first monomer is also possible. The polymerization of the alkoxyallene initiated by [(π-allyl)Ni(OCOCF$_3$)]$_2$/PPh$_3$ is followed by addition of CuI that scavenges PPh$_3$, which causes further polymerization of phenylallene to afford the block copolymer (Eq. 32). The block

copolymer of n-octyloxyallene with phenylallene is prepared by a single feed of these monomers to a solution of [(π-allyl)Ni(OCOCF$_3$)]$_2$. Since n-octyloxyallene polymerizes much more rapidly than phenylallene under these conditions, the polymerization of n-octyloxyallene takes place initially until full consumption of the monomer, which is followed by slow polymerization of phenylallene to produce the corresponding block copolymer (Eq. 33) [131].

Living polymerization of 1-allenyloxymethylnaphthalene is catalyzed by [(π-allyl)Ni(OCOCF$_3$)]$_2$/PPh$_3$, giving a polymer with naphthalene moieties [132]. Polymers with naphthalene moieties show a higher T_g than 1-allenyloxymethylbenzene polymer.

4.1.3
Polymerization of Functionalized Allenes

Since the π-allylnickel complexes are tolerant to polar amide and ester groups, the catalysts mentioned in the previous sections could polymerize allene derivatives with functional groups. The polymerization of N-allenyl-γ-butyrolactam, having a five-membered cyclic amide group, is catalyzed smoothly by [(π-allyl)Ni(OCOCF$_3$)]$_2$/PPh$_3$ to produce polymer with a narrow molecular weight distribution (M_w/M_n=1.05) (Eq. 34) [133]. The molecular weight of the

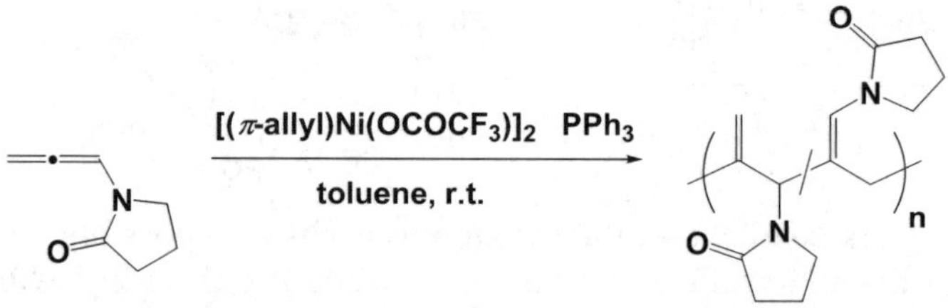

eq 34

polymer increases proportionally to the [monomer]/[Ni] ratio, suggesting living polymerization. Addition of PPh$_3$ to the catalyst is essential to obtain the stable propagating end; the living polymer becomes less active gradually after consumption of the monomer in the absence of added PPh$_3$. Living polymerization of N-allenyl-δ-valerolactam, N-allenyl-ε-caprolactam, and N-allenyl-N-methylacetamide is catalyzed by [(π-allyl)NiI]$_2$/PPh$_3$ to produce the corresponding polymers in high yields. [(π-allyl)Ni(OCOCF$_3$)]$_2$/PPh$_3$, the catalyst of the polymerization of N-allenyl-γ-valerolactam, is less suited for the polymerization of these monomers than the iodonickel catalyst.

Allenes with ester, arylthioether, and imide groups are also polymerized by [(π-allyl)Ni(OCOCF$_3$)]$_2$/PPh$_3$ to afford polymers with controlled molecular weights [121b]. The polymerization of nonsubstituted allene and Si-containing alkyl- and alkoxyallenes was reported [134, 135]. Polyallene with a pendant sugar group is obtained by the Ni-catalyzed polymerization of 1,2;5,6-di-O-isopropylidene-3-O-(allenyl)-α-D-glucofuranoside (Eq. 35) [136]. The CD spectrum of the monomer shows a positive Cotton effect at 273 nm, whereas that of the polymer has a much stronger negative Cotton effect at 229 nm, indicating that the polymer has a chiral main chain conformation.

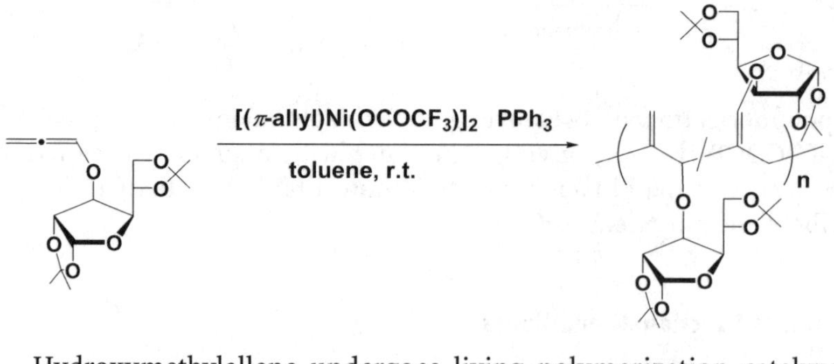

eq 35

Hydroxymethylallene undergoes living polymerization catalyzed by the π-allylnickel complex([Eq. 36) [137]. This preparation of polymer with hydroxyl groups and regulated molecular weight does not require protection of the OH group before the reaction.

4.2
Copolymerization of Allenes with 1,3-Butadiene, Propyne, and Isocyanides

Novak found the polymerization of allene catalyzed by [(π-allyl)Ni(OCOCF$_3$)]$_2$ (M_w/M_n=1.23) and its copolymerization with propyne [138]. Although [(π-allyl)Ni(OCOCF$_3$)]$_2$ does not initiate the homopolymerization of propyne, the copolymerization of allene with propyne is catalyzed by the complex to give the copolymer which has alignment of the allene-derived unit, –CH$_2$–C(=CH$_2$)–, and the propyne-derived unit, -CH=C(-CH$_3$)– (Eq. 37) [139]. [(π-Allyl)Ni(OCOCF$_3$)]$_2$ also initiates smooth polymerization of 1,3-butadiene [65] and isocyanides [140].

Endo and Tomita reported the copolymerization of allene with isocyanide [135, 141] and with 1,3-butadiene [142] to form AB-type block copolymers. The Pt-catalyzed hydrosilylation reaction of poly(allene-*b*-isocyanide) occurs selectively at the C=CH$_2$ group of the allene monomer unit to produce the polymer with silyl pendant groups (Eq. 38) [135]. The allene macromonomer having a poly(ethylene glycol) moiety forms block and graft copolymers of poly(ethylene glycol) and poly(isocyanide) [129, 143] (Eq. 39).

eq 39

4.3
Polymerization Catalyzed by Co and Rh Complexes

Although $Co_2(CO)_8$ was reported to promote the polymerization of allenes [119a], the mechanism of the initiation reaction and structure of the active species have remained as unclarified issues. Nakamura reported the reaction of allene with excess $Co_2(CO)_8$ to give $[(C_3H_4)Co(CO)_3]_2$ which is the plausible initiator of the polymerization [144]. Recently, Ungváry isolated the dinuclear π-allylcobalt complexes, $(CO)_3Co\{CH_2C(CMe_2)-C(=O)-C(CMe_2)CH_2\}Co(CO)_3$ and $(CO)_3Co\{CH_2C(CMe_2)-CH_2C(=CMe_2)-C(=O)-C(CMe_2)CH_2\}Co(CO)_3$, from the reaction of $Co_2(CO)_8$ with 3-methyl-1,2-butadiene and investigated the kinetics of their formation [145]. X-ray and NMR analyses showed the presence of stereoisomers of the above complexes [146].

Polymerization of arylallenes catalyzed by $RhH(PPh_3)_4$ produces a mixture of the 2,3-polymer with high molecular weight ($>10^6$) and oligomers [147]. The results are attributed to low initiation efficiency of the polymerization, although the growth of the polymer takes place rapidly without any chain transfer.

5
Polymerization and Copolymerization of Methylenecyclopropanes

Methylenecyclopropane with a high strain energy has been regarded as a useful synthetic equivalent of butadiene and trimethylenemethane, and used extensively in synthetic organic reactions [148]. The reaction of methylenecyclopropanes with transition metal complexes gives various reaction products, such as trimethylenemethane metal complexes [149] and allyl-metal complexes [148b], as well as dienes [150] and substituted cyclopropanes [151]. The facile ring opening of the cyclic compounds upon con-

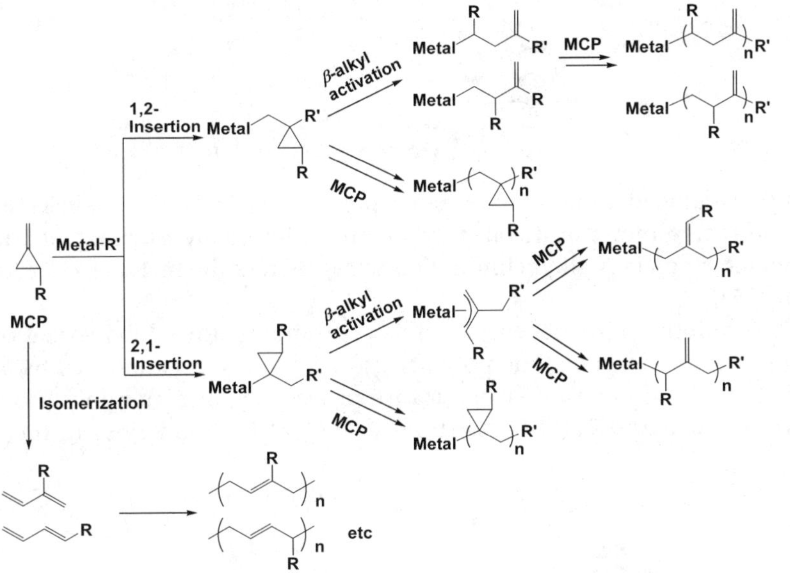

Scheme 14 Polymers expected from methylenecyclopropanes

tact with the transition metal complexes suggests its potential utility as a monomer of coordination polymerization that may be accompanied by C–C bond cleavage of the strained three-membered ring. Scheme 14 depicts possible polymer structures from substituted methylenecyclopropanes. The polymers produced are expected to have a high degree of unsaturation and/or ring strain in the backbone, which would make the polymers amenable to further functionalization.

Coordination polymerization of methylenecyclobutane was first reported in 1969 by using Ziegler–Natta catalysts [152], and then other catalysts such as $Ti(CH_2Ph)_4$ and $RhCl_3$ [153].

5.1
Ring-Opening Polymerization of Methylenecycloalkanes Catalyzed by Early Transition Metal Complexes

Marks reported ring-opening polymerization of methylenecyclopropane and methylenecyclobutane promoted by metallocenes of Zr and Lu. Cationic zirconocene complex initiates ring-opening polymerization of methylenecyclobutane to give a polymer having the $=CH_2$ substituent on every fourth carbon of the polyethylene chain (Eq. 40) [154, 155]. Polymerization using ^{13}C-labeled monomer revealed polymer growth via 1,2-insertion of the monomer into the metal–alkyl bond, followed by opening of the cyclobutane ring via β-alkyl elimination to regenerate the metal–alkyl complex [154, 155].

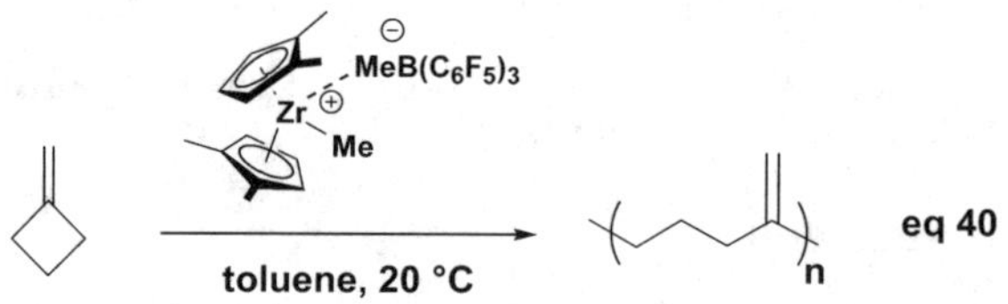

The cationic zirconocene complex promotes the polymerization of methylenecyclopropane. The polymer growth, however, is followed by an intramolecular "zipping-up" ring closure process to form the polyspirane as shown in Scheme 15 [156]. Insertion of the C=CH$_2$ bond close to the poly-

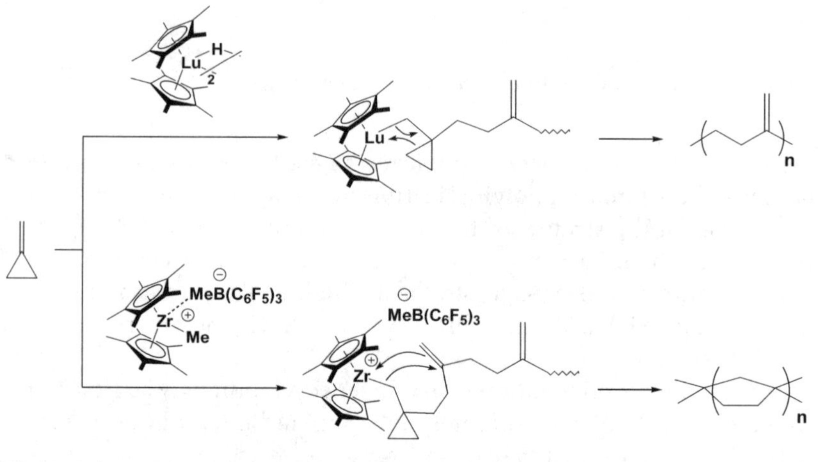

Scheme 15 Polymerization of methylenecyclopropane by lutenocene and cationic zirconocene complexes

mer end into the Zr–CH$_2$ bond forms a five-membered ring and a new Zr–CH$_2$ bond. Repetition of this ring formation via intramolecular insertion of the unsaturated bond into the Zr–C bond leads to the polymer with spiro structural units. On the other hand, lutenocene complex promotes the ring-opening polymerization of methylenecyclopropane to give a polyethylene having the *exo*-methylene group on every third carbon of the polymer chain [157]. Samarocene and lanthanocene promote dimerization of methylenecyclopropane chemoselectively to produce 1,2-dimethylene-3-methylcyclopentane [155].

The copolymerization of methylenecyclobutane and methylenecyclopropane with ethylene produces polyethylene having *exo*-methylene groups with lower density than the homopolymer of methylenecycloalkanes.

5.2 Ring-Opening Polymerization of Methylenecyclopropanes Catalyzed by Pd Complexes

A cationic π-allylpalladium complex with a diimine ligand, [(π-allyl)Pd(4-Me-C_6H_4-N=CH-CH=N-C_6H_4-4-Me)](BF$_4$) initiates ring-opening polymerization of 2-phenyl-1-methylenecyclopropane at 80 °C (Eq. 41) [158, 159].

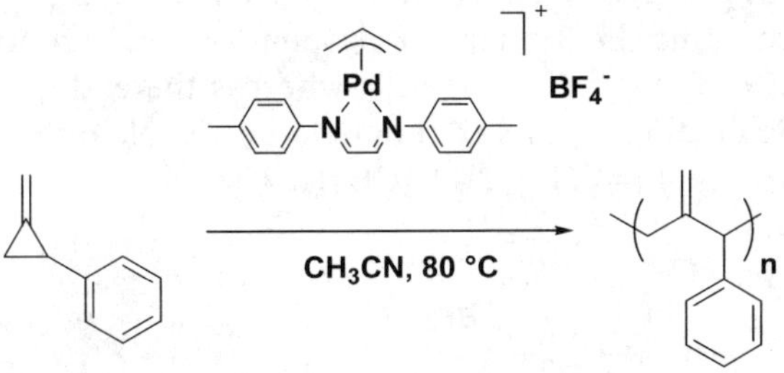

eq 41

The polymer of ^{13}C-labeled 2-phenyl-1-methylenecyclopropane by the same catalyst contains the isotope-labeled carbon exclusively at the *exo*-methylene group. The polymer consists of the 1-phenyl-2-*exo*-methylenepropan-1,3-diyl structure, exclusively. The ^{13}C{^{1}H} NMR signals of the *ipso* carbon of the phenyl group are split into four due to the four triads, *rr*, *rm*, *mr*, and *mm*, based on comparison of the spectrum with that of poly(styrene-*co*-CO) with isotactic, syndiotactic, and atactic arrangement of the monomer units [160] (Scheme 16). Thus, the polymer produced from 2-phenyl-1-

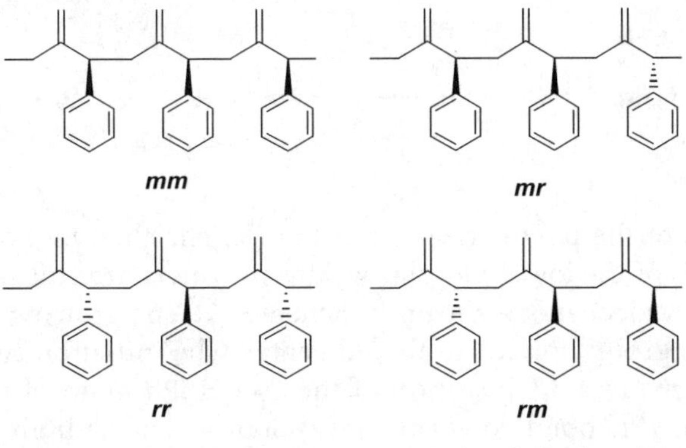

Scheme 16 Schematic drawing of the four triads of poly(2-phenyl-1-methylenecyclopropane)

methylenecyclopropane has a well-regulated head-to-tail sequence and atactic stereochemistry.

Palladium complexes with chelating diimine ligands and cationic π-allylpalladium complexes catalyze the ring-opening polymerization of 2-aryl-1-methylenecyclopropanes. The neutral palladium complex, PdCl(Me)(4-Me-C_6H_4-N=CH-CH=N-C_6H_4-4-Me), also catalyzes the polymerization, but more slowly than the reaction with the cationic Pd catalyst. The highest molecular weight is M_n=11,000 obtained from the polymerization catalyzed by $PdCl_2$(4-Me-C_6H_4-N=CH-CH=N-C_6H_4-4-Me)/AgBF$_4$.

The polymer with low molecular weight, prepared from the reaction at a small [monomer]/[Pd] ratio (20:1), shows the $^{13}C\{^1H\}$ NMR signals of the carbons of the initiating polymer end. The polymer from 1-phenyl-π-allylpalladium complex contains the 3-phenyl-1-propen-3-yl end group (Eq. 42) (δ141.7 [CH_2=], 114.3 [=CH–], 47.8 [CHPh]), whereas the end group of the polymer from π-allylpalladium complex contains the signals due to the allyl group (δ138.4 [CH_2=], 114.4 [=CH–], 34.1 [CH_2]).

eq 42

These results indicate that the π-allylic ligand bonded to Pd of the catalyst is incorporated in the initiating polymer end. Analogous insertion of 1,3-diene into the Pd–(1-methyl-π-allyl) bond was reported to form a C–C bond between the diene and the methyl-substituted carbon of the π-allylic ligand, although it is sterically less favorable than the alternative insertion of the C=C bond into the Pd–CH_2 bond (Eq. 43) [161].

eq 43

The results of the polymerization of the ^{13}C-enriched monomer and end group analyses of the low molecular weight polymers are consistent with the polymerization mechanism shown in Scheme 17. The growing polymer end is the π-allylic group bonded to the Pd center. The initiation step of the polymerization involves 2,1-insertion of the C=CH–Ph bond of the monomer into the Pd–CHPh bond to form the cyclopropyl palladium intermediate with the phenyl substituent at the γ-position. The intermediate undergoes

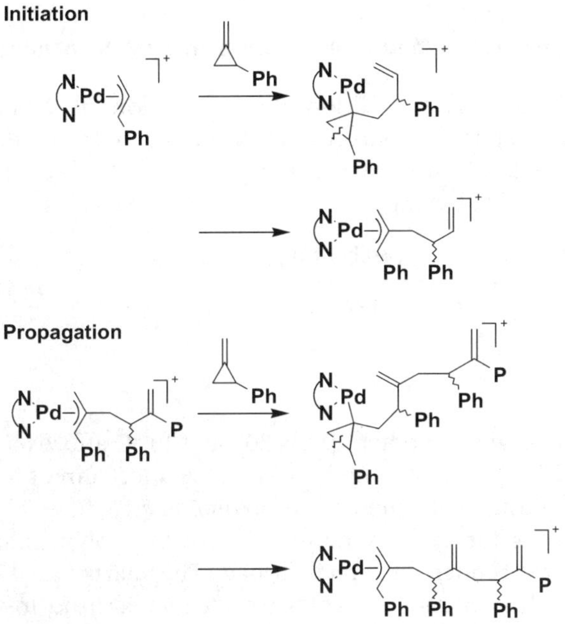

Scheme 17 Plausible mechanism of the polymerization of 2-phenyl-1-methylenecyclopropane by cationic π-allyl Pd complexes (P: polymer chain)

scission of the CH_2–CH_2 bond of the three-membered ring (distal bond) via β-alkyl elimination to generate a new π-allylic ligand with the Ph group at the terminal carbon. β-Alkyl elimination is common in the addition polymerization of propylene promoted by early transition metal complex [162]. The polypropylene whose terminal carbon is attached to the Me group, formed via 2,1-insertion of the monomer, undergoes chain transfer by β-methyl elimination to generate polymer with a vinyl end group. The C–C bond cleavage in the polymerization of 2-aryl-1-methylenecyclopropane is facilitated by release of the ring strain of the three-membered ring by the reaction. The polymer growth also involves 2,1-insertion of the monomer into the benzylic carbon of the Pd–π-allyl bond of the growing polymer and subsequent C–C bond cleavage of the three-membered ring, similarly to the initiating step of the polymerization. Repetition of the reactions produces the polymer with high head-to-tail selectivity. This mechanism is in contrast to the polymerization of methylenecyclopropane by Zr and Lu complexes, in which the polymerization proceeds via 1,2-insertion of the monomer into the metal–carbon bond followed by cleavage of the C–C bond at the proximal position (=C–CH_2–) via a β-alkyl elimination.

5.3
Addition Polymerization of Methylenecyclopropanes by Ni Complexes

[(π-Allyl)NiBr]$_2$ promotes addition polymerization of 2-phenyl-1-methylenecyclopropane at room temperature to give a polymer that contains a cyclopropylidene group in every repeating unit (Eq. 44) [163, 164]. The reaction

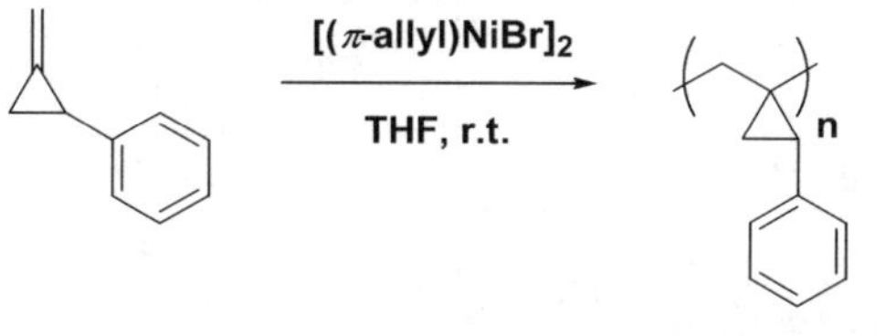

eq 44

with a molar ratio of [monomer]/[Ni]=200 at −40 °C affords a polymer with M_n=29,000 and M_w/M_n=1.59. The ^{13}C{^{1}H} NMR spectrum of the polymer revealed repeating units with the cyclopropylidene structure. The existence of four stereoisomers for each monomer unit of the polymer and rigidity of the polymer chain render the signals broad. The sharp signal of the quaternary carbon of the polymer indicates well-regulated head-to-tail linkage of the monomer units. The reaction of 2-phenyl-1-methylenecyclopropane with [(π-allyl)NiBr]$_2$ in a low molar ratio and quenching of the polymer by D$_2$O leads to the –CH$_2$D end of the oligomer. These data indicate that the polymerization proceeds via successive 1,2-insertion of 2-phenyl-1-methylenecyclopropane into the alkyl–nickel bond of the growing polymer (Eq. 45).

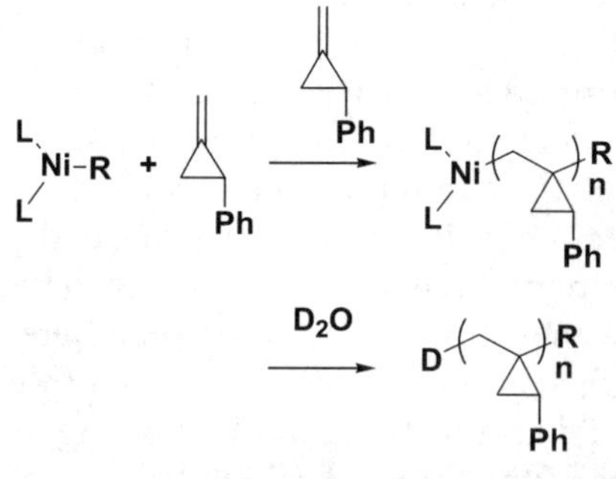

eq 45

The mechanism via 1,2-insertion is in contrast to the ring-opening polymerization of 2-phenyl-1-methylenecyclopropane by Pd complexes, where the polymerization proceeds via 2,1-insertion of the monomer into the π-allyl–Pd bond followed by β-alkyl elimination. Cyanosilylation of 2-phenyl-1-methylenecyclopropane catalyzed by Pd and Ni complexes was reported to produce the acyclic and cyclic products depending on the metals [165], which also undergo the insertion of the substrate with and without ring opening.

DSC analysis of the polymer revealed a higher T_g (178 °C) than those of many other known hydrocarbon polymers, such as thermally resistant syndiotactic polystyrene. The polymer does not decompose up to 300 °C in the TG analysis and shows good thermal stability. These results as well as the ^{13}C NMR data indicate a rigid polymer structure.

Other 2-aryl-1-methylenecyclopropanes also undergo addition polymerization in the presence of Ni catalysts to afford polymers containing cyclopropylidene groups [164]. Yields of the polymers obtained at −40 °C are high (>88%), suggesting that the polymerization is not retarded by the substituents of the phenyl ring, such as OMe and Cl. The molecular weights of the polymers increase in the order of the substituent, -OMe (M_n=3,600, M_w/M_n=1.67)<-Me (M_n=6,800, M_w/M_n=1.71)<-H (M_n=8,500, M_w/M_n=1.95)<-Cl (M_n=13,000, M_w/M_n=1.44). This is related to the electron donating or withdrawing character of the substituents or their Hammett constant (Fig. 1). The positive correlation between σ_p and the molecular weights indicates that the relative rate of propagation to that of chain transfer increases with the electron withdrawing character of the monomer substituent.

The polymerization of 2-ethoxycarbonyl-1-methylenecyclopropane catalyzed by [(π-allyl)NiBr]$_2$ at room temperature produces the polymer (Eq. 46) [164].

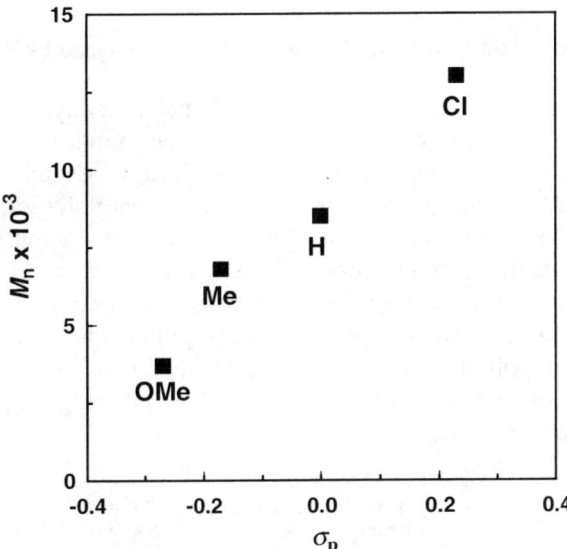

Fig. 1 Hammett plot of M_n of polymers of 2-aryl-1-methylenecyclopropanes

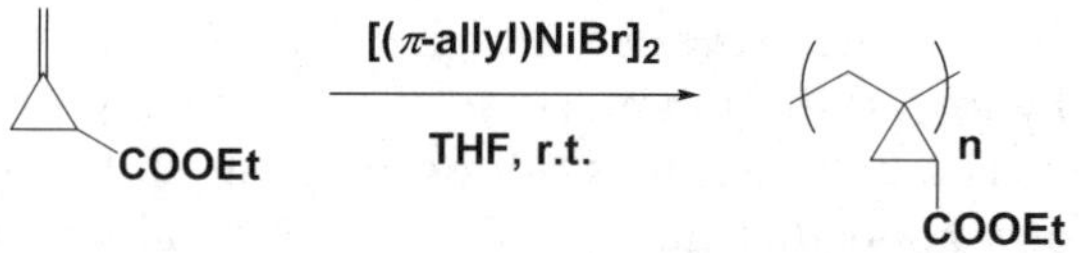

eq 46

The polymer structure is quite similar to those prepared from 2-aryl-1-methylenecyclopropanes with the Ni catalyst. The glass transition temperature is lower than that of the polymer with aryl substituents.

6
Copolymerization of Dienes and Methylenecyclopropanes with CO

Pd complex-catalyzed copolymerization of alkene and CO affords the polyketones via alternating insertion of the two monomers [166, 167]. The polymer growth involves migratory insertion of CO into the metal–carbon bond as a crucial step, which is unique to the late transition metal complexes such as Ni, Pd, Rh, and Co. The copolymerization of allenes and methylenecyclopropanes with CO has attracted much less attention than the alkene–CO copolymerization, although it would provide further functionalized polyketones due to the dual functionality of the dienes and the derivatives.

6.1
Copolymerization of Nonconjugated Dienes with CO Catalyzed by Pd Complexes

Cyclopolymerization of the nonconjugated dienes is catalyzed by early transition metal complexes to produce the unique polymer via alternating intermolecular and intramolecular insertion of the C=C bond into the metal–carbon bond. Waymouth reported that the cyclocopolymerization of 1,5-hexadiene under CO, catalyzed by $Pd(OAc)_2$/dppp/1,4-naphthaquinone/Ni $(ClO_4)_2 \cdot 6H_2O$, affords a polyketone with a cyclic structure in the repeating units [168]. The polymer contains both 5- and 6-membered rings in the polyketone backbone. Use of dipp instead of dppp as the ligand of the catalyst leads to a polymer containing the 6-membered ring only (Eq. 47). Epimerization of the polymer with 6-membered rings by 4-(dimethylamino)pyridine

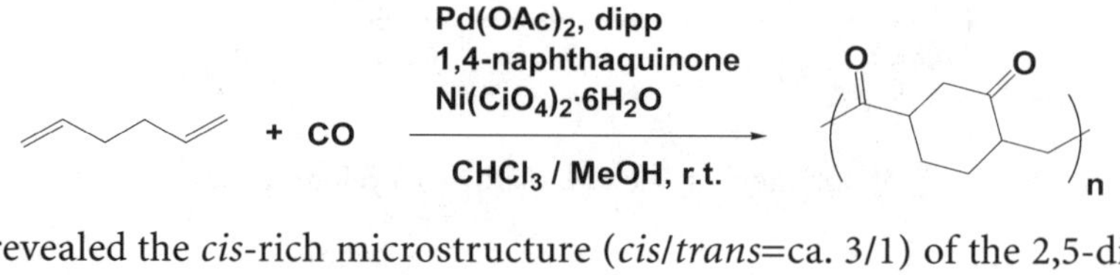

eq 47

revealed the *cis*-rich microstructure (*cis/trans*=ca. 3/1) of the 2,5-disubstituted cyclohexanone repeating units of the polyketone before epimerization.

The repeating units containing the 5-membered cyclopentanone structure are formed via an initial 2,1-insertion of 1,5-hexadiene, while the 6-membered cyclohexanone units are due to initial 1,2-insertion of the 1,5-hexadiene.

Cyclopolymerization of 1,4-pentadiene with CO in the presence of [Pd(Me)(NCMe)((R,S)-BINAPHOS)](BARF) forms the polymer with 5-membered rings only (Eq. 48) [169]. The cyclopentanone unit of the copolymer has the *cis* and *trans* structures in almost 1:1 ratio.

Abu-Surrah reported the copolymerization of several α, ω-dienes with CO catalyzed by [Pd(dppp)(NCMe)$_2$](BF$_4$)$_2$ [170]. Although copolymerization of 1,5-hexadiene with CO proceeds via complete cyclization of diene monomer, the copolymerization of 1,7-octadiene and 1,6-heptadien-4-ol gives copolymers without a ring structure and with pendant vinyl groups.

6.2
Copolymerization of Alkyl- and Arylallenes with CO Catalyzed by Pd and Rh Complexes

Pd complexes promote both the copolymerization of alkenes with CO and homopolymerization of the allene derivatives, which makes the alternating copolymerization of allenes with CO difficult. Copolymerization of allene with CO by Pd catalyst was briefly reported [171], but the following studies have not appeared in the literature until recently. Sen reported the copolymerization of 3,3-dimethylallene with CO catalyzed by [Pd(PPh$_3$)$_2$(NCMe)$_2$](BF$_4$)$_2$ to produce the polyketone $-(CH_2-C(=CMe_2)-CO)_n-$ and its use in the copolyketone synthesis (Eq. 49) [172]. In this reaction, two methyl groups of the monomer prevent homopolymerization via multiinsertion of the C=C bond of the allene into the Pd–C bond of the growing polymer. The molecular weight of the produced polyketone is limited probably due to steric hindrance of the monomer.

eq 49

The Rh complex shows lower reactivity toward insertion of unsaturated molecules into metal–carbon bonds than the Pd complex. Rh complex-promoted polymerization of ethylene, for example, is reported in a much fewer number of reports than the Pd complexes [173]. Copolymerization of norbornadiene with CO promoted by Rh complex occurs at high CO pressure [174]. Recently, substituted allenes were reported to undergo copolymerization with CO in the presence of Rh catalyst. π-Allylrhodium complex, formed by the reaction of $HRh(PPh_3)_4$ with excess (4-methoxyphenyl)allene, $Rh[\eta^3\text{-}CH(Ar)C\{C(=CHAr)CH_2C(=CHAr)CH_2CH_2CH=CHAr\}\text{-}CH_2](PPh_3)_2$ ($Ar=C_6H_4OMe\text{-}p$) (Eq. 50) [175], catalyzes alternating copolymerization of arylallenes with CO to give the polyketones, $-[-CO-C(=CHAr)-CH_2-]_n-$ (Eq. 51) [176, 177].

$$RhH(PPh_3)_4 + 4\ CH_2=C=CHAr \longrightarrow (\pi\text{-allyl})Rh(PPh_3)_2 \quad \text{eq 50}$$

$$CH_2=C=CHAr + CO \xrightarrow[\text{THF, r.t.}]{(\pi\text{-allyl})Rh(PPh_3)_2 \text{ (eq 50)}} -[-CO-C(=CHAr)-CH_2-]_n- \quad \text{eq 51}$$

The C=C double bond in the repeating unit has a Z configuration, which is confirmed by the ^1H NMR spectrum using the ROESY technique. The reaction rate does not vary depending on the concentration of the arylallene, indicating that the rate-determining step of the reaction involves insertion of CO into the Rh–allyl bond. Preferential formation of the structural unit with Z configuration of the C=C double bond is achieved via the Rh–π-allyl intermediate with the aryl ligand at the *syn* position. The living polymerization is demonstrated by (1) the ^1H NMR spectrum of the polymers showing signals of the terminal group originally contained in the Rh catalyst, (2) the increase in molecular weight (M_n) of the polymer with narrow molecular weight distribution throughout the polymerization, and (3) a linear relationship between the molecular weight and the monomer-to-catalyst ratio. Block and random copolyketones composed of two structural units, $-CO-C(=CHPh)-CH_2-$ and $-CO-C(=CHC_6H_4OMe)-CH_2-$, are prepared by using the above living polymerization. Addition of bis(allenyl)benzene to the polymerization mixture of the arylallene forms polymer with a partially cross-linked structure (Eq. 52) [178].

Substituted 1,2-dienes such as cyclohexylallene, (3-phenylpropyl)allene, 1,1-dimethylallene, 1-methyl-1-phenylallene, 4-(*tert*-butyl)phenoxyallene, and (4-allenyloxy)azobenzene undergo alternating copolymerization with CO in the presence of the Rh complex to produce unsaturated polyketones {-CO-C(=CRR')-CH$_2$-}$_n$ (R, R'=H, cyclo-C$_6$H$_{11}$; H, (CH$_2$)$_3$Ph; Me, Me; Me, Ph; H, OC$_6$H$_4$-*t*-Bu-4; H,OC$_6$H$_4$-N=N-C$_6$H$_5$) (Eq. 53) [179].

The polymers are mainly composed of the repeating units formed via insertion of the 2,3-bond of allene monomer and CO into the polymer end, while the polyketone from 3,3-dimethylallene contains the structural unit from the 1,2-polymerization of the monomer in a small ratio (<5%). Plotting the molecular weights of the polyketones during living polymerization shows a significant difference in the reaction rates of the allene monomers; the polymerization rate decreases in the order 1,1-dimethylallene>phenylallene>cyclohexylallene>(3-phenylpropyl)allene. The sterically more hindered allene derivatives undergo the copolymerization with CO more rapidly, which is explained reasonably based on the reaction mechanism involving insertion of CO into the Rh–σ-allyl bond rather than the Rh–π-allyl bond (Scheme 18). The π-allylrhodium intermediate A with the substituent at the *syn* position is equilibrated with the thermodynamically less favorable *anti* isomer. The intermediates with a bulky substituent at the π-allyl ligand or with two substituents at the same allyl carbon tend to be converted to the σ-allyl intermediate B. Insertion of CO into the Rh–C single bond to produce

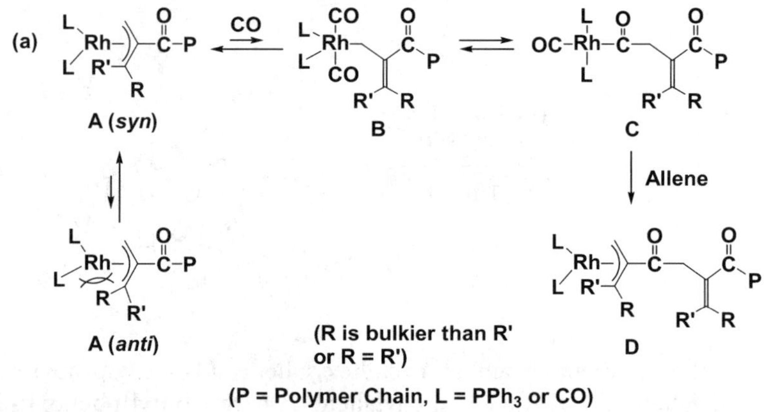

Scheme 18 Plausible mechanism of copolymerization of (a) monosubstituted and (b) 1,1-disubstituted allenes with CO by Rh complexes

the intermediate C followed by insertion of allene into the Rh–CO bond regenerates the π-allylic polymer A.

Copolymerization of mixtures of 4-(allenyloxy)azobenzene and 4-methoxyphenylallene, and of 4-(allenyloxy)azobenzene and 4-(*tert*-butyl) phenoxyallene, with CO affords random copolyketones with M_n=11,100–15,400. The allene monomer having a long-chain alkyl group connected to the azobenzene group also polymerizes with CO to produce a polyketone with liquid crystalline properties (Eq. 54) [180].

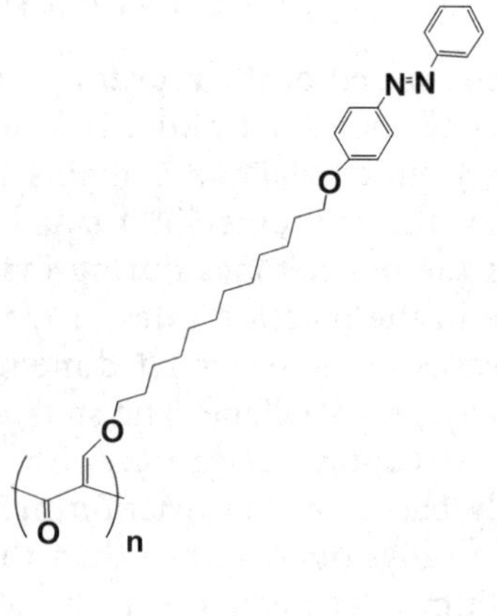

eq 54

6.3
Ring-Opening Copolymerization of Methylenecyclopropanes with CO by Pd Complexes

[Pd(dppp)(NCMe)$_2$](BF$_4$)$_2$ promotes copolymerization of methylenecyclopropane with CO [170]. The resulting copolymer was not soluble enough to be characterized by NMR spectroscopy and was characterized by IR to contain both ring-opened and cyclic microstructures from the methylenecyclopropane. Methylenecyclopropane–CO–propylene terpolymer can also be prepared by the polymerization using [Pd(dppp)(NCMe)$_2$](BF$_4$)$_2$ as the catalyst.

PdCl(Me)(bpy)/NaBARF initiates ring-opening copolymerization of 2-phenyl-1-methylenecyclopropane with CO (1 atm) at room temperature to produce a polyketone soluble in common organic solvent (Eq. 55) [181, 182].

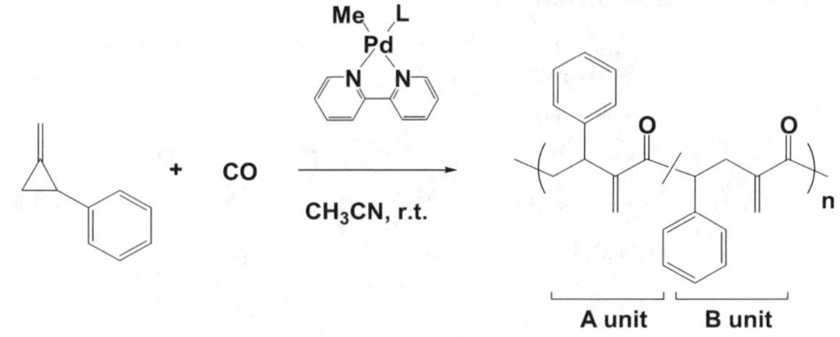

eq 55

The ^{1}H and ^{13}C{^{1}H} NMR spectra showed that the copolymer contains the ring-opened unit exclusively. The structural unit of the copolymer contains a vinylidene group, which is similar to that of the homopolymer ob-

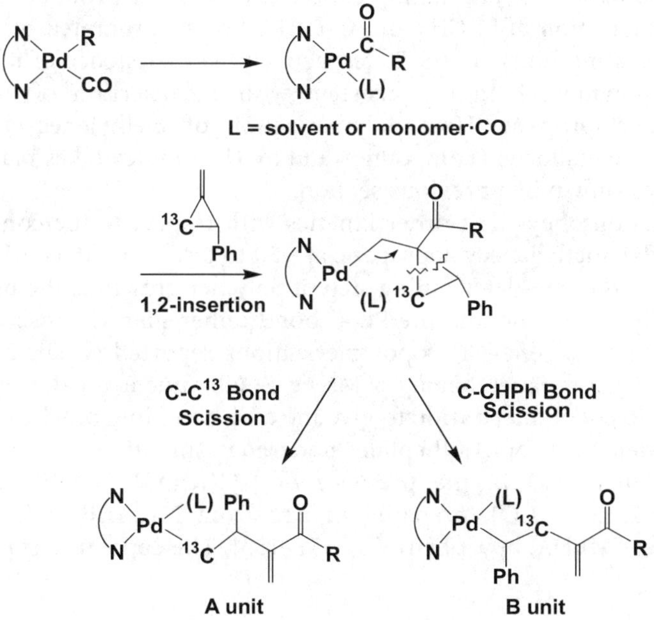

Scheme 19 Mechanism of copolymerization of 2-phenyl-1-methylenecyclopropane with CO by Pd complexes

tained by the Pd–diimine complex-catalyzed polymerization at elevated temperature. Isotope-labeled experiments, however, indicated a different insertion mode of the monomer and different end structure of the polymer between the copolymerization and homopolymerization. The copolymerization of 2-phenyl-1-methylenecyclopropane-3-^{13}C with CO introduces ^{13}C-enriched CH$_2$ carbons into the polymer chain (Eq. 56). The above labeled position of the polyketone indicates cleavage of a proximal C–C bond during the polymerization, as shown in Scheme 19.

$$\text{eq 56}$$

Insertion of CO into the Pd–alkyl bond of the growing polymer gives the acylpalladium intermediate. 1,2-Insertion of the methylenecyclopropane into the Pd–acyl bond forms an α-cyclopropylalkyl palladium intermediate that undergoes rapid β-alkyl elimination, resulting in the formation of a Pd complex having an α- or β-phenylalkyl ligand. Repetition of the above procedures accounts for the alternating copolymerization to produce the polyketone. The activation of C–CH$_2$ and C–CHPh bonds, promoted by Pd, forms the two repeating units A and B, respectively. The existence of units A and B in the polymer chain is consistent with 1,2-insertion of 2-phenyl-1-methylenecyclopropane. Homopolymerization of methylenecyclopropanes catalyzed by metallocene compounds and by Ni complex takes place via 1,2-insertion, as shown in a previous section.

The reaction obeys first-order kinetics with respect to the concentration of 2-phenyl-1-methylenecyclopropane at –30 to 0 °C. The first-order kinetics indicate that the rate-determining step of polymer growth is the insertion of methylenecyclopropane into the Pd–C bond rather than CO insertion, similar to the many alkene–CO copolymerizations reported so far. 2-Phenyl-1-methylenecyclopropanes having a Me or F substituent on the phenyl ring also undergo copolymerization to give the corresponding polyketones.

7-Methylenebicyclo[4.1.0]heptane undergoes smooth alternating copolymerization with CO in the presence of [Pd(R)(X)L]$^+$BARF$^-$ (R=Me or COMe; X=THF or CO), prepared in situ from NaBARF and PdCl(Me)L (L=diimine, TMEDA, bpy, phen) (Eq. 57) [183]. The copolymerization using diimine ligand or TMEDA ligand proceeds in a living fashion, affording a copolymer with narrow molecular weight distribution. The stereochemistry of the products is influenced significantly by the ligand as well as the solvent. For example, copolymerization in THF using Pd complexes with diimine or TMEDA ligands affords a polymer with *cis*-diisootactic structure, whereas polymerization using Pd complexes with bpy ligand in THF and CH_3CN leads to polymer with a mixed *cis*-diisotactic and -disyndiotactic structure, and polymer containing mainly *trans*-disyndiotactic structure, respectively (Scheme 20) [184].

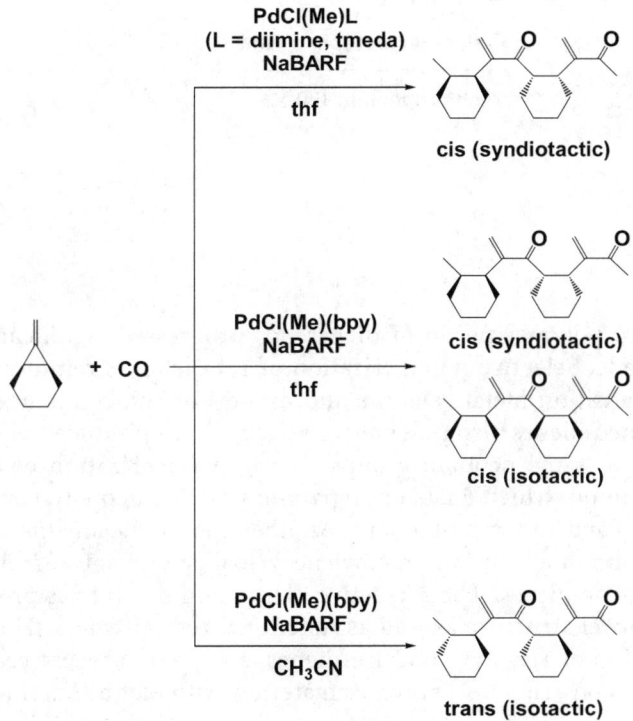

Scheme 20 Copolymerization of 7-methylenebicyclo[4.1.0]heptane with CO by Pd complexes

The bulky six-membered ring of the growing polymer end and substituents of the catalyst ligand regulate insertion of the monomer into the Pd–acyl bond to form the *cis*-polyketone with a single tacticity in THF

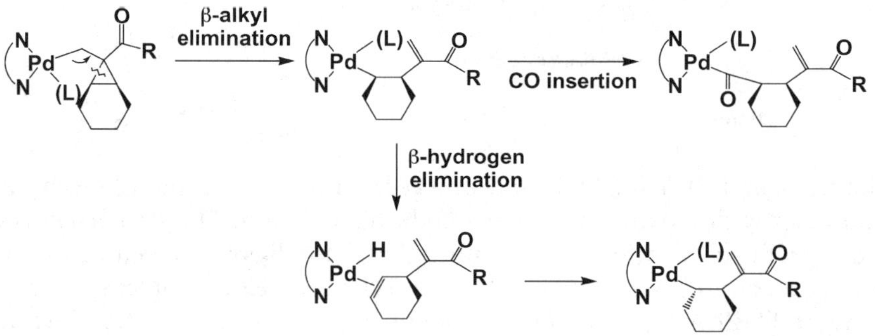

Scheme 21 Mechanism of copolymerization of 7-methylenebicyclo[4.1.0]heptane with CO by Pd complexes

(Scheme 21). The tacticity of the polyketone is kept before and after the base-promoted epimerization (Eq. 58).

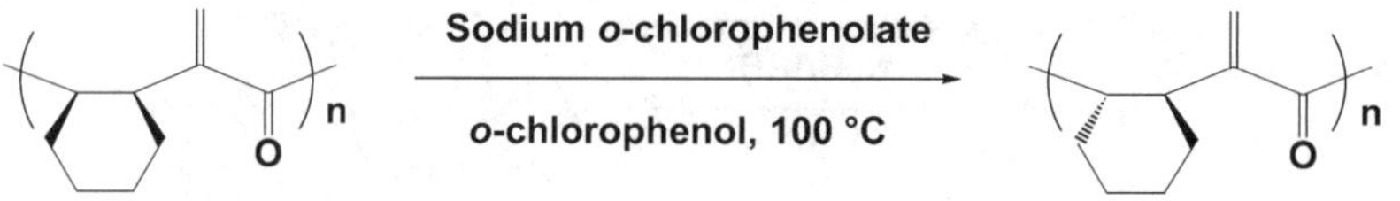

eq 58

7
Conclusion

Coordination polymerization of dienes has progressed significantly within the last decade. Selective polymerization of 1,3-dienes is reinforced by conventional transition metal catalysts and by new organolanthanide catalysts. Nonconjugated dienes also polymerize selectively to produce polymers with cyclic units or vinyl pendant groups. Living polymerization of dienes has become common, which enabled preparation of block copolymers of dienes with alkenes and other monomers. Another new topic in this field is the polymerization of allenes and methylenecycloalkanes catalyzed by late transition metal complexes. These reactive dienes and derivatives provide polymers with novel structure as well as functionalized polymers. The precision polymerization of 1,2-, 1,3-, and 1,n-dienes, achieved in recent years, will be developed to construct new polymer materials with olefin functionality.

References

1. For reviews on polymerization of dienes: (a) Porri L, Giarrusso A (1989) Conjugated diene polymerization. In: Ledwith A, Russo S, Sigewalt P (eds) Comprehensive polymer science, vol 4, Chap 5. Pergamon, Oxford, p 53; (b) Porri L, Giarrusso A, Ricci G (1991) Prog Polym Sci 16:405; (c) Watanabe H, Masuda T (1997) Diene polymerization. In: Kobayashi S (ed) Catalysis in precision polymerization. Wiley, Chichester, p 55; (d) Tindall D, Pawlow JH, Wagener KB (1998) Top Organomet Chem 1:183; (e) Lehman SE Jr, Wagener KB (2003) Catalysis in acyclic diene metathesis (ADMET) polymerization. In: Rieger B, Baugh LS, Kacker S, Striegler S (eds) Late transition metal polymerization catalysis. Wiley-VCH, Weinheim, p 193
2. (a) Yasuda H, Nakamura A (1987) Angew Chem Int Ed Engl 26:723; (b) Erker G, Krüger C, Müller G (1985) Adv Organomet Chem 24:1
3. Sinn H, Kaminsky W (1980) Adv Organomet Chem 18:99
4. Ishihara N, Seimiya T, Kuramoto M, Uoi M (1986) Macromolecules 19:2464
5. Ishihara N, Kuramoto M, Uoi M (1988) Macromolecules 21:3356
6. Oliva L, Longo P, Grassi A, Ammendola P, Pellecchia C (1990) Makromol Chem Rapid Commun 11:519
7. Ricci G, Italia S, Giarrusso A, Porri L (1993) J Organomet Chem 451:67
8. Ricci G, Italia S, Porri L (1994) Macromolecules 27:868
9. Longo P, Proto A, Oliva P, Zambelli A (1996) Macromolecules 29:5500
10. Ricci G, Bosisio C, Porri L (1996) Macromol Rapid Commun 17:781
11. Longo P, Guerra G, Grisi F, Pizzuti S, Zambelli A (1998) Macromol Chem Phys 199:149
12. Pragliola S, Cipriano M, Boccia AC, Longo P (2002) Macromol Rapid Commun 23:356
13. Miyazawa A, Kase T, Soga K (1999) J Polym Sci Part A Polym Chem 37:695
14. Miyazawa A, Kase T, Soga K (2000) Macromolecules 33:2796
15. Long P, Oliva P, Proto A, Zambelli A (1996) Gazz Chim Ital 126:377
16. Pragliola S, Forlenza E, Longo P (2001) Macromol Rapid Commun 22:783
17. Ricci G, Italia S, Comitani C, Porri L (1991) Polym Commun 32:514
18. Endo K, Yamanaka Y (2001) Macromol Chem Phys 202:201
19. Ricci G, Italia S, Porri L (l994) Macromol Chem Phys 195:1389
20. Suárez PAZ, Rosa NT, Einloft S, Souza RF, Dick YP (1998) Polym Bull 41:175
21. Suárez PAZ, Silva FM, Souza RF, Dick YP (1997) Ti complex with hematein ligand also promotes cis-1,4-polymerization of butadiene. Polym Bull 39:311
22. Ricci G, Panagia A, Porri L (1996) Polymer 37:363
23. Bradley S, Camm KD, Furtado SJ, Gott AL, McGowan PC, Podesta TJ, Thornton-Pett M (2002) Organometallics 21:3443
24. Ricci G, Battistella M, Porri L (2001) Macromolecules 34:5766
25. Ricci G, Battistella M, Bertini F, Porri L (2002) Polym Bull 48:25
26. Endo K, Uchida Y (2000) J Appl Polym Sci 78:1621
27. Ricci G, Italia S, Giarrusso A, Porri L (1993) J Organomet Chem 451:67
28. (a) Costabile C, Milano G, Cavallo L, Guerra G (2001) Macromolecules 34:7952; (b) Guerra G, Cavallo L, Corradini P, Fusco R (1997) Macromolecules 30:677; (c) Peluso A, Improta R, Zambelli A (2000) Organometallics 19:411; (d) Improta R, Peluso A (1999) Macromolecules 32:6852; (e) Peluso A, Improta R, Zambelli A (1997) Macromolecules 30:2219
29. Von Dohlen WC, Wilson TP, Caflisch EG (1964) Belg Patent 644,291

30. (a) Iovu H, Hubca G, Simionescu E, Badea E, Hurst JS (1997) Eur Polym J 33:811; (b) Srinivasa Rao GS, Upadhyay VK, Jain RC (1999) J Appl Polym Sci 71:595
31. Barbotin F, Spitz R, Boisson C (2001) Macromol Rapid Commun 22:1411
32. Nickaf JB, Burford RP, Chaplin RP (1995) J Polym Sci Part A Polym Chem 33:1125
33. (a) Kobayashi E, Hayashi N, Aoshima S, Furukawa J (1998) J Polym Sci Part A Polym Chem 36:1707; (b) Quirk RP, Kells AM, Yunlu K, Cuif JP (2000) Polymer 41:5903
34. Friebe L, Nuyken O, Windisch H, Obrecht W (2002) Macromol Chem Phys 203:1055
35. Kwag G, Lee H, Kim S (2001) Macromolecules 34:5367
36. Evans WJ, Giarikos DG, Ziller JW (2001) Organometallics 20:5751
37. Boisson C, Barbotin F, Spitz R (1999) Macromol Chem Phys 200:1163
38. Quirk RP, Kells AM (2000) Polym lnt 49:751
39. Dong W, Endo K, Masuda T (2003) Macromol Chem Phys 204:104
40. Dong W, Masuda T (2002) J Polym Sci Part A Polym Chem 40:1838
41. Natta G, Porri S, Corradini P, Zanini G, Ciampelli F (1963) J Polym Sci 51:463
42. (a) Natta G, Porri L, Stoppa G, Allegra G, Ciampelli F (1964) J Polym Sci B 1:67; (b) Natta G, Porri L, Carbonaro A, Stoppa G (1964) Makromol Chem 77:114
43 (a) Porri L, Carbonaro A (1963) Makromol Chem 60:236; (b) Natta G, Porri A, Carbonaro A, Ciampelli F, Allegra G (1962) Makromol Chem 51:229
44. Purevsuren B, Allegra G, Meille SV, Farina A, Porri L, Ricci G (1998) Polymer J 130:431
45. (a) Taube R, Maiwald S, Sieler J (1996) J Organomet Chem 513:37; (b) Taube R, Windisch H, Maiwald S, Hemling H, Schumann H (1996) J Organomet Chem 513:49; (c) Maiwald S, Weissenborn H, Sommer C, Müller G, Taube R (2001) J Organomet Chem 640:1
46. Maiwald S, Sommer C, Müller G, Taube R (2002) Macromol Chem Phys 203:1029
47. (a) Taube R, Windisch H, Weissenborn H, Hemling H, Schumann H (1997) J Organomet Chem 548:229; (b) Maiwald S, Taube R, Hemling H, Schumann H (1998) J Organomet Chem 552:195
48. Maiwald S, Sommer C, Müller G, Taube R (2001) Macromol Chem Phys 202:1446
49. For reviews on polymerization by organolanthanide complexes: (a) Hou Z, Wakatsuki Y (2002) Coord Chem Rev 231:1; (b) Hou Z, Wakatsuki Y (2002) J Organomet Chem 647:61
50. Cui L, Ba X, Teng H, Ying L, Li K, Jin Y (1998) Polym Bull 40:729
51. Kaita S, Hou Z, Wakatsuki Y (1999) Macromolecules 32:9078
52. Kaita S, Hou Z, Wakatsuki Y (2001) Macromolecules 34:1539
53. Kaita S, Hou Z, Nishiura M, Doi Y, Kurazumi J, Horiuchi AC, Wakatsuki Y (2003) Macromol Rapid Commun 24:179
54. Baudry-Barbier D, Andre N, Dormond A, Pardes C, Richard P, Visseaux M, Zhu CJ (1998) Eur J Inorg Chem 1721
55. Tardif O, Hou Z, Nishiura M, Koizumi T, Wakatsuki Y (2001) Organometallics 20:4565
56. Bönnemann H, Grade C, Kopp W, Wilke G (1971) Pure Appl Chem 23:265
57. Rinehart RE, Smith HP, Witt HS, Romeyn H Jr (1961) J Am Chem Soc 83:4864
58. Lobach MI, Lormer VA, Tsereteli IY, Kondoratenkov GP, Babitsikii BD, Klepikova VI (1971) J Polym Sci Polym Lett 9:71
59. Sato H, Yagi Y (1992) Bull Chem Soc Jpn 65:1299
60. Endo K, Uchida Y, Matsuda Y (1996) Macromol Chem Phys 197:3515
61. (a) Endo K, Yamanaka Y (1999) Macromol Rapid Commun 20:312; (b) Endo K, Matsuda Y (1999) J Polym Sci Part A Polym Chem 37:3838;

62. Natori I, Imaizumi K, Yamagishi H, Kazunori M (1998) J Polym Sci Part B Polym Phys 36:1657
63. Nakano M, Yao Q, Usuki A, Tanimura S, Matsuoka T (2000) Chem Commun 2207
64. Tanimura S, Matsuoka T, Nakano M, Usuki A (2001) J Polym Sci Part B Polym Phys 39:973
65. Hadjiandreou P, Julémont M, Teyssié P (1984) Macromolecules 17:2455
66. Winter H, Aagaard O (1998) Macromol Rapid Commun 19:345
67. Taube R, Wache S, Kehlen H (1995) J Mol Catal A Chemical 97:21
68. Taube R, Schmidt U, Gehrke JP, Böhme P, Langlotz J, Wache S (1993) Makromol Chem Macromol Symp 66:245
69. Taube R, Windisch H, Maiwald S (1995) Macromol Symp 89:393
70. (a) Tobisch S (2002) Acc Chem Res 35:96; (b) Tobisch S, Bögel H, Taube R (1996) Organometallics 15:3563; (c) Tobisch S, Bögel H, Taube R (1998) Organometallics 17:1177; (d) Tobisch S, Taube R (1999) Organometallics 18:5204; (e) Tobisch S, Taube R (1999) Organometallics 18:3045; (f) Tobisch S, Taube R (2001) Chem Eur J 7:3681; (g) Tobisch S, Taube R (2002) Chem Eur J 8:4756
71. Nath DCD, Shiono T, Ikeda T (2002) Macromol Chem Phys 203:756
72. Nath DCD, Shiono T, Ikeda T (2002) Macromol Chem Phys 203:1171
73. Endo K, Hatakeyama N (2001) J Polym Sci Part A Polym Chem 39:2793
74. Endo K, Matsuda Y (2000) Macromol Chem Phys 201:1426
75. Jang Y, Kim P, Lee H (2002) Macromolecules 35:1477
76. Cass P, Pratt K, Mann T, Laslett B, Rizzardo E, Burford R (1999) J Polym Sci Part A Polym Chem 37:3277
77. Cass P, Pratt K, Mann T, Laslett B, Rizzardo E (2001) J Polym Sci Part A Polym Chem 39:2244
78. Cass P, Pratt K, Fairhall K, Laslett B, Rizzardo E (2001) J Polym Sci Part A Polym Chem 39:2256
79. Yamamoto A, Morifuji K, Ikeda S, Sato T, Uchida Y, Misono A (1968) J Am Chem Soc 90:1878
80. (a) Yamamoto T, Yamamoto A, Ikeda S (1972) Bull Chem Soc Jpn 45:1111; (b) Yamamoto T, Yamamoto A, Ikeda S (1972) Bull Chem Soc Jpn 45:1140 (c) Yamamoto A, Shimizu T, Ikeda S (1970) Makromol Chem 136:297; (d) Yamamoto T, Yamamoto A (1978) J Polym Sci Macromol Rev 13:161
81. Bazzini C, Giarrusso A, Porri L (2002) Macromol Rapid Commun 23:922
82. (a) Pellecchia C, Proto A, Zambelli A (1992) Macromolecules 25:4450; (b) Zambelli A, Proto A, Longo P, Oliva P (1994) Macromol Chem Phys 195:2623; (c) Longo P, Proto A, Oliva P, Sessa I, Zambelli A (1997) J Polym Sci Part A Polym Chem 35:2697
83. Pragliola S, Milano G, Guerra G, Longo P (2002) J Am Chem Soc 124:3502
84. Laird JL, Edmondson MS, Riedel JA (1997) Rubber World 217:42
85. Simanke AG, Mauler RS, Galland GB (2002) J Polym Sci Part A Polym Chem 40:471
86. (a) Barbotin F, Monteil V, Llauro MF, Boisson C, Spitz R (2000) Macromolecules 33:8521; (b) Llauro MF, Monnet C, Barbotin F, Monteil V, Spitz R, Boisson C (2001) Macromolecules 34:6304
87. Visseaux M, Barbier-Baudry D, Bonnet F, Dormond A (2001) Macromol Chem Phys 202:2485
88. Bonnet F, Visseaux M, Barbier-Baudry D, Dormond A (2002) Macromolecules 35:1143
89. Bonnet F, Barbier-Baudry D, Dormond A, Visseaux M (2002) Polym lnt 51:986
90. Barbier-Baudry D, Bonnet F, Domenichini B, Dormond A, Visseaux M (2002) J Organomet Chem 647:167

91. Deming TJ, Novak BM (1993) Macromolecules 26:7089
92. Deming TJ, Novak BM (1991) Macromolecules 24:5478
93. Deming TJ, Novak BM, Ziller JW (1994) J Am Chem Soc 116:2366
94. Nath DCD, Shiono T, Ikeda T (2002) J Polym Sci Part A Polym Chem 40:3086
95. Doi Y, Tokuhiro N, Soga K (1989) Makromol Chem 190:643
96. (a) Resconi L, Waymouth RM (1990) J Am Chem Soc 112:4953; (b) Mogstad AL, Waymouth RM (1992) Macromolecules 25:2282; (c) Cavallo L, Guerra G , Corradini P, Resconi L, Waymouth RM (1993) Macromolecules 26:260; (d) Kesti MR, Waymouth RM (1992) J Am Chem Soc 114:3565
97. Schaverien CJ (1994) Organometallics 13:69
98. For reviews on enantioselective cyclopolymerization of nonconjugated dienes: Coates GW (2000) Chem Rev 100:1223
99. (a) Coates GW, Waymouth RM (1991) J Am Chem Soc 113:6270; (b) Coates GW, Waymouth RM (1992) J Mol Catal 76:189; (c) Coates GW, Waymouth RM (1993) J Am Chem Soc 115:91
100. Mitani M, Oouchi K, Hayakawa M, Yamada T, Mukaiyama T (1995) Chem Lett 905
101. Jayaratne KC, Sita LR (2000) J Am Chem Soc 122:958
102. Jayaratne KC, Keaton RJ, Henningsen DA, Sita LR (2000) J Am Chem Soc 122:10490
103. Lee DH, Yoon KB, Park JR, Lee BH (1997) Eur Polym J 33:447
104. Hackmann M, Repo T, Jany G, Rieger B (1998) Macromol Chem Phys 199:1511
105. Pietikäinen P, Väänänen T, Seppälä JV (1999) Eur Polym 135:1047; and references therein
106. Jeremic D, Wang Q, Quyoum R, Baird MC (1995) J Organomet Chem 497:143
107. Naga N, Shiono T, Ikeda T (l999) Macromol Chem Phys 200:1466
108. Naga N, Shiono T, Ikeda T (1999) Macromolecules 32:1348
109. Pietikäinen P, Seppälä JV, Ahjopalo L, Pietilä LO (2000) Eur Polymer J 36:183
110. Kim I, Shin YS, Lee JK (2000) J Polym Sci Part A Polym Chem 38:1590
111. Kokko E, Pietikäinen P, Koivunen J, Seppälä JV (2001) J Polym Sci Part A Polym Chem 39:3805
112. (a) Naga N, lmanishi Y (2002) Macromol Chem Phys 203:771; (b) Naga N, Imanishi Y (2002) Macromol Chem Phys 203:2155
113. Jin HS, Choi CH, Park ES, Lee IM, Yoon JS (2002) J Appl Polym Sci 84:1048
114. Santos JM, Ribeiro MR, Portela MF, Cramail H, Deffieux A (2001) Macromol Chem Phys 202:3043
115. Arnold M, Bornemann S, Schimmel T, Heinze T (2002) Macromol Symp 181:5
116. Hustad PD, Tian J, Coates GW (2002) J Am Chem Soc 124:3614
117. Hustad PD, Coates GW (2002) J Am Chem Soc 124:11578
118. (a) Mori H, Kono H, Terano M (2000) Macromol Chem Phys 201:543; (b) Kono H, Mori H, Terano M (2001) Macromol Chem Phys 202:1319; (c) Kono H, Ichiki T, Mori H, Nakatani H, Terano M (2001) Polym lnt 50:568
119. (a) Greenfield H, Wender I, Wotiz JH (1956) J Org Chem 21:875; (b) Baker Jr WP (1963) J Polym Sci Part A 1:655; (c) Otsuka S, Mori K, Imaizumi F (1965) J Am Chem Soc 87:3017; (d) Otsuka S, Nakamura A (1967) J Polym Sci Polym Lett 5:973; (e) Otsuka S, Mori K, Suminoe T, Imaizumi F (1967) Eur Polym J 3:73; (f) Krentsel BA, Mushina EA, Khar'kova EM, Shishkina MV (1975) Eur Polym J 11:865; (g) Leland J, Boucher J, Anderson K (1977) J Polym Sci Part A Polym Chem 15:2785; (h) Ghalamkar-Moazzam M, Jacobs TL (1978) J Polym Sci Part A Polym Chem 16:615
120. (a) Scholten JP; van der Ploeg HJ (1972) J Polym Sci Polym Chem Ed 10:3067; (b) Scholten JP, van der Ploeg HJ (1973) J Polym Sci Polym Chem Ed 11:3205
121. (a) Endo T, Tomita I (1997) Prog Polym Sci 46:79; (b) Tomita I, Takagi K, Endo T (1998) J Synth Org Chem Jpn 56:260

122. Takagi K, Tomita I, Endo T (1997) Macromolecules 30:7386
123. Endo T, Takagi K, Tomita I (1997) Tetrahedron 53:15187
124. Takagi K, Tomita I, Endo T (1997) Chem Lett 1187
125. Tomita I, Kondo Y, Takagi K, Endo T (1994) Macromolecules 27:4413
126. Tomita I, Kondo Y, Takagi K, Endo T (1995) Acta Polym 46:432
127. Takagi K, Tomita I, Nakamura Y, Endo T (1998) Macromolecules 31:2779
128. Tomita I, Abe T, Takagi K, Endo T (1995) J Polym Sci Part A Polym Chem 33:2487
129. Taguchi M, Tomita I, Yoshida Y, Endo T (2000) Macromol Chem Phys 201:1025
130. Takagi K, Tomita I, Endo T (1997) Polym Bull 39:685
131. Takagi K, Tomita I, Endo T (1998) Chem Commun 681
132. Tomita I, Ubukata M, Endo T (1998) React Funct Polym 37:27
133. Takagi K, Tomita I, Endo T (1998) Macromolecules 31:6741
134. Suzuki M, Takao T, Sakamoto N, Tomita I, Endo T (1999) Polym J 31:1021
135. Taguchi M, Tomita I, Endo T (2000) Macromol Chem Phys 201:2322
136. Wang J, Tomita I, Endo T (2001) Macromolecules 34:4294
137. Taguchi M, Tomita I, Endo T (2000) Angew Chem Int Ed 39:3667
138. Nakano M, Novak BM (1996) ACS Polym Prepr 37:200
139. Novak BM, Nakano M (1997) Polym Mater Sci Eng 76:102
140. (a) Deming TJ, Novak BM (1991) Macromolecules 24:6043; (b) Deming TJ, Novak BM (1993) J Am Chem Soc 115:9101
141. Tomita I, Taguchi M, Takagi K, EndoT (1997) J Polym Sci Part A Polym Chem 35:431
142. Taguchi M, Tomita I, Yoshida Y, Endo T (1999) J Polym Sci Part A Polym Chem 37:3916
143. Taguchi M, Tomita I, Yoshida Y, EndoT (2001) J Polym Sci Part A Polym Chem 39:495
144. (a) Nakamura A (1966) Bull Chem Soc Jpn 39:543; (b) Bowden FL, Giles R (1976) Coord Chem Rev 20:81
145. Sóvágó J, Newton MG, Mushina EA, Ungváry F (1996) J Am Chem Soc 118:9589
146. Szalontai G, Sóvágó J, Ungváry F (1999) J Organomet Chem 586:54
147. Choi JC, Osakada K, Yamaguchi I, Yamamoto T (1997) Appl Organomet Chem 11:957
148. For reviews of transition metal-catalyzed reactions of methylenecyclopropanes: (a) Binger P, Büch HM (1987) Top Curr Chem 135:77; (b) Nakamura I, Yamamoto Y (2002) Adv Synth Cat 344:111
149. Harrington PJ (1995) Transition metal allyl complexes: trimethylene methane complexes. In: Abel EW, Stone FGA, Wilkinson G (eds) Comprehensive organometallic chemistry, vol 12, Chap 8.4. Pergamon, Oxford, p 923
150. (a) Osakada K, Takimoto H, Yamamoto T (1999) J Chem Soc Dalton Trans 853; (b) Nishihara Y, Yoda C, Osakada K (2001) Organometallics 20:2124
151. (a) Nishihara Y, Itazaki M, Osakada K (2002) J Org Chem 67:6889; (b) Nishihara Y, Itazaki M, Osakada K (2002) Tetrahedron Lett 43:2059
152. (a) Pinazzi CP, Brossas J (1969) Makromol Chem 122:105; (b) Pinazzi CP, Brossas J (1971) Makromol Chem 147:15; (c) Pinazzi CP, Brossas J, Clouet G (1971) Makromol Chem 148:81
153. Rossi R, Diversi P, Porri L (1972) Macromolecules 5:247
154. Yang X, Jia L, Marks TJ (1993) J Am Chem Soc 115:3392
155. Jia L, Yang X, Seyam AM, Albert IDL, Fu PF, Yang S, Marks TJ (1996) J Am Chem Soc 118:7900
156. Jia L, Yang X, Yang S, Marks TJ (1996) J Am Chem Soc 118:1547
157. Yang X, Seyam AM, Fu PF, Marks TJ (1994) Macromolecules 27:4625
158. Takeuchi D, Kim S, Osakada K (2001) Angew Chem Int Ed 40:2685

159. Kim S, Takeuchi D, Osakada K (2003) Pd complex-catalyzed ring-opening polymerisation of 2-aryl-1-methylenecyclopropanes. In: Screttas CG, Steele BR (eds) Perspectives in organometallic chemistry. The Royal Society of Chemistry, Cambridge, p 306
160. Aeby A, Gsponer A, Consiglio G (1998) J Am Chem Soc 120:11000
161. Hughes RP, Powell J (1972) J Am Chem Soc 94:7723
162. (a) Burger BJ, Thompson ME, Cotter WD, Bercaw JB (1990) J Am Chem Soc 112:1566; (b) Takaya M, Kageyama A, Miya S, Yamazaki H (1991) Chem Lett 20:1525; (c) Resconi L, Piemontesi F, Franciscono G, Abis L, Fiorani T (1992) J Am Chem Soc 114:1025; (d) Eshuis JJW, Tan YY, Meetsma A, Teuben JH, Renkema J, Evens GG (1992) Organometallics 11:362; (e) Guo Z, Swenson DC, Jordan RF (1994) Organometallics 13:1424
163. Takeuchi D, Osakada K (2002) Chem Commun 646
164. Takeuchi D, Anada K, Osakada K (2002) Macromolecules 35:9628
165. Chatani N, Takeyasu T, Hanafusa T (1988) Tetrahedron Lett 29:3979
166. For reviews of copolymerization of alkene and CO: (a) Drent E, Budzelaar PHM (1996) Chem Rev 96:663; (b) Sen A (1993) Acc Chem Res 26:303; (c) Sperrle M, Consiglio G (1997) Chem Ber 130:1557; (d) Nozaki K, Hiyama T (1999) J Organomet Chem 576:248
167. (a) Brookhart M, Rix FC, DeSimone JM, Barborak JC (1992) J Am Chem Soc 114:5894; (b) Brookhart M, Wagner MI, Balavoine GGA, Haddou HA (1994) J Am Chem Soc 116:3641; (e) Bronco S, Consiglio G, Hutter R, Batistini A, Suter UW (1994) Macromolecules 27:4436; (d) Sesto B, Consiglio G (2001) J Am Chem Soc 123:4097; (e) Nozaki K, Sato N, Takaya H (1995) J Am Chem Soc 117:9911; (f) Nozaki K, Komaki H, Kawashima Y, Hiyama T, Matsubara T (2001) J Am Chem Soc 123:534
168. Borkowsky SL, Waymouth RM (1996) Macromolecules 29:6377
169. Nozaki K, Sato N, Nakamoto K, Takaya H (1997) Bull Chem Soc Jpn 70:659
170. Kettunen M, Abu-Surrah AS, Repo T, Leskelä M (2001) Polym Int 50:1223
171. Maatschappij BV (1988) Neth Appl 8801168 (1990) Chem Abstr 113:24686f
172. Kacker S, Sen A (1997) J Am Chem Soc 119:10028
173. (a) Wang L, Flood TC (1992) J Am Chem Soc 114:3169; (b) Wang L, Lu RS, Bau R, Flood TC (1993) J Am Chem Soc 115:6999
174. (a) Zhang SW, Takahashi S (2000) Chem Commun 315; (b) Zhang SW, Kaneko T, Takahashi S (2000) Macromolecules 33:6930
175. Choi JC, Osakada K, Yamamoto T (1998) Organometallics 17:3044
176. Osakada K, Choi JC, Yamamoto T (1997) J Am Chem Soc 119:12390
177. Choi JC, Yamaguchi I, Osakada K, Yamamoto T (1998) Macromolecules 31:8731
178. Osakada K, Takenaka Y, Choi JC, Yamaguchi I, Yamamoto T (2000) J Polym Sci Part A Polym Chem 38:1505
179. Takenaka Y, Osakada K (2001) Macromol Chem Phys 202:3571
180. Takenaka Y, Osakada K, Nakano M, Ikeda T (2003) Macromolecules 36:1414
181. Kim S, Takeuchi D, Osakada K (2002) J Am Chem Soc 124:762
182. Kim S, Takeuchi D, Osakada K (2003) Macromol Chem Phys 204:666
183. Takeuchi D, Yasuda A, Osakada K (2003) J Chem Soc Dalton Trans 2029
184. Syndiotactic and isotactic denote the stereochemistry of the polymer of the substituted alkenes such as propylene and methyl methacrylate. In Scheme 20, the stereochemical relationships of the neighboring cyclohexane groups along the polymer chain are expressed using this terminology, although the original definition of syndiotactic and isotactic does not cover such polymers. The stereochemistry of several polymers in this review is expressed according to each of the original papers

Editor: Akihiro Abe   Received: November 2003

Author Index Volumes 101-171

Author Index Volumes 1-100 see Volume 100

de, Abajo, J. and *de la Campa, J. G.*: Processable Aromatic Polyimides.Vol. 140, pp. 23-60.
Abetz, V. see Förster, S.: Vol. 166, pp. 173-210.
Adolf, D. B. see Ediger, M. D.: Vol. 116, pp. 73-110.
Aharoni, S. M. and *Edwards, S. F.*: Rigid Polymer Networks.Vol. 118, pp. 1-231.
Albertsson, A.-C., Varma, I. K.: Aliphatic Polyesters: Synthesis, Properties and Applications. Vol. 157, pp. 99-138.
Albertsson, A.-C. see Edlund, U.: Vol. 157, pp. 53-98.
Albertsson, A.-C. see Söderqvist Lindblad, M.: Vol. 157, pp. 139-161.
Albertsson, A.-C. see Stridsberg, K. M.: Vol. 157, pp. 27-51.
Albertsson, A.-C. see Al-Malaika, S.: Vol. 169, pp. 177-199.
Al-Malaika, S.: Perspectives in Stabilisation of Polyolefins. Vol. 169, pp. 121-150.
Améduri, B., Boutevin, B. and *Gramain, P.*: Synthesis of Block Copolymers by Radical Polymerization and Telomerization. Vol. 127, pp. 87-142.
Améduri, B. and *Boutevin, B.*: Synthesis and Properties of Fluorinated Telechelic Monodispersed Compounds. Vol. 102, pp. 133-170.
Amselem, S. see Domb, A. J.: Vol. 107, pp. 93-142.
Andrady, A. L.: Wavelenght Sensitivity in Polymer Photodegradation. Vol. 128, pp. 47-94.
Andreis, M. and *Koenig, J. L.*: Application of Nitrogen-15 NMR to Polymers.Vol. 124, pp. 191-238.
Angiolini, L. see Carlini, C.: Vol. 123, pp. 127-214.
Anjum, N. see Gupta, B.: Vol. 162, pp. 37-63.
Anseth, K. S., Newman, S. M. and *Bowman, C. N.*: Polymeric Dental Composites: Properties and Reaction Behavior of Multimethacrylate Dental Restorations. Vol. 122, pp. 177-218.
Antonietti, M. see Cölfen, H.: Vol. 150, pp. 67-187.
Armitage, B. A. see O'Brien, D. F.: Vol. 126, pp. 53-58.
Arndt, M. see Kaminski, W.: Vol. 127, pp. 143-187.
Arnold Jr., F. E. and *Arnold, F. E.*: Rigid-Rod Polymers and Molecular Composites. Vol. 117, pp. 257-296.
Arora, M. see Kumar, M. N. V. R.: Vol. 160, pp. 45-118.
Arshady, R.: Polymer Synthesis via Activated Esters:A New Dimension of Creativity in Macromolecular Chemistry. Vol. 111, pp. 1-42.

Bahar, I., Erman, B. and *Monnerie, L.*: Effect of Molecular Structure on Local Chain Dynamics: Analytical Approaches and Computational Methods. Vol. 116, pp. 145-206.
Ballauff, M. see Dingenouts, N.: Vol. 144, pp. 1-48.
Ballauff, M. see Holm, C.: Vol. 166, pp. 1-27.
Ballauff, M. see Rühe, J.: Vol. 165, pp. 79-150.
Baltá-Calleja, F. J., González Arche, A., Ezquerra, T. A., Santa Cruz, C., Batallón, F., Frick, B. and *López Cabarcos, E.*: Structure and Properties of Ferroelectric Copolymers of Poly(vinylidene) Fluoride. Vol. 108, pp. 1-48.
Barnes, M. D. see Otaigbe, J.U.: Vol. 154, pp. 1-86.
Barshtein, G. R. and *Sabsai, O. Y.*: Compositions with Mineralorganic Fillers.Vol. 101, pp. 1-28.

Baschnagel, J., Binder, K., Doruker, P., Gusev, A. A., Hahn, O., Kremer, K., Mattice, W. L., Müller-Plathe, F., Murat, M., Paul, W., Santos, S., Sutter, U. W., Tries, V.: Bridging the Gap Between Atomistic and Coarse-Grained Models of Polymers: Status and Perspectives. Vol. 152, pp. 41-156.
Batallán, F. see Baltá-Calleja, F. J.: Vol. 108, pp. 1-48.
Batog, A. E., Pet'ko, I.P., Penczek, P.: Aliphatic-Cycloaliphatic Epoxy Compounds and Polymers. Vol. 144, pp. 49-114.
Barton, J. see Hunkeler, D.: Vol. 112, pp. 115-134.
Bell, C. L. and *Peppas, N. A.:* Biomedical Membranes from Hydrogels and Interpolymer Complexes. Vol. 122, pp. 125-176.
Bellon-Maurel, A. see Calmon-Decriaud, A.: Vol. 135, pp. 207-226.
Bennett, D. E. see O'Brien, D. F.: Vol. 126, pp. 53-84.
Berry, G. C.: Static and Dynamic Light Scattering on Moderately Concentraded Solutions: Isotropic Solutions of Flexible and Rodlike Chains and Nematic Solutions of Rodlike Chains. Vol. 114, pp. 233-290.
Bershtein, V. A. and *Ryzhov, V. A.:* Far Infrared Spectroscopy of Polymers. Vol. 114, pp. 43-122.
Bhargava R., Wang S.-Q., Koenig J. L: FTIR Microspectroscopy of Polymeric Systems. Vol. 163, pp. 137-191.
Biesalski, M.: see Rühe, J.: Vol. 165, pp. 79-150.
Bigg, D. M.: Thermal Conductivity of Heterophase Polymer Compositions.Vol. 119, pp. 1-30.
Binder, K.: Phase Transitions in Polymer Blends and Block Copolymer Melts: Some Recent Developments. Vol. 112, pp. 115-134.
Binder, K.: Phase Transitions of Polymer Blends and Block Copolymer Melts in Thin Films. Vol. 138, pp. 1-90.
Binder, K. see Baschnagel, J.: Vol. 152, pp. 41-156.
Bird, R. B. see Curtiss, C. F.: Vol. 125, pp. 1-102.
Biswas, M. and *Mukherjee, A.:* Synthesis and Evaluation of Metal-Containing Polymers. Vol. 115, pp. 89-124.
Biswas, M. and *Sinha Ray, S.:* Recent Progress in Synthesis and Evaluation of Polymer-Montmorillonite Nanocomposites. Vol. 155, pp. 167-221.
Bogdal, D., Penczek, P., Pielichowski, J., Prociak, A.: Microwave Assisted Synthesis, Crosslinking, and Processing of Polymeric Materials. Vol. 163, pp. 193-263.
Bohrisch, J., Eisenbach, C.D., Jaeger, W., Mori H., Müller A.H.E., Rehahn, M., Schaller, C., Traser, S., Wittmeyer, P.: New Polyelectrolyte Architectures. Vol. 165, pp. 1-41.
Bolze, J. see Dingenouts, N.: Vol. 144, pp. 1-48.
Bosshard, C.: see Gubler, U.: Vol. 158, pp. 123-190.
Boutevin, B. and *Robin, J. J.:* Synthesis and Properties of Fluorinated Diols. Vol. 102. pp. 105-132.
Boutevin, B. see Amédouri, B.: Vol. 102, pp. 133-170.
Boutevin, B. see Améduri, B.: Vol. 127, pp. 87-142.
Bowman, C. N. see Anseth, K. S.: Vol. 122, pp. 177-218.
Boyd, R. H.: Prediction of Polymer Crystal Structures and Properties. Vol. 116, pp. 1-26.
Briber, R. M. see Hedrick, J. L.: Vol. 141, pp. 1-44.
Bronnikov, S. V., Vettegren, V. I. and *Frenkel, S. Y.:* Kinetics of Deformation and Relaxation in Highly Oriented Polymers. Vol. 125, pp. 103-146.
Brown, H. R. see Creton, C.: Vol. 156, pp. 53-135.
Bruza, K. J. see Kirchhoff, R. A.: Vol. 117, pp. 1-66.
Budkowski, A.: Interfacial Phenomena in Thin Polymer Films: Phase Coexistence and Segregation. Vol. 148, pp. 1-112.
Burban, J. H. see Cussler, E. L.: Vol. 110, pp. 67-80.
Burchard,W.: Solution Properties of Branched Macromolecules. Vol. 143, pp. 113-194.

Calmon-Decriaud, A., Bellon-Maurel, V., Silvestre, F.: Standard Methods for Testing the Aerobic Biodegradation of Polymeric Materials.Vol 135, pp. 207-226.
Cameron, N. R. and *Sherrington, D. C.:* High Internal Phase Emulsions (HIPEs)-Structure, Properties and Use in Polymer Preparation.Vol. 126, pp. 163-214.

de la Campa, J. G. see de Abajo, J.: Vol. 140, pp. 23-60.
Candau, F. see Hunkeler, D.: Vol. 112, pp. 115-134.
Canelas, D. A. and *DeSimone, J. M.*: Polymerizations in Liquid and Supercritical Carbon Dioxide. Vol. 133, pp. 103-140.
Canva, M., Stegeman, G. I.: Quadratic Parametric Interactions in Organic Waveguides. Vol. 158, pp. 87-121.
Capek, I.: Kinetics of the Free-Radical Emulsion Polymerization of Vinyl Chloride. Vol. 120, pp. 135-206.
Capek, I.: Radical Polymerization of Polyoxyethylene Macromonomers in Disperse Systems. Vol. 145, pp. 1-56.
Capek, I.: Radical Polymerization of Polyoxyethylene Macromonomers in Disperse Systems. Vol. 146, pp. 1-56.
Capek, I. and *Chern, C.-S.*: Radical Polymerization in Direct Mini-Emulsion Systems. Vol. 155, pp. 101-166.
Cappella, B. see Munz, M.: Vol. 164, pp. 87-210.
Carlesso, G. see Prokop, A.: Vol. 160, pp. 119-174.
Carlini, C. and *Angiolini, L.*: Polymers as Free Radical Photoinitiators. Vol. 123, pp. 127-214.
Carter, K. R. see Hedrick, J. L.: Vol. 141, pp. 1-44.
Casas-Vazquez, J. see Jou, D.: Vol. 120, pp. 207-266.
Chandrasekhar, V.: Polymer Solid Electrolytes: Synthesis and Structure. Vol 135, pp. 139-206.
Chang, J. Y. see Han, M. J.: Vol. 153, pp. 1-36.
Chang, T.: Recent Advances in Liquid Chromatography Analysis of Synthetic Polymers. Vol. 163, pp. 1-60.
Charleux, B., Faust R.: Synthesis of Branched Polymers by Cationic Polymerization. Vol. 142, pp. 1-70.
Chen, P. see Jaffe, M.: Vol. 117, pp. 297-328.
Chern, C.-S. see Capek, I.: Vol. 155, pp. 101-166.
Chevolot, Y. see Mathieu, H. J.: Vol. 162, pp. 1-35.
Choe, E.-W. see Jaffe, M.: Vol. 117, pp. 297-328.
Chow, T. S.: Glassy State Relaxation and Deformation in Polymers. Vol. 103, pp. 149-190.
Chujo, Y. see Uemura, T.: Vol. 167, pp. 81-106.
Chung, S.-J. see Lin, T.-C.: Vol. 161, pp. 157-193
Chung, T.-S. see Jaffe, M.: Vol. 117, pp. 297-328.
Cölfen, H. and *Antonietti, M.*: Field-Flow Fractionation Techniques for Polymer and Colloid Analysis. Vol. 150, pp. 67-187.
Comanita, B. see Roovers, J.: Vol. 142, pp. 179-228.
Connell, J. W. see Hergenrother, P. M.: Vol. 117, pp. 67-110.
Creton, C., Kramer, E. J., Brown, H. R., Hui, C.-Y.: Adhesion and Fracture of Interfaces Between Immiscible Polymers: From the Molecular to the Continuum Scale. Vol. 156, pp. 53-135.
Criado-Sancho, M. see Jou, D.: Vol. 120, pp. 207-266.
Curro, J. G. see Schweizer, K. S.: Vol. 116, pp. 319-378.
Curtiss, C. F. and *Bird, R. B.*: Statistical Mechanics of Transport Phenomena: Polymeric Liquid Mixtures. Vol. 125, pp. 1-102.
Cussler, E. L., Wang, K. L. and *Burban, J. H.*: Hydrogels as Separation Agents. Vol. 110, pp. 67-80.

Dalton, L. Nonlinear Optical Polymeric Materials: From Chromophore Design to Commercial Applications. Vol. 158, pp. 1-86.
Dautzenberg, H. see Holm, C.: Vol. 166, pp.113-171.
Davidson, J. M. see Prokop, A.: Vol. 160, pp.119-174.
Desai, S. M., Singh, R. P.: Surface Modification of Polyethylene. Vol. 169, pp. 231-293.
DeSimone, J. M. see Canelas D. A.: Vol. 133, pp. 103-140.
DiMari, S. see Prokop, A.: Vol. 136, pp. 1-52.
Dimonie, M. V. see Hunkeler, D.: Vol. 112, pp. 115-134.
Dingenouts, N., Bolze, J., Pötschke, D., Ballauf, M.: Analysis of Polymer Latexes by Small-Angle X-Ray Scattering. Vol. 144, pp. 1-48.

Dodd, L. R. and *Theodorou, D. N.*: Atomistic Monte Carlo Simulation and Continuum Mean Field Theory of the Structure and Equation of State Properties of Alkane and Polymer Melts. Vol. 116, pp. 249-282.
Doelker, E.: Cellulose Derivatives. Vol. 107, pp. 199-266.
Dolden, J. G.: Calculation of a Mesogenic Index with Emphasis Upon LC-Polyimides.Vol. 141, pp. 189 -245.
Domb, A. J., Amselem, S., Shah, J. and Maniar, M.: Polyanhydrides: Synthesis and Characterization. Vol. 107, pp. 93-142.
Domb, A. J. see Kumar, M. N. V. R.: Vol. 160, pp. 45118.
Doruker, P. see Baschnagel, J.: Vol. 152, pp. 41-156.
Dubois, P. see Mecerreyes, D.: Vol. 147, pp. 1-60.
Dubrovskii, S. A. see Kazanskii, K. S.: Vol. 104, pp. 97-134.
Dunkin, I. R. see Steinke, J.: Vol. 123, pp. 81-126.
Dunson, D. L. see McGrath, J. E.: Vol. 140, pp. 61-106.
Dziezok, P. see Rühe, J.: Vol. 165, pp. 79-150.

Eastmond, G. C.: Poly(ε-caprolactone) Blends. Vol. 149, pp. 59-223.
Economy, J. and *Goranov, K.*: Thermotropic Liquid Crystalline Polymers for High Performance Applications. Vol. 117, pp. 221-256.
Ediger, M. D. and *Adolf, D. B.*: Brownian Dynamics Simulations of Local Polymer Dynamics. Vol. 116, pp. 73-110.
Edlund, U. Albertsson, A.-C.: Degradable Polymer Microspheres for Controlled Drug Delivery. Vol. 157, pp. 53-98.
Edwards, S. F. see Aharoni, S. M.: Vol. 118, pp. 1-231.
Eisenbach, C. D. see Bohrisch, J.: Vol. 165, pp. 1-41.
Endo, T. see Yagci, Y.: Vol. 127, pp. 59-86.
Engelhardt, H. and *Grosche, O.*: Capillary Electrophoresis in Polymer Analysis. Vol.150, pp. 189-217.
Engelhardt, H. and Martin, H.: Characterization of Synthetic Polyelectrolytes by Capillary Electrophoretic Methods. Vol. 165, pp. 211-247.
Eriksson, P. see Jacobson, K.: Vol. 169, pp. 151-176.
Erman, B. see Bahar, I.: Vol. 116, pp. 145-206.
Eschner, M. see Spange, S.: Vol. 165, pp. 43-78.
Estel, K. see Spange, S.: Vol. 165, pp. 43-78.
Ewen, B, Richter, D.: Neutron Spin Echo Investigations on the Segmental Dynamics of Polymers in Melts, Networks and Solutions. Vol. 134, pp. 1-130.
Ezquerra, T. A. see Baltá-Calleja, F. J.: Vol. 108, pp. 1-48.

Fatkullin, N. see Kimmich, R: Vol. 170, pp. 1-113.
Faust, R. see Charleux, B: Vol. 142, pp. 1-70.
Faust, R. see Kwon, Y.: Vol. 167, pp. 107-135.
Fekete, E. see Pukánszky, B: Vol. 139, pp. 109-154.
Fendler, J. H.: Membrane-Mimetic Approach to Advanced Materials. Vol. 113, pp. 1-209.
Fetters, L. J. see Xu, Z.: Vol. 120, pp. 1-50.
Förster, S., Abetz, V., Müller, A. H. E.: Polyelectrolyte Block Copolymer Micelles. Vol. 166, pp. 173-210.
Förster, S. and *Schmidt, M.*: Polyelectrolytes in Solution. Vol. 120, pp. 51-134.
Freire, J. J.: Conformational Properties of Branched Polymers: Theory and Simulations. Vol. 143, pp. 35-112.
Frenkel, S. Y. see Bronnikov, S.V.: Vol. 125, pp. 103-146.
Frick, B. see Baltá-Calleja, F. J.: Vol. 108, pp. 1-48.
Fridman, M. L.: see Terent'eva, J. P.: Vol. 101, pp. 29-64.
Fukui, K. see Otaigbe, J. U.: Vol. 154, pp. 1-86.
Funke, W.: Microgels-Intramolecularly Crosslinked Macromolecules with a Globular Structure. Vol. 136, pp. 137-232.
Furusho, Y. see Takata, T.: Vol. 171, pp. 1-75.

Galina, H.: Mean-Field Kinetic Modeling of Polymerization: The Smoluchowski Coagulation Equation.Vol. 137, pp. 135-172.
Ganesh, K. see Kishore, K.: Vol. 121, pp. 81-122.
Gaw, K. O. and Kakimoto, M.: Polyimide-Epoxy Composites. Vol. 140, pp. 107-136.
Geckeler, K. E. see Rivas, B.: Vol. 102, pp. 171-188.
Geckeler, K. E.: Soluble Polymer Supports for Liquid-Phase Synthesis. Vol. 121, pp. 31-80.
Gedde, U. W., Mattozzi, A.: Polyethylene Morphology. Vol. 169, pp. 29-73.
Gehrke, S. H.: Synthesis, Equilibrium Swelling, Kinetics Permeability and Applications of Environmentally Responsive Gels. Vol. 110, pp. 81-144.
de Gennes, P.-G.: Flexible Polymers in Nanopores. Vol. 138, pp. 91-106.
Georgiou, S.: Laser Cleaning Methodologies of Polymer Substrates. Vol. 168, pp. 1-49.
Geuss, M. see Munz, M.: Vol. 164, pp. 87-210
Giannelis, E. P., Krishnamoorti, R., Manias, E.: Polymer-Silicate Nanocomposites: Model Systems for Confined Polymers and Polymer Brushes. Vol. 138, pp. 107-148.
Godovsky, D. Y.: Device Applications of Polymer-Nanocomposites. Vol. 153, pp. 163-205.
Godovsky, D. Y.: Electron Behavior and Magnetic Properties Polymer-Nanocomposites. Vol. 119, pp. 79-122.
González Arche, A. see Baltá-Calleja, F. J.: Vol. 108, pp. 1-48.
Goranov, K. see Economy, J.: Vol. 117, pp. 221-256.
Gramain, P. see Améduri, B.: Vol. 127, pp. 87-142.
Grest, G. S.: Normal and Shear Forces Between Polymer Brushes. Vol. 138, pp. 149-184.
Grigorescu, G, Kulicke, W.-M.: Prediction of Viscoelastic Properties and Shear Stability of Polymers in Solution. Vol. 152, p. 1-40.
Gröhn, F. see Rühe, J.: Vol. 165, pp. 79-150.
Grosberg, A. and Nechaev, S.: Polymer Topology. Vol. 106, pp. 1-30.
Grosche, O. see Engelhardt, H.: Vol. 150, pp. 189-217.
Grubbs, R., Risse, W. and Novac, B.: The Development ofWell-defined Catalysts for Ring-Opening Olefin Metathesis. Vol. 102, pp. 47-72.
Gubler, U., Bosshard, C.: Molecular Design for Third-Order Nonlinear Optics. Vol. 158, pp. 123-190.
van Gunsteren, W. F. see Gusev, A. A.: Vol. 116, pp. 207-248.
Gupta, B., Anjum, N.: Plasma and Radiation-Induced Graft Modification of Polymers for Biomedical Applications. Vol. 162, pp. 37-63.
Gusev, A. A., Müller-Plathe, F., van Gunsteren, W. F. and Suter, U. W.: Dynamics of Small Molecules in Bulk Polymers. Vol. 116, pp. 207-248.
Gusev, A. A. see Baschnagel, J.: Vol. 152, pp. 41-156.
Guillot, J. see Hunkeler, D.: Vol. 112, pp. 115-134.
Guyot, A. and Tauer, K.: Reactive Surfactants in Emulsion Polymerization. Vol. 111, pp. 43-66.

Hadjichristidis, N., Pispas, S., Pitsikalis, M., Iatrou, H., Vlahos, C.: Asymmetric Star Polymers Synthesis and Properties. Vol. 142, pp. 71-128.
Hadjichristidis, N. see Xu, Z.: Vol. 120, pp. 1-50.
Hadjichristidis, N. see Pitsikalis, M.: Vol. 135, pp. 1-138.
Hahn, O. see Baschnagel, J.: Vol. 152, pp. 41-156.
Hakkarainen, M.: Aliphatic Polyesters: Abiotic and Biotic Degradation and Degradation Products. Vol. 157, pp. 1-26.
Hakkarainen, M., Albertsson, A.-C.: Environmental Degradation of Polyethylene. Vol. 169, pp. 177-199.
Hall, H. K. see Penelle, J.: Vol. 102, pp. 73-104.
Hamley, I.W.: Crystallization in Block Copolymers. Vol. 148, pp. 113-138.
Hammouda, B.: SANS from Homogeneous Polymer Mixtures: A Unified Overview. Vol. 106, pp. 87-134.
Han, M. J. and Chang, J. Y.: Polynucleotide Analogues. Vol. 153, pp. 1-36.
Harada, A.: Design and Construction of Supramolecular Architectures Consisting of Cyclodextrins and Polymers. Vol. 133, pp. 141-192.
Haralson, M. A. see Prokop, A.: Vol. 136, pp. 1-52.

Hassan, C. M. and *Peppas, N. A.*: Structure and Applications of Poly(vinyl alcohol) Hydrogels Produced by Conventional Crosslinking or by Freezing/Thawing Methods. Vol. 153, pp. 37-65.
Hawker, C. J.: Dentritic and Hyperbranched Macromolecules Precisely Controlled Macromolecular Architectures. Vol. 147, pp. 113-160.
Hawker, C. J. see Hedrick, J. L.: Vol. 141, pp. 1-44.
He, G. S. see Lin, T.-C.: Vol. 161, pp. 157-193.
Hedrick, J. L., Carter, K. R., Labadie, J. W., Miller, R. D., Volksen, W., Hawker, C. J., Yoon, D. Y., Russell, T. P., McGrath, J. E., Briber, R. M.: Nanoporous Polyimides. Vol. 141, pp. 1-44.
Hedrick, J. L., Labadie, J. W., Volksen, W. and *Hilborn, J. G.*: Nanoscopically Engineered Polyimides. Vol. 147, pp. 61-112.
Hedrick, J. L. see Hergenrother, P. M.: Vol. 117, pp. 67-110.
Hedrick, J. L. see Kiefer, J.: Vol. 147, pp. 161-247.
Hedrick, J. L. see McGrath, J. E.: Vol. 140, pp. 61-106.
Heinrich, G. and *Klüppel, M.*: Recent Advances in the Theory of Filler Networking in Elastomers. Vol. 160, pp. 1-44.
Heller, J.: Poly (Ortho Esters). Vol. 107, pp. 41-92.
Helm, C. A.: see Möhwald, H.: Vol. 165, pp. 151-175.
Hemielec, A. A. see Hunkeler, D.: Vol. 112, pp. 115-134.
Hergenrother, P. M., Connell, J. W., Labadie, J. W. and *Hedrick, J. L.*: Poly(arylene ether)s Containing Heterocyclic Units. Vol. 117, pp. 67-110.
Hernández-Barajas, J. see Wandrey, C.: Vol. 145, pp. 123-182.
Hervet, H. see Léger, L.: Vol. 138, pp. 185-226.
Hilborn, J. G. see Hedrick, J. L.: Vol. 147, pp. 61-112.
Hilborn, J. G. see Kiefer, J.: Vol. 147, pp. 161-247.
Hiramatsu, N. see Matsushige, M.: Vol. 125, pp. 147-186.
Hirasa, O. see Suzuki, M.: Vol. 110, pp. 241-262.
Hirotsu, S.: Coexistence of Phases and the Nature of First-Order Transition in Poly-N-isopropylacrylamide Gels. Vol. 110, pp. 1-26.
Höcker, H. see Klee, D.: Vol. 149, pp. 1-57.
Holm, C., Hofmann, T., Joanny, J. F., Kremer, K., Netz, R. R., Reineker, P., Seidel, C., Vilgis, T. A., Winkler, R. G.: Polyelectrolyte Theory. Vol. 166, pp. 67-111.
Holm, C., Rehahn, M., Oppermann, W., Ballauff, M.: Stiff-Chain Polyelectrolytes. Vol. 166, pp. 1-27.
Hornsby, P.: Rheology, Compoundind and Processing of Filled Thermoplastics. Vol. 139, pp. 155-216.
Houbenov, N. see Rühe, J.: Vol. 165, pp. 79-150.
Huber, K. see Volk, N.: Vol. 166, pp. 29-65.
Hugenberg, N. see Rühe, J.: Vol. 165, pp. 79-150.
Hui, C.-Y. see Creton, C.: Vol. 156, pp. 53-135.
Hult, A., Johansson, M., Malmström, E.: Hyperbranched Polymers.Vol. 143, pp. 1-34.
Hunkeler, D., Candau, F., Pichot, C., Hemielec, A. E., Xie, T. Y., Barton, J., Vaskova, V., Guillot, J., Dimonie, M. V., Reichert, K. H.: Heterophase Polymerization: A Physical and Kinetic Comparision and Categorization. Vol. 112, pp. 115-134.
Hunkeler, D. see Macko, T.: Vol. 163, pp. 61-136.
Hunkeler, D. see Prokop, A.: Vol. 136, pp. 1-52; 53-74.
Hunkeler, D see Wandrey, C.: Vol. 145, pp. 123-182.

Iatrou, H. see Hadjichristidis, N.: Vol. 142, pp. 71-128.
Ichikawa, T. see Yoshida, H.: Vol. 105, pp. 3-36.
Ihara, E. see Yasuda, H.: Vol. 133, pp. 53-102.
Ikada, Y. see Uyama,Y.: Vol. 137, pp. 1-40.
Ikehara, T. see Jinnuai, H.: Vol. 170, pp. 115-167.
Ilavsky, M.: Effect on Phase Transition on Swelling and Mechanical Behavior of Synthetic Hydrogels. Vol. 109, pp. 173-206.
Imai, Y.: Rapid Synthesis of Polyimides from Nylon-Salt Monomers. Vol. 140, pp. 1-23.

Inomata, H. see *Saito, S.*: Vol. 106, pp. 207-232.
Inoue, S. see *Sugimoto, H.*: Vol. 146, pp. 39-120.
Irie, M.: Stimuli-Responsive Poly(N-isopropylacrylamide), Photo- and Chemical-Induced Phase Transitions. Vol. 110, pp. 49-66.
Ise, N. see *Matsuoka, H.*: Vol. 114, pp. 187-232.
Ito, K., Kawaguchi, S.: Poly(macronomers), Homo- and Copolymerization. Vol. 142, pp. 129-178.
Ito, Y. see *Suginome, M.*: Vol. 171, pp. 77-136.
Ivanov, A. E. see *Zubov, V. P.*: Vol. 104, pp. 135-176.

Jacob, S. and *Kennedy, J.*: Synthesis, Characterization and Properties of OCTA-ARM Polyisobutylene-Based Star Polymers. Vol. 146, pp. 1-38.
Jacobson, K., Eriksson, P., Reitberger, T., Stenberg, B.: Chemiluminescence as a Tool for Polyolefin. Vol. 169, pp. 151-176.
Jaeger, W. see *Bohrisch, J.*: Vol. 165, pp. 1-41.
Jaffe, M., Chen, P., Choe, E.-W., Chung, T.-S. and *Makhija, S.*: High Performance Polymer Blends. Vol. 117, pp. 297-328.
Jancar, J.: Structure-Property Relationships in Thermoplastic Matrices. Vol. 139, pp. 1-66.
Jen, A. K-Y. see *Kajzar, F.*: Vol. 161, pp. 1-85.
Jerome, R. see *Mecerreyes, D.*: Vol. 147, pp. 1-60.
Jiang, M., Li, M., Xiang, M. and *Zhou, H.*: Interpolymer Complexation and Miscibility and Enhancement by Hydrogen Bonding. Vol. 146, pp. 121-194.
Jin, J. see *Shim, H.-K.*: Vol. 158, pp. 191-241.
Jinnai, H., Nishikawa, Y., Ikehara, T. and *Nishi, T.*: Emerging Technologies for the 3D Analysis of Polymer Structures. Vol. 170, pp. 115-167.
Jo, W. H. and *Yang, J. S.*: Molecular Simulation Approaches for Multiphase Polymer Systems. Vol. 156, pp. 1-52.
Joanny, J.-F. see *Holm, C.*: Vol. 166, pp. 67-111.
Joanny, J.-F. see *Thünemann, A. F.*: Vol. 166, pp. 113-171.
Johannsmann, D. see *Rühe, J.*: Vol. 165, pp. 79-150.
Johansson, M. see *Hult, A.*: Vol. 143, pp. 1-34.
Joos-Müller, B. see *Funke, W.*: Vol. 136, pp. 137-232.
Jou, D., Casas-Vazquez, J. and *Criado-Sancho, M.*: Thermodynamics of Polymer Solutions under Flow: Phase Separation and Polymer Degradation. Vol. 120, pp. 207-266.

Kaetsu, I.: Radiation Synthesis of Polymeric Materials for Biomedical and Biochemical Applications. Vol. 105, pp. 81-98.
Kaji, K. see *Kanaya, T.*: Vol. 154, pp. 87-141.
Kajzar, F., Lee, K.-S., Jen, A. K.-Y.: Polymeric Materials and their Orientation Techniques for Second-Order Nonlinear Optics. Vol. 161, pp. 1-85.
Kakimoto, M. see *Gaw, K. O.*: Vol. 140, pp. 107-136.
Kaminski, W. and *Arndt, M.*: Metallocenes for Polymer Catalysis. Vol. 127, pp. 143-187.
Kammer, H. W., Kressler, H. and *Kummerloewe, C.*: Phase Behavior of Polymer Blends - Effects of Thermodynamics and Rheology. Vol. 106, pp. 31-86.
Kanaya, T. and *Kaji, K.*: Dynamcis in the Glassy State and Near the Glass Transition of Amorphous Polymers as Studied by Neutron Scattering. Vol. 154, pp. 87-141.
Kandyrin, L. B. and *Kuleznev, V. N.*: The Dependence of Viscosity on the Composition of Concentrated Dispersions and the Free Volume Concept of Disperse Systems. Vol. 103, pp. 103-148.
Kaneko, M. see *Ramaraj, R.*: Vol. 123, pp. 215-242.
Kang, E. T., Neoh, K. G. and *Tan, K. L.*: X-Ray Photoelectron Spectroscopic Studies of Electroactive Polymers. Vol. 106, pp. 135-190.
Karlsson, S. see *Söderqvist Lindblad, M.*: Vol. 157, pp. 139-161.
Karlsson, S.: Recycled Polyolefins. Material Properties and Means for Quality Determination. Vol. 169, pp. 201-229.
Kato, K. see *Uyama, Y.*: Vol. 137, pp. 1-40.
Kautek, W. see *Krüger, J.*: Vol. 168, pp. 247-290.
Kawaguchi, S. see *Ito, K.*: Vol. 142, p 129-178.

Kawata, S. see Sun, H-B: Vol. 170, pp. 169-273.
Kazanskii, K. S. and *Dubrovskii, S. A.*: Chemistry and Physics of Agricultural Hydrogels. Vol. 104, pp. 97-134.
Kennedy, J. P. see Jacob, S.: Vol. 146, pp. 1-38.
Kennedy, J. P. see Majoros, I.: Vol. 112, pp. 1-113.
Khokhlov, A., Starodybtzev, S. and *Vasilevskaya, V.*: Conformational Transitions of Polymer Gels: Theory and Experiment. Vol. 109, pp. 121-172.
Kiefer, J., Hedrick J. L. and *Hiborn, J. G.*: Macroporous Thermosets by Chemically Induced Phase Separation. Vol. 147, pp. 161-247.
Kihara, N. see Takata, T.: Vol. 171, pp. 1-75.
Kilian, H. G. and *Pieper, T.*: Packing of Chain Segments. A Method for Describing X-Ray Patterns of Crystalline, Liquid Crystalline and Non-Crystalline Polymers. Vol. 108, pp. 49-90.
Kim, J. see Quirk, R.P.: Vol. 153, pp. 67-162.
Kim, K.-S. see Lin, T.-C.: Vol. 161, pp. 157-193.
Kimmich, R., Fatkullin, N.: Polymer Chain Dynamics and NMR. Vol. 170, pp. 1–113.
Kippelen, B. and *Peyghambarian, N.*: Photorefractive Polymers and their Applications. Vol. 161, pp. 87-156.
Kishore, K. and *Ganesh, K.*: Polymers Containing Disulfide, Tetrasulfide, Diselenide and Ditelluride Linkages in the Main Chain. Vol. 121, pp. 81-122.
Kitamaru, R.: Phase Structure of Polyethylene and Other Crystalline Polymers by Solid-State 13C/MNR. Vol. 137, pp 41-102.
Klee, D. and *Höcker, H.*: Polymers for Biomedical Applications: Improvement of the Interface Compatibility. Vol. 149, pp. 1-57.
Klier, J. see Scranton, A. B.: Vol. 122, pp. 1-54.
v. Klitzing, R. and *Tieke, B.*: Polyelectrolyte Membranes. Vol. 165, pp. 177-210.
Klüppel, M.: The Role of Disorder in Filler Reinforcement of Elastomers on Various Length Scales. Vol. 164, pp. 1-86
Klüppel, M. see Heinrich, G.: Vol. 160, pp 1-44.
Knuuttila, H., Lehtinen, A., Nummila-Pakarinen, A.: Advanced Polyethylene Technologies – Controlled Material Properties. Vol. 169, pp. 13-27.
Kobayashi, S., Shoda, S. and *Uyama, H.*: Enzymatic Polymerization and Oligomerization. Vol. 121, pp. 1-30.
Köhler, W. and *Schäfer, R.*: Polymer Analysis by Thermal-Diffusion Forced Rayleigh Scattering. Vol. 151, pp. 1-59.
Koenig, J. L. see Bhargava, R.: Vol. 163, pp. 137-191.
Koenig, J. L. see Andreis, M.: Vol. 124, pp. 191-238.
Koike, T.: Viscoelastic Behavior of Epoxy Resins Before Crosslinking. Vol. 148, pp. 139-188.
Kokko, E. see Löfgren, B.: Vol. 169, pp. 1-12.
Kokufuta, E.: Novel Applications for Stimulus-Sensitive Polymer Gels in the Preparation of Functional Immobilized Biocatalysts. Vol. 110, pp. 157-178.
Konno, M. see Saito, S.: Vol. 109, pp. 207-232.
Konradi, R. see Rühe, J.: Vol. 165, pp. 79-150.
Kopecek, J. see Putnam, D.: Vol. 122, pp. 55-124.
Koßmehl, G. see Schopf, G.: Vol. 129, pp. 1-145.
Kozlov, E. see Prokop, A.: Vol. 160, pp. 119-174.
Kramer, E. J. see Creton, C.: Vol. 156, pp. 53-135.
Kremer, K. see Baschnagel, J.: Vol. 152, pp. 41-156.
Kremer, K. see Holm, C.: Vol. 166, pp. 67-111.
Kressler, J. see Kammer, H. W.: Vol. 106, pp. 31-86.
Kricheldorf, H. R.: Liquid-Cristalline Polyimides. Vol. 141, pp. 83-188.
Krishnamoorti, R. see Giannelis, E. P.: Vol. 138, pp. 107-148.
Kirchhoff, R. A. and *Bruza, K. J.*: Polymers from Benzocyclobutenes. Vol. 117, pp. 1-66.
Krüger, J. and *Kautek, W.*: Ultrashort Pulse Laser Interaction with Dielectrics and Polymers, Vol. 168, pp. 247-290.
Kuchanov, S. I.: Modern Aspects of Quantitative Theory of Free-Radical Copolymerization. Vol. 103, pp. 1-102.

Kuchanov, S. I.: Principles of Quantitive Description of Chemical Structure of Synthetic Polymers. Vol. 152, p. 157-202.
Kudaibergennow, S. E.: Recent Advances in Studying of Synthetic Polyampholytes in Solutions. Vol. 144, pp. 115-198.
Kuleznev, V. N. see Kandyrin, L. B.: Vol. 103, pp. 103-148.
Kulichkhin, S. G. see Malkin, A. Y.: Vol. 101, pp. 217-258.
Kulicke, W.-M. see Grigorescu, G.: Vol. 152, p. 1-40.
Kumar, M. N. V. R., Kumar, N., Domb, A. J. and *Arora, M.:* Pharmaceutical Polymeric Controlled Drug Delivery Systems. Vol. 160, pp. 45-118.
Kumar, N. see Kumar M. N. V. R.: Vol. 160, pp. 45-118.
Kummerloewe, C. see Kammer, H. W.: Vol. 106, pp. 31-86.
Kuznetsova, N. P. see Samsonov, G.V.: Vol. 104, pp. 1-50.
Kwon, Y. and *Faust, R.:* Synthesis of Polyisobutylene-Based Block Copolymers with Precisely Controlled Architecture by Living Cationic Polymerization. Vol. 167, pp. 107-135.

Labadie, J. W. see Hergenrother, P. M.: Vol. 117, pp. 67-110.
Labadie, J. W. see Hedrick, J. L.: Vol. 141, pp. 1-44.
Labadie, J. W. see Hedrick, J. L.: Vol. 147, pp. 61-112.
Lamparski, H. G. see O'Brien, D. F.: Vol. 126, pp. 53-84.
Laschewsky, A.: Molecular Concepts, Self-Organisation and Properties of Polysoaps. Vol. 124, pp. 1-86.
Laso, M. see Leontidis, E.: Vol. 116, pp. 283-318.
Lazár, M. and *Rychl, R.:* Oxidation of Hydrocarbon Polymers. Vol. 102, pp. 189-222.
Lechowicz, J. see Galina, H.: Vol. 137, pp. 135-172.
Léger, L., Raphaël, E., Hervet, H.: Surface-Anchored Polymer Chains: Their Role in Adhesion and Friction. Vol. 138, pp. 185-226.
Lenz, R. W.: Biodegradable Polymers. Vol. 107, pp. 1-40.
Leontidis, E., de Pablo, J. J., Laso, M. and *Suter, U. W.:* A Critical Evaluation of Novel Algorithms for the Off-Lattice Monte Carlo Simulation of Condensed Polymer Phases. Vol. 116, pp. 283-318.
Lee, B. see Quirk, R. P.: Vol. 153, pp. 67-162.
Lee, K.-S. see Kajzar, F.: Vol. 161, pp. 1-85.
Lee, Y. see Quirk, R. P: Vol. 153, pp. 67-162.
Lehtinen, A. see Knuuttila, H.: Vol. 169, pp. 13-27.
Leónard, D. see Mathieu, H. J.: Vol. 162, pp. 1-35.
Lesec, J. see Viovy, J.-L.: Vol. 114, pp. 1-42.
Li, M. see Jiang, M.: Vol. 146, pp. 121-194.
Liang, G. L. see Sumpter, B. G.: Vol. 116, pp. 27-72.
Lienert, K.-W.: Poly(ester-imide)s for Industrial Use. Vol. 141, pp. 45-82.
Lin, J. and *Sherrington, D. C.:* Recent Developments in the Synthesis, Thermostability and Liquid Crystal Properties of Aromatic Polyamides. Vol. 111, pp. 177-220.
Lin, T.-C., Chung, S.-J., Kim, K.-S., Wang, X., He, G. S., Swiatkiewicz, J., Pudavar, H. E. and *Prasad, P. N.:* Organics and Polymers with High Two-Photon Activities and their Applications. Vol. 161, pp. 157-193.
Lippert, T.: Laser Application of Polymers. Vol. 168, pp. 51-246.
Liu, Y. see Söderqvist Lindblad, M.: Vol. 157, pp. 139161
López Cabarcos, E. see Baltá-Calleja, F. J.: Vol. 108, pp. 1-48.
Löfgren, B., Kokko, E., Seppälä, J.: Specific Structures Enabled by Metallocene Catalysis in Polyethenes. Vol. 169, pp. 1-12.
Löwen, H. see Thünemann, A. F.: Vol. 166, pp. 113-171.

Macko, T. and *Hunkeler, D.:* Liquid Chromatography under Critical and Limiting Conditions: A Survey of Experimental Systems for Synthetic Polymers. Vol. 163, pp. 61-136.
Majoros, I., Nagy, A. and *Kennedy, J. P.:* Conventional and Living Carbocationic Polymerizations United. I.A Comprehensive Model and New Diagnostic Method to Probe the Mechanism of Homopolymerizations. Vol. 112, pp. 1-113.

Makhija, S. see Jaffe, M.: Vol. 117, pp. 297-328.
Malmström, E. see Hult, A.: Vol. 143, pp. 1-34.
Malkin, A. Y. and *Kulichkhin, S. G.*: Rheokinetics of Curing. Vol. 101, pp. 217-258.
Maniar, M. see Domb, A. J.: Vol. 107, pp. 93-142.
Manias, E. see Giannelis, E. P.: Vol. 138, pp. 107-148.
Martin, H. see Engelhardt, H.: Vol. 165, pp. 211-247.
Mashima, K., Nakayama, Y. and *Nakamura, A.*: Recent Trends in Polymerization of a-Olefins Catalyzed by Organometallic Complexes of Early Transition Metals.Vol. 133, pp. 1-52.
Mathew, D. see Reghunadhan Nair, C.P.: Vol. 155, pp. 1-99.
Mathieu, H. J., Chevolot, Y, Ruiz-Taylor, L. and Leónard, D.: Engineering and Characterization of Polymer Surfaces for Biomedical Applications. Vol. 162, pp. 1-35.
Matsumoto, A.: Free-Radical Crosslinking Polymerization and Copolymerization of Multivinyl Compounds. Vol. 123, pp. 41-80.
Matsumoto, A. see Otsu, T.: Vol. 136, pp. 75-138.
Matsuoka, H. and *Ise, N.*: Small-Angle and Ultra-Small Angle Scattering Study of the Ordered Structure in Polyelectrolyte Solutions and Colloidal Dispersions. Vol. 114, pp. 187-232.
Matsushige, K., Hiramatsu, N. and *Okabe, H.*: Ultrasonic Spectroscopy for Polymeric Materials. Vol. 125, pp. 147-186.
Mattice, W. L. see Rehahn, M.: Vol. 131/132, pp. 1-475.
Mattice, W. L. see Baschnagel, J.: Vol. 152, pp. 41-156.
Mattozzi, A. see Gedde, U. W.: Vol. 169, pp. 29-73.
Mays, W. see Xu, Z.: Vol. 120, pp. 1-50.
Mays, J. W. see Pitsikalis, M.: Vol. 135, pp. 1-138.
McGrath, J. E. see Hedrick, J. L.: Vol. 141, pp. 1-44.
McGrath, J. E., Dunson, D. L., Hedrick, J. L.: Synthesis and Characterization of Segmented Polyimide-Polyorganosiloxane Copolymers. Vol. 140, pp. 61-106.
McLeish, T. C. B., Milner, S. T.: Entangled Dynamics and Melt Flow of Branched Polymers. Vol. 143, pp. 195-256.
Mecerreyes, D., Dubois, P. and *Jerome, R.*: Novel Macromolecular Architectures Based on Aliphatic Polyesters: Relevance of the Coordination-Insertion Ring-Opening Polymerization. Vol. 147, pp. 1-60.
Mecham, S. J. see McGrath, J. E.: Vol. 140, pp. 61-106.
Menzel, H. see Möhwald, H.: Vol. 165, pp. 151-175.
Meyer, T. see Spange, S.: Vol. 165, pp. 43-78.
Mikos, A. G. see Thomson, R. C.: Vol. 122, pp. 245-274.
Milner, S. T. see McLeish, T. C. B.: Vol. 143, pp. 195-256.
Mison, P. and *Sillion, B.*: Thermosetting Oligomers Containing Maleimides and Nadiimides End-Groups. Vol. 140, pp. 137-180.
Miyasaka, K.: PVA-Iodine Complexes: Formation, Structure and Properties. Vol. 108. pp. 91-130.
Miller, R. D. see Hedrick, J. L.: Vol. 141, pp. 1-44.
Minko, S. see Rühe, J.: Vol. 165, pp. 79-150.
Möhwald, H., Menzel, H., Helm, C. A., Stamm, M.: Lipid and Polyampholyte Monolayers to Study Polyelectrolyte Interactions and Structure at Interfaces. Vol. 165, pp. 151-175.
Monnerie, L. see Bahar, I.: Vol. 116, pp. 145-206.
Mori, H. see Bohrisch, J.: Vol. 165, pp. 1-41.
Morishima, Y.: Photoinduced Electron Transfer in Amphiphilic Polyelectrolyte Systems. Vol. 104, pp. 51-96.
Morton M. see Quirk, R. P: Vol. 153, pp. 67-162.
Motornov, M. see Rühe, J.: Vol. 165, pp. 79-150.
Mours, M. see Winter, H. H.: Vol. 134, pp. 165-234.
Müllen, K. see Scherf, U.: Vol. 123, pp. 1-40.
Müller, A.H.E. see Bohrisch, J.: Vol. 165, pp. 1-41.
Müller, A.H.E. see Förster, S.: Vol. 166, pp. 173-210.
Müller, M. see Thünemann, A. F.: Vol. 166, pp. 113-171.
Müller-Plathe, F. see Gusev, A. A.: Vol. 116, pp. 207-248.

Müller-Plathe, F. see Baschnagel, J.: Vol. 152, p. 41-156.
Mukerherjee, A. see Biswas, M.: Vol. 115, pp. 89-124.
Munz, M., Cappella, B., Sturm, H., Geuss, M., Schulz, E.: Materials Contrasts and Nanolithography Techniques in Scanning Force Microscopy (SFM) and their Application to Polymers and Polymer Composites. Vol. 164, pp. 87-210
Murat, M. see Baschnagel, J.: Vol. 152, p. 41-156.
Mylnikov, V.: Photoconducting Polymers. Vol. 115, pp. 1-88.

Nagy, A. see Majoros, I.: Vol. 112, pp. 1-11.
Naka, K. see Uemura, T.: Vol. 167, pp. 81-106.
Nakamura, A. see Mashima, K.: Vol. 133, pp. 1-52.
Nakayama, Y. see Mashima, K.: Vol. 133, pp. 1-52.
Narasinham, B., Peppas, N. A.: The Physics of Polymer Dissolution: Modeling Approaches and Experimental Behavior. Vol. 128, pp. 157-208.
Nechaev, S. see Grosberg, A.: Vol. 106, pp. 1-30.
Neoh, K. G. see Kang, E. T.: Vol. 106, pp. 135-190.
Netz, R.R. see Holm, C.: Vol. 166, pp. 67-111.
Netz, R.R. see Rühe, J.: Vol. 165, pp. 79-150.
Newman, S. M. see Anseth, K. S.: Vol. 122, pp. 177-218.
Nijenhuis, K. te: Thermoreversible Networks. Vol. 130, pp. 1-252.
Ninan, K. N. see Reghunadhan Nair, C.P.: Vol. 155, pp. 1-99.
Nishi, T. see Jinnai, H.: Vol. 170, pp. 115–167.
Nishikawa, Y. see Jinnai, H.: Vol. 170, pp. 115–167.
Noid, D. W. see Otaigbe, J. U.: Vol. 154, pp. 1-86.
Noid, D. W. see Sumpter, B. G.: Vol. 116, pp. 27-72.
Novac, B. see Grubbs, R.: Vol. 102, pp. 47-72.
Novikov, V. V. see Privalko, V. P.: Vol. 119, pp. 31-78.
Nummila-Pakarinen, A. see Knuuttila, H.: Vol. 169, pp. 13-27.

O'Brien, D. F., Armitage, B. A., Bennett, D. E. and *Lamparski, H. G.:* Polymerization and Domain Formation in Lipid Assemblies. Vol. 126, pp. 53-84.
Ogasawara, M.: Application of Pulse Radiolysis to the Study of Polymers and Polymerizations. Vol.105, pp. 37-80.
Okabe, H. see Matsushige, K.: Vol. 125, pp. 147-186.
Okada, M.: Ring-Opening Polymerization of Bicyclic and Spiro Compounds. Reactivities and Polymerization Mechanisms. Vol. 102, pp. 1-46.
Okano, T.: Molecular Design of Temperature-Responsive Polymers as Intelligent Materials. Vol. 110, pp. 179-198.
Okay, O. see Funke, W.: Vol. 136, pp. 137-232.
Onuki, A.: Theory of Phase Transition in Polymer Gels. Vol. 109, pp. 63-120.
Oppermann W. see Holm, C.: Vol. 166, pp. 1-27.
Oppermann W. see Volk, N.: Vol. 166, pp. 29-65.
Osad'ko, I. S.: Selective Spectroscopy of Chromophore Doped Polymers and Glasses. Vol. 114, pp. 123-186.
Osakada, K., Takeuchi, D.: Coordination Polymerization of Dienes, Allenes, and Methylenecycloalkanes. Vol. 171, pp. 137-194.
Otaigbe, J. U., Barnes, M. D., Fukui, K., Sumpter, B. G., Noid, D. W.: Generation, Characterization, and Modeling of Polymer Micro- and Nano-Particles. Vol. 154, pp. 1-86.
Otsu, T., Matsumoto, A.: Controlled Synthesis of Polymers Using the Iniferter Technique: Developments in Living Radical Polymerization. Vol. 136, pp. 75-138.

de Pablo, J. J. see Leontidis, E.: Vol. 116, pp. 283-318.
Padias, A. B. see Penelle, J.: Vol. 102, pp. 73-104.
Pascault, J.-P. see Williams, R. J. J.: Vol. 128, pp. 95-156.
Pasch, H.: Analysis of Complex Polymers by Interaction Chromatography. Vol. 128, pp. 1-46.
Pasch, H.: Hyphenated Techniques in Liquid Chromatography of Polymers. Vol. 150, pp. 1-66.

Paul, W. see Baschnagel, J.: Vol. 152, p. 41-156.
Penczek, P. see Batog, A. E.: Vol. 144, pp. 49-114.
Penczek, P. see Bogdal, D.: Vol. 163, pp. 193-263.
Penelle, J., Hall, H. K., Padias, A. B. and *Tanaka, H.*: Captodative Olefins in Polymer Chemistry. Vol. 102, pp. 73-104.
Peppas, N. A. see Bell, C. L.: Vol. 122, pp. 125-176.
Peppas, N. A. see Hassan, C. M.: Vol. 153, pp. 37-65
Peppas, N. A. see Narasimhan, B.: Vol. 128, pp. 157-208.
Pet'ko, I. P. see Batog, A. E.: Vol. 144, pp. 49-114.
Pheyghambarian, N. see Kippelen, B.: Vol. 161, pp. 87-156.
Pichot, C. see Hunkeler, D.: Vol. 112, pp. 115-134.
Pielichowski, J. see Bogdal, D.: Vol. 163, pp. 193-263.
Pieper, T. see Kilian, H. G.: Vol. 108, pp. 49-90.
Pispas, S. see Pitsikalis, M.: Vol. 135, pp. 1-138.
Pispas, S. see Hadjichristidis: Vol. 142, pp. 71-128.
Pitsikalis, M., Pispas, S., Mays, J. W., Hadjichristidis, N.: Nonlinear Block Copolymer Architectures. Vol. 135, pp. 1-138.
Pitsikalis, M. see Hadjichristidis: Vol. 142, pp. 71-128.
Pleul, D. see Spange, S.: Vol. 165, pp. 43-78.
Plummer, C. J. G.: Microdeformation and Fracture in Bulk Polyolefins. Vol. 169, pp. 75-119.
Pötschke, D. see Dingenouts, N.: Vol 144, pp. 1-48.
Pokrovskii, V. N.: The Mesoscopic Theory of the Slow Relaxation of Linear Macromolecules. Vol. 154, pp. 143-219.
Pospíšil, J.: Functionalized Oligomers and Polymers as Stabilizers for Conventional Polymers. Vol. 101, pp. 65-168.
Pospíšil, J.: Aromatic and Heterocyclic Amines in Polymer Stabilization. Vol. 124, pp. 87-190.
Powers, A. C. see Prokop, A.: Vol. 136, pp. 53-74.
Prasad, P. N. see Lin, T.-C.: Vol. 161, pp. 157-193.
Priddy, D. B.: Recent Advances in Styrene Polymerization.Vol. 111, pp. 67-114.
Priddy, D. B.: Thermal Discoloration Chemistry of Styrene-co-Acrylonitrile. Vol. 121, pp. 123-154.
Privalko, V. P. and *Novikov, V. V.*: Model Treatments of the Heat Conductivity of Heterogeneous Polymers.Vol. 119, pp 31-78.
Prociak, A see Bogdal, D.: Vol. 163, pp. 193-263
Prokop, A., Hunkeler, D., Powers, A. C., Whitesell, R. R., Wang, T. G.: Water Soluble Polymers for Immunoisolation II: Evaluation of Multicomponent Microencapsulation Systems. Vol. 136, pp. 53-74.
Prokop, A., Hunkeler, D., DiMari, S., Haralson, M. A., Wang, T. G.: Water Soluble Polymers for Immunoisolation I: Complex Coacervation and Cytotoxicity. Vol. 136, pp. 1-52.
Prokop, A., Kozlov, E., Carlesso, G and *Davidsen, J. M.*: Hydrogel-Based Colloidal Polymeric System for Protein and Drug Delivery: Physical and Chemical Characterization, Permeability Control and Applications. Vol. 160, pp. 119-174.
Pruitt, L. A.: The Effects of Radiation on the Structural and Mechanical Properties of Medical Polymers. Vol. 162, pp. 65-95.
Pudavar, H. E. see Lin, T.-C.: Vol. 161, pp. 157-193.
Pukánszky, B. and *Fekete, E.*: Adhesion and Surface Modification. Vol. 139, pp. 109 -154.
Putnam, D. and *Kopecek, J.*: Polymer Conjugates with Anticancer Acitivity. Vol. 122, pp. 55-124.

Quirk, R. P. and *Yoo, T., Lee, Y., M., Kim, J.* and *Lee, B.*: Applications of 1,1-Diphenylethylene Chemistry in Anionic Synthesis of Polymers with Controlled Structures. Vol. 153, pp. 67-162.

Ramaraj, R. and *Kaneko, M.*: Metal Complex in Polymer Membrane as a Model for Photosynthetic Oxygen Evolving Center. Vol. 123, pp. 215-242.
Rangarajan, B. see Scranton, A. B.: Vol. 122, pp. 1-54.
Ranucci, E. see Söderqvist Lindblad, M.: Vol. 157, pp. 139-161.
Raphaël, E. see Léger, L.: Vol. 138, pp. 185-226.

Reddinger, J. L. and *Reynolds, J. R.:* Molecular Engineering of p-Conjugated Polymers. Vol. 145, pp. 57-122.
Reghunadhan Nair, C. P., Mathew, D. and *Ninan, K. N.,* : Cyanate Ester Resins, Recent Developments. Vol. 155, pp. 1-99.
Reichert, K. H. see Hunkeler, D.: Vol. 112, pp. 115-134.
Rehahn, M., Mattice, W. L., Suter, U. W.: Rotational Isomeric State Models in Macromolecular Systems. Vol. 131/132, pp. 1-475.
Rehahn, M. see Bohrisch, J.: Vol. 165, pp. 1-41.
Rehahn, M. see Holm, C.: Vol. 166, pp. 1-27.
Reineker, P. see Holm, C.: Vol. 166, pp. 67-111.
Reitberger, T. see Jacobson, K.: Vol. 169, pp. 151-176.
Reynolds, J. R. see Reddinger, J. L.: Vol. 145, pp. 57-122.
Richter, D. see Ewen, B.: Vol. 134, pp.1-130.
Risse, W. see Grubbs, R.: Vol. 102, pp. 47-72.
Rivas, B. L. and *Geckeler, K. E.:* Synthesis and Metal Complexation of Poly(ethyleneimine) and Derivatives.Vol. 102, pp. 171-188.
Robin, J.J.: The Use of Ozone in the Synthesis of New Polymers and the Modification of Polymers. Vol. 167, pp. 35-79.
Robin, J. J. see Boutevin, B.: Vol. 102, pp. 105-132.
Roe, R.-J.: MD Simulation Study of Glass Transition and Short Time Dynamics in Polymer Liquids. Vol. 116, pp. 111-114.
Roovers, J., Comanita, B.: Dendrimers and Dendrimer-Polymer Hybrids. Vol. 142, pp 179-228.
Rothon, R. N.: Mineral Fillers in Thermoplastics: Filler Manufacture and Characterisation.Vol. 139, pp. 67-108.
Rozenberg, B. A. see Williams, R. J. J.: Vol. 128, pp. 95-156.
Rühe, J., Ballauff, M., Biesalski, M., Dziezok, P., Gröhn, F., Johannsmann, D., Houbenov, N., Hugenberg, N., Konradi, R., Minko, S., Motornov, M., Netz, R. R., Schmidt, M., Seidel, C., Stamm, M., Stephan, T., Usov, D. and *Zhang, H.:* Polyelectrolyte Brushes. Vol. 165, pp. 79-150.
Ruckenstein, E.: Concentrated Emulsion Polymerization. Vol. 127, pp. 1-58.
Ruiz-Taylor, L. see Mathieu, H. J.: Vol. 162, pp. 1-35.
Rusanov, A. L.: Novel Bis (Naphtalic Anhydrides) and Their Polyheteroarylenes with Improved Processability. Vol. 111, pp. 115-176.
Russel, T. P. see Hedrick, J. L.: Vol. 141, pp. 1-44.
Rychlý, J. see Lazár, M.: Vol. 102, pp. 189-222.
Ryner, M. see Stridsberg, K. M.: Vol. 157, pp. 2751.
Ryzhov, V. A. see Bershtein, V. A.: Vol. 114, pp. 43-122.

Sabsai, O. Y. see Barshtein, G. R.: Vol. 101, pp. 1-28.
Saburov, V. V. see Zubov, V. P.: Vol. 104, pp. 135-176.
Saito, S., Konno, M. and *Inomata, H.:* Volume Phase Transition of N-Alkylacrylamide Gels. Vol. 109, pp. 207-232.
Samsonov, G. V. and *Kuznetsova, N. P.:* Crosslinked Polyelectrolytes in Biology. Vol. 104, pp. 1-50.
Santa Cruz, C. see Baltá-Calleja, F. J.: Vol. 108, pp. 1-48.
Santos, S. see Baschnagel, J.: Vol. 152, p. 41-156.
Sato, T. and *Teramoto, A.:* Concentrated Solutions of Liquid-Christalline Polymers. Vol. 126, pp. 85-162.
Schaller, C. see Bohrisch, J.: Vol. 165, pp. 1-41.
Schäfer R. see Köhler, W.: Vol. 151, pp. 1-59.
Scherf, U. and *Müllen, K.:* The Synthesis of Ladder Polymers.Vol. 123, pp. 1-40.
Schmidt, M. see Förster, S.: Vol. 120, pp. 51-134.
Schmidt, M. see Rühe, J.: Vol. 165, pp. 79-150.
Schmidt, M. see Volk, N.: Vol. 166, pp. 29-65.
Scholz, M.: Effects of Ion Radiation on Cells and Tissues. Vol. 162, pp. 97-158.
Schopf, G. and *Koßmehl, G.:* Polythiophenes - Electrically Conductive Polymers. Vol. 129, pp. 1-145.
Schulz, E. see Munz, M.: Vol. 164, pp. 97-210.

Seppälä, J. see Löfgren, B.: Vol. 169, pp. 1-12.
Sturm, H. see Munz, M.: Vol. 164, pp. 87-210.
Schweizer, K. S.: Prism Theory of the Structure, Thermodynamics, and Phase Transitions of Polymer Liquids and Alloys. Vol. 116, pp. 319-378.
Scranton, A. B., Rangarajan, B. and *Klier, J.:* Biomedical Applications of Polyelectrolytes. Vol. 122, pp. 1-54.
Sefton, M. V. and *Stevenson, W. T. K.:* Microencapsulation of Live Animal Cells Using Polycrylates. Vol.107, pp. 143-198.
Seidel, C. see Holm, C.: Vol. 166, pp. 67-111.
Seidel, C. see Rühe, J.: Vol. 165, pp. 79-150.
Shamanin, V. V.: Bases of the Axiomatic Theory of Addition Polymerization. Vol. 112, pp. 135-180.
Sheiko, S. S.: Imaging of Polymers Using Scanning Force Microscopy: From Superstructures to Individual Molecules. Vol. 151, pp. 61-174.
Sherrington, D. C. see Cameron, N. R.,Vol. 126, pp. 163-214.
Sherrington, D. C. see Lin, J.: Vol. 111, pp. 177-220.
Sherrington, D. C. see Steinke, J.: Vol. 123, pp. 81-126.
Shibayama, M. see Tanaka, T.: Vol. 109, pp. 1-62.
Shiga, T.: Deformation and Viscoelastic Behavior of Polymer Gels in Electric Fields. Vol. 134, pp. 131-164.
Shim, H.-K., Jin, J.: Light-Emitting Characteristics of Conjugated Polymers. Vol. 158, pp. 191-241.
Shoda, S. see Kobayashi, S.: Vol. 121, pp. 1-30.
Siegel, R. A.: Hydrophobic Weak Polyelectrolyte Gels: Studies of Swelling Equilibria and Kinetics. Vol. 109, pp. 233-268.
Silvestre, F. see Calmon-Decriaud, A.: Vol. 207, pp. 207-226.
Sillion, B. see Mison, P.: Vol. 140, pp. 137-180.
Simon, F. see Spange, S.: Vol. 165, pp. 43-78.
Singh, R. P. see Sivaram, S.: Vol. 101, pp. 169-216.
Singh, R. P. see Desai, S. M.: Vol. 169, pp. 231-293.
Sinha Ray, S. see Biswas, M: Vol. 155, pp. 167-221.
Sivaram, S. and *Singh, R. P.:* Degradation and Stabilization of Ethylene-Propylene Copolymers and Their Blends: A Critical Review. Vol. 101, pp. 169-216.
Söderqvist Lindblad, M., Liu, Y., Albertsson, A.-C., Ranucci, E., Karlsson, S.: Polymer from Renewable Resources.Vol. 157, pp. 139–161
Spange, S., Meyer, T., Voigt, I., Eschner, M., Estel, K., Pleul, D. and *Simon, F.:* Poly(Vinylformamide-co-Vinylamine)/Inorganic Oxid Hybrid Materials. Vol. 165, pp. 43-78.
Stamm, M. see Möhwald, H.: Vol. 165, pp. 151-175.
Stamm, M. see Rühe, J.: Vol. 165, pp. 79-150.
Starodybtzev, S. see Khokhlov, A.: Vol. 109, pp. 121-172.
Stegeman, G. I. see Canva, M.: Vol. 158, pp. 87-121.
Steinke, J., Sherrington, D. C. and *Dunkin, I. R.:* Imprinting of Synthetic Polymers Using Molecular Templates. Vol. 123, pp. 81-126.
Stenberg, B. see Jacobson, K.: Vol. 169, pp. 151-176.
Stenzenberger, H. D.: Addition Polyimides. Vol. 117, pp. 165-220.
Stephan, T. see Rühe, J.: Vol. 165, pp. 79-150.
Stevenson,W. T. K. see Sefton, M. V.: Vol. 107, pp. 143-198.
Stridsberg, K. M., Ryner, M., Albertsson, A.-C.: Controlled Ring-Opening Polymerization: Polymers with Designed Macromoleculars Architecture. Vol. 157, pp. 2751.
Sturm, H. see Munz, M.: Vol. 164, pp. 87–210.
Suematsu, K.: Recent Progress of Gel Theory: Ring, Excluded Volume, and Dimension. Vol. 156, pp. 136-214.
Sugimoto, H. and *Inoue, S.:* Polymerization by Metalloporphyrin and Related Complexes. Vol. 146, pp. 39-120.
Suginome, M., Ito, Y.: Transition Metal-Mediated Polymerization of Isocyanides. Vol. 171, pp. 77-136.

Sumpter, B. G., Noid, D. W., Liang, G. L. and *Wunderlich, B.*: Atomistic Dynamics of Macromolecular Crystals. Vol. 116, pp. 27-72.
Sumpter, B. G. see *Otaigbe, J.U.*: Vol. 154, pp. 1-86.
Sun, H.-B., Kawata, S.: Two-Photon Photopolymerization and 3D Lithographic Microfabrication. Vol. 170, pp. 169-273.
Suter, U. W. see *Gusev, A. A.*: Vol. 116, pp. 207-248.
Suter, U. W. see *Leontidis, E.*: Vol. 116, pp. 283-318.
Suter, U. W. see *Rehahn, M.*: Vol. 131/132, pp. 1-475.
Suter, U. W. see *Baschnagel, J.*: Vol. 152, p. 41-156.
Suzuki, A.: Phase Transition in Gels of Sub-Millimeter Size Induced by Interaction with Stimuli. Vol. 110, pp. 199-240.
Suzuki, A. and *Hirasa, O.*: An Approach to Artifical Muscle by Polymer Gels due to Micro-Phase Separation. Vol. 110, pp. 241-262.
Swiatkiewicz, J. see *Lin, T.-C.*: Vol. 161, pp. 157-193.

Tagawa, S.: Radiation Effects on Ion Beams on Polymers.Vol. 105, pp. 99-116.
Takata, T., Kihara, N., Furusho, Y.: Polyrotaxanes and Polycatenanes: Recent Advances in Syntheses and Applications of Polymers Comprising of Interlocked Structures. Vol. 171, pp. 1-75.
Takeuchi, D. see *Osakada, K.*: Vol. 171, pp. 137-194.
Tan, K. L. see *Kang, E. T.*: Vol. 106, pp. 135-190.
Tanaka, H. and *Shibayama, M.*: Phase Transition and Related Phenomena of Polymer Gels.Vol. 109, pp. 1-62.
Tanaka, T. see *Penelle, J.*: Vol. 102, pp. 73-104.
Tauer, K. see *Guyot, A.*: Vol. 111, pp. 43-66.
Teramoto, A. see *Sato, T.*: Vol. 126, pp. 85-162.
Terent'eva, J. P. and *Fridman, M. L.*: Compositions Based on Aminoresins. Vol. 101, pp. 29-64.
Theodorou, D. N. see *Dodd, L. R.*: Vol. 116, pp. 249-282.
Thomson, R. C., Wake, M. C., Yaszemski, M. J. and *Mikos, A. G.*: Biodegradable Polymer Scaffolds to Regenerate Organs. Vol. 122, pp. 245-274.
Thünemann, A. F., Müller, M., Dautzenberg, H., Joanny, J.-F., Löwen, H.: Polyelectrolyte complexes. Vol. 166, pp. 113-171.
Tieke, B. see *v. Klitzing, R.*: Vol. 165, pp. 177-210.
Tokita, M.: Friction Between Polymer Networks of Gels and Solvent. Vol. 110, pp. 27-48.
Traser, S. see *Bohrisch, J.*: Vol. 165, pp. 1-41.
Tries, V. see *Baschnagel, J.*: Vol. 152, p. 41-156.
Tsuruta, T.: Contemporary Topics in Polymeric Materials for Biomedical Applications. Vol. 126, pp. 1-52.

Uemura, T., Naka, K. and *Chujo, Y.*: Functional Macromolecules with Electron-Donating Dithiafulvene Unit. Vol. 167, pp. 81-106.
Usov, D. see *Rühe, J.*: Vol. 165, pp. 79-150.
Uyama, H. see *Kobayashi, S.*: Vol. 121, pp. 1-30.
Uyama, Y: Surface Modification of Polymers by Grafting. Vol. 137, pp. 1-40.

Varma, I. K. see *Albertsson, A.-C.*: Vol. 157, pp. 99-138.
Vasilevskaya, V. see *Khokhlov, A.*: Vol. 109, pp. 121-172.
Vaskova, V. see *Hunkeler, D.*: Vol.: 112, pp. 115-134.
Verdugo, P.: Polymer Gel Phase Transition in Condensation-Decondensation of Secretory Products. Vol. 110, pp. 145-156.
Vettegren, V. I. see *Bronnikov, S. V.*: Vol. 125, pp. 103-146.
Vilgis, T. A. see *Holm, C.*: Vol. 166, pp. 67-111.
Viovy, J.-L. and *Lesec, J.*: Separation of Macromolecules in Gels: Permeation Chromatography and Electrophoresis. Vol. 114, pp. 1-42.
Vlahos, C. see *Hadjichristidis, N.*: Vol. 142, pp. 71-128.
Voigt, I. see *Spange, S.*: Vol. 165, pp. 43-78.

Volk, N., Vollmer, D., Schmidt, M., Oppermann, W., Huber, K.: Conformation and Phase Diagrams of Flexible Polyelectrolytes. Vol. 166, pp. 29-65.
Volksen, W.: Condensation Polyimides: Synthesis, Solution Behavior, and Imidization Characteristics. Vol. 117, pp. 111-164.
Volksen, W. see Hedrick, J. L.: Vol. 141, pp. 1-44.
Volksen, W. see Hedrick, J. L.: Vol. 147, pp. 61-112.
Vollmer, D. see Volk N.: Vol. 166, pp. 29-65.

Wake, M. C. see Thomson, R. C.: Vol. 122, pp. 245-274.
Wandrey C., Hernández-Barajas, J. and *Hunkeler, D.*: Diallyldimethylammonium Chloride and its Polymers. Vol. 145, pp. 123-182.
Wang, K. L. see Cussler, E. L.: Vol. 110, pp. 67-80.
Wang, S.-Q.: Molecular Transitions and Dynamics at Polymer/Wall Interfaces: Origins of Flow Instabilities and Wall Slip. Vol. 138, pp. 227-276.
Wang, S.-Q. see Bhargava, R.: Vol. 163, pp. 137-191.
Wang, T. G. see Prokop, A.: Vol. 136, pp.1-52; 53-74.
Wang, X. see Lin, T.-C.: Vol. 161, pp. 157-193.
Webster, O.W.: Group Transfer Polymerization: Mechanism and Comparison with Other Methods of Controlled Polymerization of Acrylic Monomers. Vol. 167, pp. 1-34.
Whitesell, R. R. see Prokop, A.: Vol. 136, pp. 53-74.
Williams, R. J. J., Rozenberg, B. A., Pascault, J.-P.: Reaction Induced Phase Separation in Modified Thermosetting Polymers. Vol. 128, pp. 95-156.
Winkler, R. G. see Holm, C.: Vol. 166, pp. 67-111.
Winter, H. H., Mours, M.: Rheology of Polymers Near Liquid-Solid Transitions. Vol. 134, pp. 165-234.
Wittmeyer, P. see Bohrisch, J.: Vol. 165, pp. 1-41.
Wu, C.: Laser Light Scattering Characterization of Special Intractable Macromolecules in Solution. Vol 137, pp. 103-134.
Wunderlich, B. see Sumpter, B. G.: Vol. 116, pp. 27-72.

Xiang, M. see Jiang, M.: Vol. 146, pp. 121-194.
Xie, T. Y. see Hunkeler, D.: Vol. 112, pp. 115-134.
Xu, Z., Hadjichristidis, N., Fetters, L. J. and *Mays, J. W.*: Structure/Chain-Flexibility Relationships of Polymers. Vol. 120, pp. 1-50.

Yagci, Y. and *Endo, T.*: N-Benzyl and N-Alkoxy Pyridium Salts as Thermal and Photochemical Initiators for Cationic Polymerization. Vol. 127, pp. 59-86.
Yannas, I. V.: Tissue Regeneration Templates Based on Collagen-Glycosaminoglycan Copolymers. Vol. 122, pp. 219-244.
Yang, J. S. see Jo, W. H.: Vol. 156, pp. 1-52.
Yamaoka, H.: Polymer Materials for Fusion Reactors. Vol. 105, pp. 117-144.
Yasuda, H. and *Ihara, E.*: Rare Earth Metal-Initiated Living Polymerizations of Polar and Nonpolar Monomers. Vol. 133, pp. 53-102.
Yaszemski, M. J. see Thomson, R. C.: Vol. 122, pp. 245-274.
Yoo, T. see Quirk, R. P.: Vol. 153, pp. 67-162.
Yoon, D. Y. see Hedrick, J. L.: Vol. 141, pp. 1-44.
Yoshida, H. and *Ichikawa, T.*: Electron Spin Studies of Free Radicals in Irradiated Polymers. Vol. 105, pp. 3-36.

Zhang, H. see Rühe, J.: Vol. 165, pp. 79-150.
Zhang, Y.: Synchrotron Radiation Direct Photo Etching of Polymers. Vol. 168, pp. 291-340.
Zhou, H. see Jiang, M.: Vol. 146, pp. 121-194.
Zubov, V. P., Ivanov, A. E. and *Saburov, V. V.* : Polymer-Coated Adsorbents for the Separation of Biopolymers and Particles. Vol. 104, pp. 135-176.

Subject Index

ADMET polymerization 139
Alkenes, nonconjugated dienes 163
Alkoxyallenes 166
Alkylallenes 165
Allenes (1,2-dienes) 165
–, coordination polymerization 137
–, functionalized 169
–, (n-octyloxy)allene 167
Amide polymer, crosslinked 55
Amide-type macrocycle 42
Amino acids 19
Antenna molecules 29
Aryl isocyanides, organorhodium complexes 96
– –, Pt-Pd 94
2-Aryl-1-methylenecyclopropane 176, 177
Arylallenes 165

Bi(quinoxalinyl)palladium 120
4,4'-Bipyridinium salt 11
Bis[2]catenanes 58
Bis(dibenzo-24-crown-8 ether) 53
Bis(3,5-dinitrobenzoyl)-poly(ethylene glycol) 26
Butadiene 90, 138
–, allenes 171
–, $CoCl_2$/MAO 155
–, $FeEt_2(bpy)_2$/MAO 156
–, late transition metals 152
–, $Nd(OCOR)_3$ 148
–, Ni complexes 153
–, organolanthanide complexes 150
Butadiene-isocyanide 91
$tert$-Butyl isocyanide 108

Carbene, stable 78
Catenanes 2

[2]Catenanes 58
rac-[CH_2(3-$tert$-butyl-1-indenyl)$_2$]ZrCl$_2$ 156
Charge-transfer (CT) interaction 11
Chiral poisoning 106
Cholesterol 47
Chymotrypsin 112
$Co_2(CO)_8$ 172
$CoCl_2$/MAO, butadiene 155
Coordination polymerization 137
Cp_2Ni 82
$CpNi(CO)_2$ 82
$CpTiCl_3$/MAO 143
$Cr(acac)_3$ 146
$Cr(CO)_5(py)$ 146
$Cr(NCPh)_6$ 146
$CrCl_2(dmpe)$/MAO 146
Crosslinking, topological 3
Crown ethers 9, 10, 44
Cucurbituril 8, 33, 48
–, oligorotaxanes 35
Cyclodextrins 8, 19, 44
1,5-Cyclohexadiene 161
Cyclohexyl isocyanide 81
Cyclophane 40
–, 4,4'-bipyridinium 41

Dibenzo-24-crown-8 (DB24C8) 13-18, 44
Dibenzo-30-crown-10 (DB30C10) 15
Di-$tert$-butylacetylenedicarboxylate 14
Diene polymerization, lanthanides 148
Dienes, conjugated, copolymerization 156
–, –, polymerization 139
–, coordination polymerization 137, 139
–, $Cr(acac)_3$ 146

–, nonconjugated, alkenes 163
–, –, Et(Ind)$_2$ZrCl$_2$/MAO 162
–, –, polymerization 159
–, V(acac)$_3$/MAO 145
Dienes/methylenecyclopropanes 180
1,2-Diisocyanobenzenes 77, 80, 126
–, polymerization 118
1,2-Dimethylene-3-methylcyclopentane 174

Et(Ind)$_2$ZrCl$_2$/MAO, ethylene/propylene 162
2-Ethoxycarbonyl-1-methylenecyclopropane 179

FEB 27
FeEt$_2$(bpy)$_2$/MAO, butadiene 156
Frequency-domain electric birefringence (FEB) 27

Gel materials, crosslinks 57
Glucopyranosyl isocyanide 115

Half-titanocene 139
1,6-Heptadiene 163
1,5-Hexadiene, cyclization polymerization 159
–, phenoxyimine-Ti 164
1-Hexene 161

Iminocarbene, N-substituted 77
o-Iodobenzoic acid 128
Isocyanides 77
–, allenes 171
–, palladium-catalyzed 93
–, peptide-based 111
1,2-Isocyanobenzenes 121

K–S 7

Lanthanocenes 138
Ln(OCOR)$_3$ 148
Lutenocene, methylenecycloalkane 173

Methoxyallene 166
4-Methoxyphenylallene 184
4-Methyl-1,3-pentadiene 143
2-Methyl-1,5-hexadiene 159
N-Methyl-2-pyrrolidone (NMP) 27
Methylene diphenyldiisocyanate 9

Methylenebicyclo[4.1.0]heptane 186
Methylenecycloalkanes 160
–, coordination polymerization 137
Methylenecyclopropanes 172
Mobility, rotaxanes/catenanes 2
Molecular abacus 27
Molecular necklaces 34
Molecular tube 32
Molecular wires, insulated 27
Monoisocyanide, polymerization 81

Nanotubes 32
[NbO(C$_{16}$H$_{11}$O$_6^-$)(C$_2$O$_4^{2-}$)] 145
Nd(OCOR)$_3$ 148
Necklaces, molecular 34
Ni(acac)$_2$ 86
Ni(naph)$_2$ 153
Ni(OCOR)$_2$/MAO 153

Oligo[n]catenanes 62-65
Oligoether-macrocycles 8
Oligoquinoxalines 125
Oligoquinoxalinylpalladium 124, 129
Oligorotaxanes 12
Olympiadane 62
Organorhodium complexes, aryl isocyanides 96

Palladium complexes, cyclohexyl isocyanide 92
Paraquat-type cationic host 8
[Pd(dppp)(NCMe)$_2$](BF$_4$)$_2$ 184
PdCl(Me)(bpy)/NaBARF 185
PEG 21
PEG-1N$_2$ 26
(Z)-1,3-Pentadiene 143
Peptide-based polymers 111
Phenanthroline-copper(I), oligorotaxanes 38
2-Phenyl-1-methylenecyclopropane 174, 178
Phenylallene 165
1-Phenylethyl isocyanide 90
Platinum-palladium, isocyanides 94
Poly(allene-b-isocyanide) 171
Poly(benzimidazole), side chain-type polyrotaxane 44
Poly(1,3-butadiene) 140
–, isomeric, pathways 142
Poly(N-tert-butyliminomethylene) 81

Subject Index

Poly(*tert*-butyl isocyanide)   110
Poly[n]catenane   62
Poly[2]catenane   58
Poly(*N-tert*-cyclohexyliminomethylene)   81
Poly(dimethylsiloxane)   21
Poly(ether sulfone)
Poly(ethyl isocyanide)   86
Poly(ethylene glycol)   21
– bisamine   32
Poly(1,5-hexadiene)   160
Poly(hexamethyleneimine)   37
Poly(iminoundecamethylene)   23
Poly(isocyanide)s   77
–, peptide-based   111
–, racemic   82
Poly(ϵ-lysine)   31
Poly(methoxyallene)   166
Poly(methyl methacrylate)   168
Poly(oligo)catenanes   1
Poly(oligo)rotaxanes   1
Poly(propylene glycol)   21
Poly(quinoxaline-2,3-diyl)s   77
–, non-racemic   122
–, racemic   119
Poly[2]rotaxane   49
Poly[3]rotaxane   51
Poly(styrene-*co*-CO)   175
Poly(vinyl isocyanide)s   116
Polycatenane network   66
Polycatenanes   1
Polyketones   138, 184, 186
Polyquinoxalines   125
Polyrotaxanes   1
–, biodegradable   30
–, crosslinked   55
–, stimuli-responsive   31
–, synthesis   22
Polystyrene   168

Polyurethane, crosslinked   56
Polyurethane-cyclodextrin   20
PPG   21
Propyne, allenes   171
Pseudopolyrotaxanes   9, 49
–, doubly-stranded   26
–, synthesis   19

Random copolyketones   184
RhH(PPh$_3$)$_4$ 172
Rhodium, aryl isocyanides   97
Rotaxanes   2

Screw-sense induction, Ni-catalyzed isocyanides   99
Slipping approach   15
Spin trap   90
Stilbene, cyclodextrin   25

Tetrakis(isocyanide)nickel(II)   107
Tetrathiafulvalenes   8
Tetra(ethylene glycol)   9
Ti(CH$_2$Ph)$_4$/MAO   144
Titanocene   143
Topological bond   2
Topological crosslinking   3
Topological polymers   8

V(acac)$_3$/MAO   145

Wheel component   4
Wires, molecular, insulated   27

Yttrium, 1,5-hexadiene cyclopolymerization   160

Zirconocene   143
–, methylenecycloalkane   173

Printing: Saladruck, Berlin
Binding: Stein+Lehmann, Berlin